HERMAN DALY'S ECONOMICS FOR A FULL WORLD

As the first biography of Professor Herman Daly, this book provides an in-depth account of one of the leading thinkers and most widely read writers on economics, environment and sustainability.

Herman Daly's economics for a full world, based on his steady-state economics, has been widely acknowledged through numerous prestigious international awards and prizes. Drawing on extensive interviews with Daly and in-depth analysis of his publications and debates, Peter Victor presents a unique insight into Daly's life from childhood to the present day, describing his intellectual development, inspirations and influence. Much of the book is devoted to a comprehensive account of Daly's foundational contributions to ecological economics. It describes how his insights and proposals have been received by economists and non-economists and the extraordinary relevance of Daly's full world economics to solving the economic problems of today and tomorrow.

Innovative and timely, this book will be of great interest to students, scholars, researchers, activists and policy makers concerned with economics, environment and sustainability.

Peter A. Victor is Professor Emeritus and Senior Scholar in the Faculty of Environmental Studies at York University, Canada.

'Herman was instrumental in shaping my views on all the work that I now do... Truly a kind, gentle man of enormous courage and wisdom. I revere him and cherish the times I have had with him. He deserves the Nobel, but so much more.'

L. Hunter Lovins, *President of Natural Capitalism Solutions, Longmont, Colorado, USA*

'Looking for alternative approaches to the economics of natural resources, I found a light in Herman Daly's publications.'

Alejandra Saravia, *Researcher and Assistant Professor, Universidad Mayor de San Simon, Cochabamba, Bolivia*

'Herman deserves a biography as he has inspired so many who question an endless growth economy!'

Haydn Washington, *Environmental scientist, writer and activist, PANGEA Research Unit, UNSW, Australia*

'He deserves to be recognized as one of the great economists of our time.'

Anders Hayden, *Associate Professor, Department of Political Science, Dalhousie University, Canada*

'This man is a giant in an age of midgets. The best of the kind of southern gentlemen. A visionary in an age blinded by its own arrogance; drunk on a superficial idea of progress. We should all thank god for sending this man among us.'

Peter Brown, *Professor, Bieler School of Environment/Natural Resource Sciences/ Geography, McGill University, Montreal, Canada*

'As a snooty physical scientist, I dismissed economics as shallow. Daly has converted me to appreciating how deep a subject it is.'

Robert Herendeen, *Gund Affiliate, Adjunct Professor, Rubenstein School for the Environment and Natural Resources, University of Vermont, USA*

'I think Herman is a visionary and genius, yet among the kindest, most humble academics I know. He is brilliant at using metaphors to explain difficult concepts and has a clarity of thought rarely seen among economists.'

Joshua Farley, *Professor, Rubenstein School of Environmental and Natural Resources, The University of Vermont*

'When I did have the opportunity to meet him, it was a pleasure to discover that he is a gracious, warm, brilliant, and humble person--a true reflection of his writing.'

Richard Heinberg, *Senior Fellow, Post-Carbon Institute*

HERMAN DALY'S ECONOMICS FOR A FULL WORLD

His Life and Ideas

Peter A. Victor

Routledge
Taylor & Francis Group

LONDON AND NEW YORK

from Routledge

First published 2022
by Routledge
2 Park Square, Milton Park, Abingdon, Oxon OX14 4RN

and by Routledge
605 Third Avenue, New York, NY 10158

Routledge is an imprint of the Taylor & Francis Group, an informa business

British Library Cataloguing-in-Publication Data
A catalogue record for this book is available from the British Library

Library of Congress Cataloging-in-Publication Data
Names: Victor, Peter A., 1946– author.
Title: Herman Daly's economics for a full world : his life and ideas / Peter A. Victor.
Description: Abingdon, Oxon; New York, NY : Routledge, 2022. |
Includes bibliographical references and index.
Identifiers: LCCN 2021018010 (print) | LCCN 2021018011 (ebook)
Subjects: LCSH: Daly, Herman E. | Economists–United States–Biography. |
Economic development–Environmental aspects. |
Economic development–Moral and ethical aspects. | Environmental economics.
Classification: LCC HB119.D35 .V53 2022 (print) |
LCC HB119.D35 (ebook) | DDC 330.092 [B]–dc23
LC record available at https://lccn.loc.gov/2021018010
LC ebook record available at https://lccn.loc.gov/2021018011

ISBN: 978-0-367-55694-5 (hbk)
ISBN: 978-0-367-55695-2 (pbk)
ISBN: 978-1-003-09474-6 (ebk)

DOI: 10.4324/9781003094746

Typeset in Bembo
by Newgen Publishing UK

To the Gothic Group for years of conversation,
camaraderie, and fulsome refreshment of mind and body

CONTENTS

FIGURES

FOREWORD BY HERMAN DALY

Anyone who has tried for a long time to understand and explain something that is both important and difficult, but very unpopular, will know what a pleasure it is to find that someone else has not only understood the same ideas, but has used and clarified them. Such is the pleasure I first had in reading Peter Victor's classic study, *Managing Without Growth*, and now have had again in reading his generous biography of me. I should emphasize that this is a biography, not an autobiography. Reading it has taught me things about myself. The organization, selection of topics and writing are Peter's, not mine. It is mainly a story of ideas related to the economic transition from a world relatively empty of humans to a world over-full of us and our baggage, rather than the story of my personal life. However, personal history is included for explaining the background, partnerships, conflicts and origins of ideas and influences. Although I supplied Peter with whatever information he asked for, and that I could recall, I confess that I minimized sharing those aspects of my life which in retrospect seem stupid or shameful, so in that respect I probably biased his account in my favor. In any case, the 'tell-all true confessions' kind of biography is best reserved for famous movie stars and politicians, not economics professors. Nevertheless, the academic world of ideas and policies is not without its share of drama, conflict and intrigue.

Economic growth is the number one goal of nearly all economists, politicians and governments. To argue against it on the grounds that so-called 'economic' growth has now become *uneconomic*, because it increases environmental and social costs by more than production benefits, is to poke a big hornets' nest with a short stick. It rudely upsets a very large and comfortable consensus. Without growth how do we reduce poverty? By redistribution and sharing. Without growth and the consequent automatic demographic transition what will limit population expansion? Consciously reasoned demographic policy. Without growth how will we pay the huge, accumulated costs of environmental repair? By reducing current resource depletion to allow natural systems to recover, and by stopping the mining and burning of climate-ruining carbon-based fuels. Redistribution, population policy and reduced resource consumption – each by itself is considered a political anathema. Of course, the growthists were furious to be challenged by all three at once! They defensively doubled down on the call for more GDP growth, failing to recognize that growth in GDP at the current margin no longer makes us richer. Growth within our finite and entropic world now makes us poorer by adding more to illth than to wealth – very bad economics!

How many times did I get stung by angry hornets? Which teachers inspired me? Who were and are my partners? Did I escape? Who helped me? Who did I help? My life has not been dull, even though I have never fought in a war, or been put in prison, or elected to political office.

Since the rest of this book is about me and my past it would be repetitious to dwell on that in this foreword. Instead, I will speculate a bit on the future. My students, at the end of a semester, after hearing my arguments for a steady-state economy (SSE) would often say, 'OK, Professor Daly, your premises *seem* true and we can't *yet* see any flaws in your logic, but you know as well as we do that Congress will never vote for the institutions and policies of a SSE because, even in the unlikely event that they personally favored a SSE, average citizens would vote them out of office. So should we not be studying something else? Are you sure you have not wasted our time?'

A challenging question. How to answer? Two replies, if not really answers, come to me.

First is that after the present system crashes in failure there will be not only an interest, but an immediate need to rebuild something more sustainable and just. If the basic ideas of a steady-state or ecological economy have at least been outlined and are on the shelf, that will be a big help in reconstruction. Big changes usually require a big crisis to make them politically possible. And the likelihood of a big enough crisis increases daily, as seen on TV and reported in the newspapers. So let's get ready!

Second is that improbable events, novel unexpected things, really do happen. There are big social examples like the fall of empires, and the abolition of slavery. But there are also individual revolutions in people's minds. To give a relevant personal example, I used to be a neoclassical growth economist. I hoped that my contribution to the world would be to help increase the growth rate of GDP, especially in the poor regions of Latin America, but in wealthy countries too. But experience, arguments, and evidence changed my mind, and I became an ecological economist who advocates a steady-state economy (SSE) with redistribution and qualitative improvement instead of perpetual growth. Might not the same happen to other economists? Indeed, is it not now happening, although slowly? Why won't the same evidence and logic that has convinced me (and a number of others) eventually convince many more? Are we ecological economists unique? I am certainly not, and the events recounted in this book are testimony to the fact that ideas, evidence, and arguments can change the mind of at least one ordinary growth economist. And if one, why not two, and if two, why not many? Yes, I know that there are large financial and intellectual vested interests in growth, and the fight will continue. But every day I run into good people who I can tell are smarter than me. Given time and exposure to the facts and arguments they too will likely change their minds, and will clarify and renew the classical economists' early vision of a steady-state economy, an economy that improves without growing. May this book help them to hurry up, because the matter is urgent!

Herman Daly
February 2021
Midlothian, Virginia, USA

PREFACE

One sunny afternoon in 1979 the phone rang in the Toronto office of Victor & Burrell, my recently established consulting company. I was greeted by a voice speaking with a deep Southern drawl. 'I'm Herman Daly', the voice said. 'I would like your permission to include your paper "Economics and the Challenge of Environmental Issues" in the second edition of *Toward A Steady-State Economy*.' I had read the first edition of Daly's book published in 1973 and had browsed his 1977 book *Steady-State Economics*. I had also included a review of his 1968 paper 'On Economics as a Life Science' in my Ph.D. dissertation so I knew his work quite well. I was surprised and flattered by his request and agreed without hesitation.

In the years that followed I read a number of Herman's publications, but it wasn't until we met in person that I appreciated his warmth, intelligence and humility. We had arranged to have lunch in the Watergate Hotel in Washington DC with Robert Goodland, his boss at the World Bank. They sounded me out about joining them in the Bank. Working with them was a very attractive prospect but leaving Canada with my wife and two young daughters was not, so I politely declined. Subsequently, Herman and I met at a few conferences but by far the most memorable meeting came in the later 1990s when, as Dean of the Faculty of Environmental Studies at York University, I invited Herman to Toronto to deliver a public lecture and to meet with students and faculty in my university. Everything was arranged but shortly before Herman was due to arrive the faculty union called a strike and effectively closed down the campus. Herman's visit went ahead but, in respect of the picket line, he delivered his very well-received public lecture at the nearby Regional Offices of Environment Canada. What impressed me even more was that for two days Herman welcomed all comers to a meeting room provided by Environment Canada. Tirelessly, patiently and in a disarmingly honest manner he answered whatever questions were put to him to the obvious delight of those who asked them.

About five years ago I began to think that someone should write Herman's biography. After casting around without success to find a suitable author, my wife Maria looked at me and said, 'you should do it.' I was retired, had just completed the second edition of *Managing without Growth. Slower by Design, not Disaster* and was ready for something new. I approached Herman with the idea and receiving his blessing travelled to Midlothian, Virginia where I spent several days interviewing him about his life and ideas. This was followed by my immersion in his writings and those of his critics, conversations with people who have worked with him, and an online questionnaire that generated many of the quotes that you will find in this book showing his far-reaching influence. You will also find many quotes from Herman since this is his story.

The work of most economists, and of social thinkers more generally, becomes less relevant as time passes. Circumstances change, as do ways of explaining them. In the case of a few, the opposite happens. Their work becomes increasingly timely, providing answers to questions they were asking before the rest of the world woke up to their significance. Herman Daly is among these few.

ACKNOWLEDGEMENTS

The first person I must thank is Maria Paez Victor, my wife of fifty years, who urged me to take on what turned out to be an enormously rewarding task and kept up her support and gave advice throughout the two years it took me to write this biography.

I also received encouragement from the members of the Gothic Group, to whom this book is dedicated. For the past several years this group of my past and present students has met at my house to discuss all sorts of interesting ideas and issues relating to ecological economics. When I mentioned that I was thinking of writing a biography of Herman Daly they thought it was well worth doing and gave me suggestions about how to approach it. Later on, when I was deep into my research, Herman joined us via Zoom for a memorable evening of discussion and debate. We now consider him an honorary member of the Gothic Group.

I want to thank all of the 90+ respondents to the questionnaire that I posted on the internet, many of whom made a considerable effort to answer my questions and some who provided additional material that I would not otherwise have found. I have quoted from the responses to the questionnaire, edited for brevity, throughout the book. Although I could not include something from everyone, I want to thank them all for providing such useful information.

In addition to the questionnaire, I interviewed Peter Brown and Stuart Scott in person. Both have known Herman for long time and think very highly of him. Our conversations gave me an added appreciation of Herman the man. Unfortunately, I was unable to take up offers from several other of Herman's friends and colleagues to talk in more detail about Herman, but I thank them nonetheless for their willingness to do so.

A few people who I do want to acknowledge by name for the particular help they gave me are:

Bob Costanza for providing a photograph taken at the first meeting of the International Society for Ecological Economics and copies of several papers he co-authored with Herman; Brian Czech for helping publicize the Questionnaire; Josh Farley with whom I first discussed the possibility of a Daly biography and who agreed wholeheartedly with the idea; John Gowdy for urging me to address Herman's controversial views on evolution; Judith Grant for her advice on writing a biography; Tim Jackson, with whom I have enjoyed a remarkable collaboration and friendship for over a decade, who encouraged me to write this biography; Chris and Karen Junker, Herman's son-in-law and daughter, for applying their photographic and scanning skills to improve the photographs; Fridolin Krausmann for providing information on Herman's early influence on the Viennese School of Social Ecology; Deepak Malghan for offering the transcripts of his interviews of Herman in 2008; Dennis Meadows for confirming Herman's recollection of their first meeting; Inge Røpke for providing copies of several of her papers; Vera Rosenbluth for her advice on how to interview Herman; Giandomenico Scarpelli who gave me with his extremely useful, detailed bibliography of Herman's writings, and my daughter Carmen Victor who drew one of the illustrations in the book.

I also wish to acknowledge the excellent service of the Rubenstein Library at Duke University who provided me with copies of correspondence between Herman and Georgescu-Roegen dating back to the 1960s.

I owe a special thanks to Peter Timmerman and Ed Hanna. Peter read drafts of the chapters offering criticism and suggestions derived from his outstanding breadth and depth of knowledge of the humanities, social and natural sciences. I called on Ed, my go to person for all matters of environmental science, to read the chapter on Economics as a Life Science. He readily obliged and to my great relief found no egregious errors.

Finally, I am deeply grateful to Herman and his wife Marcia. In November 2018 they welcomed me into their home in Midlothian where Herman recounted stories of his life and we discussed his ideas and experiences over a long and eventful career. We followed this with numerous emails and video conversations. At my request Herman checked drafts of each chapter for factual errors and misinterpretations, but we were always very clear that while the subject matter is his life and ideas, their presentation and how there are discussed in this biography, are my responsibility alone.

PROLOGUE

In 2014 at the age of 76, Herman Daly was awarded the Blue Planet Prize. The prize is given to individuals or organizations 'that make outstanding achievements in scientific research and its application, and in so doing help to solve global environmental problems'. The selection committee said that 'Professor Herman Daly redefined "steady state economics" through the concept of sustainability by incorporating such factors as the environment, local communities, quality of life, and ethics into economic theory, which lead to building a foundation of ecological economics. He has been questioning whether economic growth brings happiness to humans and has been issuing warnings to society, which tends to overemphasize economic growth. As a consequence, he has had a significant international influence'.[1]

When Herman was born in 1938 the life expectancy of an American male was just over 65 years. He is now 82 and going strong, despite having suffered a very serious attack of polio as a child that left him without a left arm for his entire adult life. Herman's sound constitution was matched with an extraordinary mental capacity, profound ethical and religious convictions, and a willingness to tell it like it is to any and all who would listen. From Herman's birth to the present (2021) the American population increased from 130 million to 330 million. The global population almost doubled twice, which is unique in human history and will likely remain so unless the science fiction writers turn out to be correct and humans populate the solar system and beyond.

Had it been just an increase in numbers of people the impact on the planet and the countless other species who live on it would still have been significant. As the top predator of the global food chain, it could hardly have been different. But of course, the numbers of people are only part of the story. The other part is the massive and extremely unequal increase in human production, consumption and wastes, depleting, as Herman would say, the limited sources of 'low entropy' energy and materials. It is the combination of more people consuming more resources, generating more wastes and destroying the natural habitat of other species, that led Daly to the conclusion that the world has transitioned from 'empty' to 'full'.

Daly's proposal for dealing with these changed circumstances starts with a distinction between qualitative development and quantitative growth. All countries should seek development, but growth should be restricted to poor ones where its benefits are most obvious. Even then, he says, growth in these countries can only be temporary without overwhelming the capacity of Earth's resources and ecological systems. The time for growth could be extended if developed economies were to practice material and energy degrowth and reduce their demands on the biosphere in general.

Herman's perspective on growth raises questions about the optimal scale of an economy which according to Herman should be 'defined in terms other than efficiency, namely ecological sustainability' (Daly and Cobb 1989, 1994, p. 145). In Herman's terms sustainable development is development that is first and foremost ecologically sustainable. This is not to say that he is unconcerned with the economic and social dimensions of sustainable development that others have emphasized, only that ecological sustainability is fundamental if the natural conditions for life are not imperiled.

The vast number of citations to his work, the numerous testimonials, and the many prizes he has received[2] reveal a level and scope of influence that few contemporary economists can match. But who is Herman Daly? What are these ideas of his that have become increasingly relevant as the world has changed? How did he arrive at them and what has been the reaction to them and to him from more conventional thinkers? What is his life story and what were the influences on his thinking? And most important of all, what can we learn from him that will help us cope well with the threats and opportunities of the twenty-first century? The purpose of this biography is to provide answers to these questions in the hope that others will learn from and build on his life and ideas for an economics fit for purpose in a full world.

Notes

1 2014 Blue Planet Prize: Announcement of Prize Winners.
2 Doctor of Science, Honoris Causa, Muhlenberg College, May 1992.
 Grawemeyer Award for Ideas Improving World Order, 1991 (For the Common Good), Kenneth Boulding Memorial Award for Contributions to Ecological Economics, 1994. Conservation Biology Society Award, 1995. Royal Netherlands Academy of Arts and Sciences, Heineken Prize in Environmental Sciences, 1996. Honorary Right Livelihood Award, Stockholm, Sweden, 1996. Sophie Prize for Environment and Development, Oslo, Norway 1999. Association of American Geographers 'Honorary Geographer' for 1999. Doctor of Humane Letters, Honoris Causa, Allegheny College, May 2000. Leontief Prize for Advancing the Frontiers of Economic Thought, 2001, awarded by Tufts University. Medal of the Presidency of the Italian Republic, October 2002. National Council for Science and the Environment, Lifetime Achievement Award, January 2010. Club of Rome (International Member, elected 2011).

1

OUT OF TEXAS[1]

The early years

I was very fortunate in my 'choice' of parents

Daly

Imagine if you can what it must have been like for a young boy just shy of his eighth birthday to wake up one sunny morning with a high fever, stiff everywhere and unable to move his left arm. He thought he was going to die. The night before he had gone to bed with a crick in his neck and a slight pain in his arm, nothing unusual for a boy who was used to the bumps and scrapes of a childhood spent as much as possible outdoors. In the morning he knew something was terribly wrong – not just by how he felt but from the look in the eyes of his worried parents who said, 'oh my God what's this?' They feared the worst – polio – which, because of its partiality for young victims, was known by its more revealing and disturbing name, infantile paralysis.

The young boy in question, one Herman Edward Daly, was on a trip to Houston from Beaumont where he and his parents, Ed and Mildred, lived. They were visiting his grandmother and cousins for some happy family time together. With this unexpected turn of events their plans changed. Mildred and Ed drove their young son to the John Sealy hospital in Galveston. On the way they had to take a ferry at a place called Point Bolivar. Herman lay in the backseat of the car in a daze, but he heard his father tell a policeman, 'Look, I've got a sick boy. I have to go to the head of the line – can you take me?' At the hospital Herman was given a spinal tap, polio was diagnosed, and for the next six weeks he endured well-intentioned treatments and therapy that in the end proved of little use. Herman never regained the use of his left arm and the mobility of his right shoulder remained limited throughout his life. In the 1940s the etiology of polio was unknown, and treatments were based on

DOI: 10.4324/9781003094746-1

what seemed to help but lacked a scientific basis. An Australian nurse named Sister Kinney had claimed great success using hot packs on afflicted limbs so in the hospital they wrapped Herman's immobile arm in extremely hot packs that smelled bad. A therapist would then stretch his arm to try to prevent it from getting stiff.

During this unwelcome period of hospitalization so early in life, Herman's parents and grandmother would visit on weekends, leaving him to himself in the week with a radio for companionship. His favourite shows featured comedians Jack Benny and Fred Allen. Herman would memorize their jokes and retell them to the nurses and therapists, earning him a reputation as a funny kid. These lighter moments helped him allay the fear of being encased in an 'iron lung', a coffin-like metal enclosure that enabled a polio patient to breathe. With only their head exposed they faced the ceiling and viewed the world through an angled mirror placed above their eyes. There were so many iron lungs at the John Sealy hospital that they had to place some along the corridor outside Herman's room. All day and all night he heard the swoosh, swoosh of mechanical breathing, thankful that he was not in one of them.

After about six weeks when there was nothing more the hospital could do for him, Herman was sent home to make the best of it. For years his mother tried all sorts of remedies without success, anything she thought that might stimulate the muscles of his withered arm. A saddle maker named I.J. Ableman, a friend of Herman's father, designed a Sam Brown leather belt that kept the useless appendage from flopping around. This enabled Herman to lead a fairly normal life.

Before his encounter with polio which would remain a factor throughout his life, Herman had been a typical boy growing up in southern Texas. He was born on 21 July 1938 at Wright's Clinic on North Main Street in Houston, Texas. He was the first child of Edward (Ed) Daly and Mildred Julia Herrmann who, in naming their son Herman Daly, echoed his Irish–German ancestry. Introduced by mutual friends, Ed and Mildred fell in love and were married in Mildred's mother's house in Houston in May 1934 right in the middle of the Great Depression. Ed was from an Irish Catholic family and Mildred a German Protestant one, so it was an interfaith marriage that was less a problem on the Protestant side of Herman's family than on the Catholic side. Ed was essentially disowned by his family for marrying outside the Catholic religion. This division in the family distressed Herman and left a bitter taste in his mouth regarding the Catholic church.

The circumstances of the Daly family's life in Beaumont, Texas in the 1940s were modest. Herman's father Ed ran a BF Goodrich service station in Beaumont which was a substantial town about 80 miles from Houston where the Daly family lived. Then, at the outbreak of the Second World War, Ed opened his own store, proudly named: Daly's Home and Auto Appliances. Ed had been badly injured in a motorcycle accident, so he was excused military service and was able to mind the store. The store prospered with the advent of the Second World War, which reinvigorated the US economy after the desperate years of the 1930s. Ed did well enough for he and Mildred to buy their own house on McFaddin Street, which Herman moved into with his parents and his sister, Denis (Deni) Lynn Daly, five years his junior.

One day, quite unexpectedly, Ed received an offer for his store from a wealthy Beaumont patriarch who wanted it for his son. Thinking that he would never see that much money again Ed agreed to the sale and used the money to open a tire store in Beaumont. Unfortunately, the tire business did not flourish so Ed became a travelling salesman for a wholesale firm selling appliances and hardware. This meant that he was often away from home, leaving Herman and his sister Deni's upbringing largely up to their mother Mildred.

After Mildred graduated from high school she worked happily as a secretary until she married Ed. Throughout her life, she wanted to have a paid job and not only be a housewife, but in those times, women were discouraged from working outside the home, so it was not to be. Like so many other women denied opportunities open to men, Mildred had to content herself with the considerable responsibilities of raising her children and running the family home.

In 1944, a year after Deni was born, Herman entered first grade at Averill Elementary School. He was a sturdy boy with blond hair and a ready smile. At just six years of age, he rode his bicycle two miles to school with a basket on the handlebars to carry his lunch box. The road passed through a wooded area and sometimes Herman would see a snake slithering across the road in front of him. In later years he would have avoided the snake, but not as a little kid. Quite the opposite, in fact. Putting his feet on the handlebars, Herman would coast across the snake, feeling a small bump and a strong sense of fulfilment at inflicting harm on what he thought was such a dangerous and evil creature… and getting away with it. 'Even good little boys have a mean streak,' said Herman many years later, looking back with regret at this childhood memory.

By this time the USA had already been at war for three years, having declared war on Japan, Germany and Italy after Japan bombed the US naval base at Pearl Harbour in Hawaii. Herman, like other children across the USA, brought dimes to school to buy savings stamps to help finance the war effort. He also helped his mother with shopping. During the war when some foods were rationed, his mother would give him money and ration tickets to buy milk and margarine. And as was common in countries on both sides of the conflict Mildred planted a 'victory garden' to free up food for the soldiers. For his part in the war effort, Herman collected paper and tin cans in his wagon, pulling it barefooted along sun-baked streets, taking advantage of every opportunity to walk on the grass and cool off his scorched feet.

Children can be quite mean to people who are different. At school Herman was sometimes taunted because of his name. He was called 'Herman the German' which could have upset him, but he didn't take it very seriously. The children would all talk about the war, especially those whose fathers were away fighting. Herman's closest relative in the armed forces was his foster uncle who was in the Navy which made Herman proud. Like most white children in his town, Herman would go regularly to the movies, there being no TV, where they would see newsreels about the war. The stories were told in a way intended to boost morale and the children liked them but what they really wanted to see were the Westerns with stars like Gene Autry and Roy Rogers, the singing cowboys as they were known.

Summers in Beaumont were hot and humid due to the moisture that flows inland from the Gulf of Mexico. Little boys wore short pants, no shoes and no shirt.

There was no air conditioning (except in movie theatres) so kids often had heat rash, which was treated with pink calamine lotion and an electric fan. Compared with today, children were given much more freedom to roam without adult supervision and Herman was no exception. There were no cell phones or any other electronic means by which parents could keep track of their children so when Herman was growing up it was a time of liberty that Americans celebrate but which fewer and fewer children nowadays get to experience. Of course, such freedom came with risks and the children suffered the consequences – which were usually minor cuts and bruises, though not always. In later years, looking back at his childhood Herman thought that he was left alone, absolutely alone, to excess: 'I could easily have gotten killed many times.'

In the early years of his life Herman was heavily influenced by religion and his religious convictions remained strong throughout his life. His mother was responsible for his religious education so Herman was raised a Protestant, regularly attending services with her in the Evangelical and Reformed Church.[2] A notable member of that denomination was Reinhold Niebuhr, one of the most influential Christian theologians of the twentieth century. Niebuhr's political and religious philosophy are brilliantly encapsulated in the Serenity Prayer attributed to him: 'Father, give us courage to change what must be altered, serenity to accept what cannot be helped, and the insight to know the one from the other.' This approach to life served Herman well throughout his career when he challenged mainstream thinking in economics, believing it could be changed by reasoned argument and persuasion.

Seeing how his father was treated for marrying outside the Catholic faith, Herman developed a distaste for the Catholic church. This was softened by the close relationship he developed with the youngest of his father's five sisters who became a nun and a teacher at a Catholic girls' school. Later in life she helped Herman and Deni care for their aging mother, deepening his respect for her even further. As an adult Herman did not see much difference among the various protestant churches and described himself as a 'general purpose protestant'.

Not only did Herman learn religion from his parents, he also learned the importance of prudence in one's own affairs. 'Don't waste your money and work hard. Think about the future. Nobody is going to give you anything. You are going to have to do it yourself.' This personal ethic, derived from his parents' experiences in the Great Depression, was taught to Herman, particularly by his father. It contradicted the more outward looking, inclusive and charitable perspective he learned at church. Herman was not the only American to seek a reconciliation of the Christian teachings of goodwill towards all with the ethic of individualism and personal responsibility. Both are in evidence in much of his work (see, for example, Daly and Cobb 1989 and 1994).

In 1947, when Herman was nine, the family moved from Beaumont to La Porte Texas near the Gulf coast, across the street from Herman's beloved Grandma Clara. Of all his grandparents, Mildred's mother had the greatest influence on Herman. Clara Kohlhauff Herrman was very loving, kind and openhearted. She would feed tramps that came to her door, giving them some little job to do in exchange to protect their dignity. This made a big impression on Herman, teaching him that not only is charity important but so is the manner in which it is given. Grandma Clara also fostered two brothers from an orphanage and raised them as if they were her own. To Herman they were uncles, and he was very fond of them.

In the same year that Herman's family moved, there was an enormous explosion and massive fire at Texas City, a refinery port town about 20 miles from Herman's home. It was one of the largest non-nuclear explosions in history. Herman saw the sky turn a smoky orange and as an inquisitive nine year old, wondered what had caused it. The memory of it stayed with him throughout his life. Apparently, a fire started on a French-registered ship that was in the port carrying a cargo of ammonium nitrate which exploded.[3] Fires spread to other ships and to nearby oil-storage tanks, causing more explosions. Hundreds of people died, including all but one member of the Texas City fire department.

Living in La Porte placed the Daly family more centrally in father Ed's selling territory and reduced his travelling time. At the time it was believed that water and swimming could help polio victims, probably because of films of President Roosevelt, another polio victim, swimming at Warm Springs Georgia. Herman spent most of his time reading, beachcombing, and teaching himself to swim from a book borrowed from the library. On a calm day he would practise and each day, alone in the bay, he would go out a little further, on his own and without supervision: 'out there in the whole damn bay all by myself.' At night he often sat alone on the end of the community pier enjoying the solitude and wondering about the future. Sometimes he would sleep overnight in an army-surplus jungle hammock with a plastic roof and mosquito net, feeling very cozy and keeping dry even when it rained. Though only 11, he had a 22 rifle and was allowed to hunt in the nearby woods. This was all part of life for a young boy in Texas in the 1940s, but Herman wasn't a stupid kid. He knew that guns were dangerous and that he could hurt himself or someone else. He became scared and decided that shooting was not really for him.

The school system in La Porte was not very good. Herman was a keen student but was not learning much from his classes so he read a lot on his own. His test results in reading ranked at the high-school level even though he was in fourth or fifth grade, but in maths he was only average. Herman was not disappointed when, after two years at the La Porte Elementary School, the family moved to Houston where Herman's father had again opened a small store called American Hardware Co., on South Main Street. The school system was better in Houston and there was a park close to his new home with tennis courts and a swimming pool. Herman enjoyed them both. He became quite proficient as an athlete and, despite having only one useful arm, made it into the school tennis team. Unfortunately, although Houston's school system was an improvement over La Porte, Herman's experience at Pershing Junior High was not good. He did well in Latin but learned little of other subjects except from what he read on his own. Later he came to resent the fact that he could have learned a lot more mathematics, but it was often taught by people who really had no feel for the subject.

From the age of 11 right through grade school and into college Herman worked in his father's store. Starting at $6.50 a day he learned to serve customers and ring up sales on the cash register. Later, at 14, his duties included driving a pick-up truck one-handed all over Houston to get orders from wholesalers and make deliveries to industrial customers. One regular stop was a small chemical company where Herman witnessed Larry, a Mexican employee, being overworked and breathing toxic vapours around the vats all day. His eyes were always blood shot and he had a cough and looked tired. Years later, reading Marx on the exploitation of labour, the memory of Larry made it all the more real to Herman.

During these early years in Herman's life Texas was still a segregated State. African Americans and, to a lesser degree, Mexican Americans, were kept separate from white Texans in public places and facilities including schools and universities, and in private business establishments such as restaurants and barber shops. This meant that Herman had very little contact with anyone who wasn't white. One important exception was in his father's hardware store where the customers were from all races and backgrounds. Herman admired his father. He thought him wise, thoughtful and perceptive and Herman was greatly influenced by him, especially in his dealings with people. Ed Daly, who left school after grade eight, was very accepting of any kind of person. His rule was to treat everyone very courteously especially in his store. He would say 'their money is just as green as anybody else's'. This may sound crass today but at the time, in segregated and racist Texas, it was very progressive. It set an example of an inclusive attitude towards people that reinforced what Herman had learned at Sunday school as a five year old. When he sang 'Jesus loves the little children, red and yellow, black and white, they are precious in His sight' the words really meant something to him, yet he could see that this was not the way of the world, at least not in Texas. Herman simply didn't understand why he couldn't play with black children. This taught him the life lesson that the way things are was not necessarily the way they should be – a lesson he did not learn at school.

Despite his own modest education, or perhaps because of it, Herman's father encouraged his children to get a good education. At the same time, he had a rather critical view of educated people. He loved to tell the story of Joe, a PhD student that he employed for a summer in his hardware store. He would say, 'I told him one day, Joe, it looks like it's gonna rain. I want you to go upstairs into the storage room there and close the window.' Joe goes up and comes back down. Later on, Ed Daly goes up to the storage room. It's still raining and there was an open window and the rain was coming in. He comes down and says, 'Joe I thought I told you to close the window. Why didn't you close it?' 'Oh, I did.' 'Well, I went up there and it was raining in.' 'Oh, but you told me to close the window and there were two that were open, and I closed one of them.' Herman's father replied, 'Joe, how did you know which one I meant?' He loved to tell that story if only to show that higher education is not everything.

Herman's childhood beyond the age of 8 had been dominated by attempts to regain the use of his arm. His mother was always eager to help. She tried anything that would come along – like chiropractic, hot baths and electric shock treatment to stimulate the muscles – but all to no avail. The adolescent Herman thought it was taking up too much time and energy and he rebelled against it. He also realized at a very young age that some things really are impossible and that it was wrong, even dangerous, to assume otherwise. There really was no cure for his medical condition. The highly touted folk remedies had all failed to bring life back to his withered arm and he had received the best that conventional medicine could offer. Herman learned from this that when you come up against an impossibility it is best to recognize it and switch your energy to good things that are still possible. This line of thinking may well have influenced Herman years later when thinking about economic growth. He realized that unlimited economic growth on a finite planet was impossible and that a steady-state economy was not only feasible but, under the right conditions, could be highly beneficial.

In the summer of 1953, before starting high school, Herman talked to his father about amputation. To Herman, his left arm was a deadweight and he wanted to be rid of it. His father was very understanding but realized that this was not a decision to be made in haste. He said to Herman, 'Think about it for two months and then we'll talk again. If you still feel that way, we'll go talk to a doctor'. The two months passed, and Herman became even more determined to be free of his useless arm. His father, true to his word, took Herman to see a doctor who said, 'Yes, there's no way this arm will come back … it is atrophied to skin and bone and motor nerves are dead. We can do this we can do that and we can do the other.' So, without any sort of ethical oversight of the kind might happen today, Herman had the arm removed. At the time, as Herman remembered, it was popular for people to make their own decisions.

He recuperated at his grandmother's little 'bay house' where he could watch the tankers and freighters in Galveston Bay through the new telescope that he had bought from his earnings in Ed's shop. And he prepared, with a feeling of excitement, for high school and an improved sense of self-worth in the firm belief that the God who had made him still loved him.

After his convalescence, Herman made a 60-year personal 'truce' with polio. He felt freer than ever after the operation and was not significantly limited by the consequences of the disease until well into his retirement. Polio does not return, but it leaves a calling card of post-polio syndrome ('late sequelae', as the physicians now elegantly call it), consisting of fatigue, weakness and progressive loss of some muscle functions. But 60 years is a generous truce, and the 'late sequelae', not officially recognized until the 1980s, are usually gradual as they have been for Herman.

High school

In 1953 Herman joined Lamar High School in Houston. The school was named after Mirabeau B. Lamar, a leader in the Texas Revolution and the second President of the Republic of Texas before it joined the Union in 1846 as the 28th state.

Herman thoroughly enjoyed his three years at Lamar High. He had matured physically and emotionally, and after the removal of his arm had come to a clearer understanding of his situation. He felt like a better person, well prepared to enjoy life more and that included school, where he did well in English and science but not as well in mathematics, which remained his most challenging subject. He continued to read widely and began developing into a good writer, so much so that he thought for a time that he might earn a living as a novelist.

The social scene at Lamar High School also suited Herman. He liked his teachers and classmates, and he was eager to make friends. His faith in a loving God was strong, studying was fun, and girls liked him. For Herman, the formula for happiness as a teenager was good health, good parents and a good high school.

Today, the existential crisis that deeply disturbs so many young people is the climate crisis. In the 1950s and 60s it was the Cold War, which threatened – at any moment – to become a hot war fought with nuclear weapons causing mass annihilation. It was a period of geo-political tension between the Soviet Union-led Eastern Bloc and the US-led Western Bloc. The Cold War came to end, or perhaps an interregnum, in 1991 with the collapse of the Soviet Union. In the United States, the threat of Communism was a cause of great concern, real and imagined. Starting in Junior high and continuing in high school, children were shown propaganda movies with a red octopus enveloping the world and strangling everything. Even then Herman thought it was over the top. Maybe it will happen, maybe it won't, but he could tell propaganda when he saw it. More serious and more disturbing to Herman than these movies were the drills for a nuclear attack. In schools across the United States children had to practise diving under their desks and covering the back of their necks, as if that would help. Like all the kids, Herman knew it was silly. He had seen what had happened to Hiroshima and Nagasaki on the newsreels, but the drills, pathetic as they were, were something to do. Fortunately, they were never put to the test.

On a lighter note, Herman spent many hours listening to the radio. Radios in those days were not the small portable transistor radios that were invented in the 1950s, becoming widespread in the 1960s. They were large pieces of furniture placed prominently in the living rooms of families that could afford them and they gathered round to listen. The radio stimulated the imagination, more so than TV which was still in its infancy. With radio, the listener has to supply the visual images to accompany the words. A special favourite of Herman's childhood was *The Lone Ranger*. He often ate his dinner listening to the radio, where the masked cowboy on his horse Silver and his trusty friend Tonto were having yet another thrilling adventure. Herman liked listening to scary mysteries that kept him awake at night. In his later teens Herman used the radio to listen to all kinds of music. He started with country music, which was very popular in Texas in the 1950s, and as he got older graduated to classical music and developed a particular liking for music by Beethoven, Mozart and Schubert. He also liked Cajun music and, as a sign of things to come, developed a particular affection for the wonderful rhythms of music from Brazil.

Herman continued to work in his father's hardware store throughout High School and college. From his earnings he bought an old Ford and used it to drive friends to and from school for five dollars a week. He also became quite the handyman. There was a little shed behind the store where he threaded pipe and pipe fittings and learned other skills. But it was the variety of people who came to the store that impressed him most. There was the maintenance man, a Polish Jew from Temple Beth Israel, who Herman enjoyed talking to and there was another maintenance man from the Houston Art Museum, who was interesting. Herman made several friends among the Mexicans and Central American customers. All of this helped the teenage Herman broaden his view of the world beyond the boundaries of Texas and Texan values.

Mexico

At 18, the summer of his high-school graduation, Herman and his best friend Bobby drove his father's pickup truck all the way from Houston down through the mountains to Acapulco, Mexico, a journey of 2,000 kilometres. They stayed in small hotels or motels along the way. In Acapulco Herman realized a childhood dream denied him by his father of going deep sea fishing. With considerable help from the Mexican captain he caught a large sailfish and had his picture taken with it. The experience did not satisfy Herman and he never wanted to repeat it. His father had been right. A sailfish is too beautiful a creature to kill.

The trip through Mexico was a big adventure for two 18-year olds who had seen very little of the world. This was the first time they had gone out on their own, very far from home into a different country where they didn't really speak the language. Not that Herman went totally unprepared. Typical of him, when he wanted to learn something, he would look for a book. 'The source of all wisdom is to bury yourself in a book and read it.' That was his theory. To prepare for Mexico he found a Spanish grammar book and during lunch hour every day when he was working for his father in the summer, he would sit in a truck in the parking lot and work through a chapter. To learn the pronunciation, he relied on 'living language' records and was able to say simple things in Spanish.

One of the people Herman befriended in his father's store was a Mexican named Arnoldo. He was the black sheep of his family. Arnaldo hardly spoke English so he and Herman got together and helped each other learn the other's language. When Herman told Arnaldo they were going to Mexico, he said, 'Here's a letter, take this to my mother who lives in Tampico'. Herman discovered that Arnaldo, who was scraping a living in the United States as a maintenance man, came from a family that owned the Ford distributorship in Tampico and was pretty well off. This surprised Herman, who knew Arnaldo's difficult circumstances in Houston, circumstances that only got worse when later his Mexican wife left him after getting what she really wanted from the marriage – a green card.

The trip to Acapulco opened Herman's eyes to the larger world as only travel can do. From Texas, he had seen the borderlands where the USA and Mexico were joined. They seemed to combine the worst of both cultures. 'But as you got down into Mexico and Mexico City that just blew my mind. Wow, this is a whole different civilization, a whole different culture and it deserves respect, more then it got in Texas.' Herman was very taken with Mexico City finding it to be a huge modern metropolis that seemed to work well. It was quite a revelation to him. Even more impressive were the pre-Columbian ruins in Teotihuacan with its pyramids and broad avenue. Teotihuacan had been the largest and most populated urban centre in Mesoamerica almost 1000 years before the Aztecs.

Such a trip today – by two guys barely turned 18 driving a truck to Acapulco and back – would be a lot more dangerous than it was then, if it were even possible. For Herman, it was a life-changing experience. For one thing, it piqued his interest in the Spanish language so after he returned, he took some formal Spanish lessons and became more proficient. This paid off handsomely a few years later when he met and married Marcia, a beautiful young woman from Brazil. In Brazil they speak Portuguese, not Spanish, but the languages are close enough so that Herman had a head start in learning Portuguese and communicating with Marcia.

College

By the time Herman graduated from high school he realized that he would not be able to make a living as a carpenter or a plumber, or a prize fighter or some other athletic hero. He would have to do something more intellectual. He had been impressed by the medical doctors that he had met and thought that he could be a physician. Herman liked the idea of doing something to help people and he enjoyed science so medicine looked like a good fit. He carried that idea with him as he set off to the Rice Institute for the Advancement of Literature, Science and Art in 1956 to start college.

The Rice Institute, which became Rice University in 1960, was established in 1912. Its origins are the stuff of crime fiction. William Marsh Rice was a successful businessman from Massachusetts who made a fortune in real estate, railroads and cotton trading in Texas. He decided to leave most of his wealth to a new educational Institute in Houston to be established after his death. The Institute, which would be named after him, would have no tuition fees, would be 'a competitive institution of the highest grade', and would only admit white students (The JBH Foundation 1996).

On the morning of 23 September 1900 Rice was found dead by his valet Charles F. Jones. At 84 years of age Rice was presumed to have died in his sleep. Soon after, an alert bank teller noticed a spelling error in a check for a large amount made out to Rice's New York City lawyer, Albert T. Patrick. Patrick claimed that Rice had changed his will to leave most of his fortune to him and not to a new educational institute. An investigation by the District Attorney of New York revealed that the valet had given Rice chloroform while he slept, and that the will was a fake. Lawyer Patrick and valet Jones were arrested. Jones escaped prosecution by cooperating with the district attorney and testifying against Patrick, who was found guilty of conspiring to steal Rice's fortune and of murder. Justice was served and Rice's considerable estate was used to establish the Rice Institute, just as William Rice had wished.

Sixty years later, when Herman entered the Rice Institute, it still had no tuition fees and there were no black students at Rice. Some years after that, both the no tuition fee and whites only conditions in the will and in the University's charter were rescinded. Segregation, which Herman opposed, became illegal and since the University needed more money, tuition fees were introduced. When Herman was at Rice it was not a very welcoming place for women either. In his first-year chemistry course, Herman had a female lab partner who was very good at chemistry, a subject that she loved. Herman asked her: 'Why don't you major in it?' She replied, 'Well they won't let me.' The University actively discouraged women taking science on the grounds that it was physically too hard for a woman to be a scientist. In those days, especially in the South, women were expected to become teachers or nurses, and not much else if they were to work outside the home.

On the basis of his high score on the Rice entrance exam Herman gained admittance to an augmented academic programme where he was exempted from first year English. He took advantage of this opportunity by enrolling in a freshman course on philosophy, science and maths. It was unusual at the time for an American college to provide tutorials but Herman was fortunate. He had a weekly tutorial with an English professor named Connor, who gave assignments for discussion. This was how Herman first read writers such as Joseph Conrad. In this style of teaching long practised in Oxford and Cambridge Universities in the UK, Professor Connor would pose philosophical questions to which Herman was expected to respond, defend his answer or sometimes be forced to back off from an untenable position. It was a kind of educational experience quite distinct from attending lectures and which cannot be replicated online.

In the summer after his freshman year, Herman had another opportunity to travel outside the USA, this time to Europe to participate in a World Council of Churches Youth Work Camp in northern Italy where he helped build a community centre together with Christian students from many nations. He spent part of the summer with his friend John, also at Rice, riding the trains in Europe and sometimes hitchhiking. It was a formative experience for both of them. They saw bombsites left over from World War II and, even though it was only 12 years after the war ended, when the USA and Germany had been enemies, the friendly treatment they received from young Germans made a lasting impression.

It was in his second year at Rice that Herman took his first course in economics. Rice had hired Edgar Edwards from Princeton to chair the Department of Economics and he was followed by two more economists from Princeton, Dwight Brothers and Gaston Rimlinger. Brothers' course introduced Herman to economics through a survey of the history of economic thought. On his own initiative, Herman supplemented the assigned text by reading Robert Heilbroner's highly readable *The Worldly Philosophers* (Heilbroner 1953), the second best-selling economics textbook of all time. He also read books by other leading economists.

Among those who influenced Herman and who, to this day like Heilbroner, are very well worth reading were John Kenneth Galbraith and Karl Polanyi. In Galbraith's *The Affluent Society* (Galbraith 1958) Herman was struck by the contrast that Galbraith made between private affluence and public squalor in the USA. Galbraith also coined the term 'conventional wisdom' to describe commonly held views which often go unquestioned, something that Herman, with his penchant for questioning economic orthodoxy, would not be guilty of in his own work. In Polanyi's seminal account in *The Great Transformation* (Polanyi 1944) of the development of capitalism he explained how land and labour were commodified, meaning they became bought and sold much like anything else in a capitalist economy. In doing so it changed the relationships among people, often to their detriment. These writers gave Herman a different, more critical perspective on economics compared with what he was exposed to in the popular textbooks of the day by Paul Samuleson (*Economics*), Kenneth Boulding (*Economic Analysis*) and Charles Kindleberger (*International Economics*).

These days most students rely on the internet to search the literature and more often than not, read what they find online. In Herman's day students and professors had to visit their university library and search the library shelves, or 'stacks' as they are called, for items of interest. When Herman was looking for a copy of *The Great Transformation* in the stacks at Rice University, he mistakenly pulled out another book with the name Polanyi on the spine. This was *Personal Knowledge*, a book about science and philosophy, by Michael Polanyi, Karl's younger brother. Herman borrowed both Polanyi books. *Personal Knowledge* included a chapter on entropy, a topic about which Herman would learn much more and write about in years to come.

With his interest in philosophy stimulated by what he was reading and learning at Rice, Herman joined an informal discussion group that met in the basement of the Rice library called the 'Agora'. Someone, usually a faculty member or a graduate student, would present a paper and then it would be discussed. Herman presented a book review of *Dr Zhivago* by Boris Pasternak. In his review, and in the discussion that followed, Herman sided with Pasternak, whose writings he thought had the ring of truth in his criticism of the Russian revolution, in contrast to the more sympathetic views of others in the group.

When it came time to choose a major in third year Herman was still thinking about medicine as a career. He had done well in the biology course taught by Dr Davies who had been a student of Julian Huxley. Herman realized that surgery was out, but he could be a physician. He knew that even physicians had to study surgery so he wondered whether he would be able to get through medical school. In search of an answer, he made his way to the Baylor Medical School close to Rice to meet the Dean, who said: 'Well you took biology at Rice. Did you take the lab there?' 'Yes.' 'Did you get through the lab?' 'Yes.' 'Well, you'll be alright then.' Herman was not convinced. He was concerned that he was not good enough in math to major in science and in any case, he also liked the humanities and didn't want to give up one for the other. Economics, at least as represented in the history of economic thought, looked like a chance to combine the two. It also offered the promise of a decent living. He thought that by studying economics he would have one foot in the humanities and one foot in science. As he later discovered, that turned out not to be true. Herman has argued throughout his career that economics, especially in its dominant neoclassical version, has both feet in the air: disconnected from the Earth and disconnected from ethics.

Herman's sophomore mistake of thinking that economics was grounded in science and the humanities, gave him his mission in life: to put one foot of economics in science and the other in the humanities. He had glimpsed this in the history of economic thought but it had somehow been lost in modern economics. This is truly tragic since, as Herman was later to argue, it makes economics in its neoclassical version seriously inadequate for addressing connections between the economy and the environment, and for answering questions of social and environmental justice. In the first decades of the twenty-first century, with problems such as climate change, the loss of biodiversity, energy supply, distribution and use, mass-migration, and social and environmental injustice to contend with, this is a most serious deficiency. As Herman would later argue, our ideas were developed when the world was essentially 'empty', in the sense that humanity was not yet a force of global change. Now the world is full of people and our artifacts, far from unequally shared of course which only complicates the task of adjusting our ideas and lives to these sharply changed circumstances. Fortunately, in his economics for a full world, Herman has given us many ideas and insights that can help make the kind of transformations in thinking and acting that are so crucial to humanity as we proceed further into the twenty-first century.

Notes

1 To write Chapters 1 and 2 on Herman's life, I have relied heavily on interviews I conducted with him on 13–15 November 2018, his own unpublished Memoire written in August 2019, an interview he gave Benjamin Kunkel (Kunkel 2018), and various emails and Zoom calls. Much of the text has been edited from these sources and is in Herman's own words.

2 In 1957 the Evangelical and Reformed Church merged with the General Council of the Congregational Christian Churches to become the United Church of Christ.

3 This was the same chemical that exploded in Beirut in 2020, demolishing a considerable part of the Lebanese capital.

PHOTO 1 1943 at 5 Houston, little guy, big hat

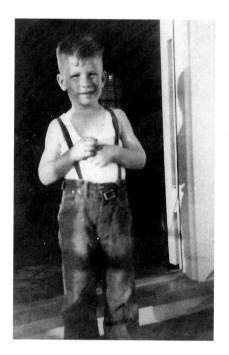

PHOTO 2 1944 at 6 Beaumont, free spirited urchin

PHOTO 3 1956 at 18 Acapulco, Herman and Bobby Byrne caught sailfish with help

PHOTO 4 1962 at 24 Vanderbilt University, graduate class in Economic Development, back row, second from right

PHOTO 5 1963 at 25 Montevideo, in taxi after wedding in Anglican church

PHOTO 6 1974 at 22 Baton Rouge, father and son

PHOTO 7 1980 at 42 Rio di Janeiro, on sabbatical, Karen, Terri, Marcia, Herman

PHOTO 8 1981 at 43 Baton Rouge, pensive pose for interview

PHOTO 9 1983 at 45 Brazil, teaching as Fulbright lecturer

PHOTO 10 1990 at Wye Island workshop. Herman Daly 2nd row 2nd from left. Also mentioned in the book: Garrett Hardin, 2nd row, 3rd from left, Kenneth Boulding, 2nd row 5th from left, Richard Norgaard, back row far right, Joan Martinez-Alier, 2nd row, far right, Bob Costanza, front row 3rd from right. Full list of names available on ISEE website. (Photo provided by Bob Costanza)

PHOTO 11 1993 at 55 Stockholm, recipient of the Right Livelihood Award

PHOTO 12 1993 at 55 Potomac River, with Robert and Jonmin Goodland

PHOTO 13 1997 at 39 Pensacola Beach, mother and son sharing a memory

PHOTO 14 1997 at 60 Richmond, happy parents of the bride at Karen and Chris' wedding

PHOTO 15 1999 at 61 Maryland, graduation day at University of Maryland

PHOTO 16 2010 at 72 Maryland, University of Maryland

PHOTO 17 2011 at 73 Atlanta, Thanksgiving at Terri's house

PHOTO 18 2014 at 76 Tokyo, receiving the Blue Planet Prize

PHOTO 19 2018 at 80 Midlothian, Herman and Marica

PHOTO 20 2018 at 80 Midlothian, Herman and Peter, two ecological economists and friends

2

HERMAN THE ECONOMIST

> In retrospect I see that my sophomore error has become my lifetime profes-
> sional project – namely to reestablish the foundations of economics in both
> physical science and ethics. But that developed gradually. My immediate aim
> was to work in the area of Latin American economic development.
>
> *Daly*

Unlike lawyers, accountants, medical doctors, engineers and other professionals,
there is no official body that accredits economists and there are no specific
formal requirements needed to call oneself an economist. In the 1960s it was not
uncommon for professors of economics to have a Masters' degree; some even only
had a Bachelors' degree. This has changed and the expectation today is that newly
appointed assistant professors of economics will have at least a Ph.D. and likely a
couple of years of post-doctoral research as well. Economists hired outside academia may also have reached this advanced level of education, but it is less common.
Herman's path to becoming an economist followed what is now the norm – intensive post-graduate studies leading to a Ph.D. – but there were bumps along the way,
an indication perhaps of what was to come later.

Post-graduate studies at Rice and Vanderbilt

When Herman was contemplating graduate studies in economics his parents
were supportive but left him, as they had done in earlier years, to make his own
decisions. They said, 'If you need us come see us. If something's bleeding, come
home.' Herman's trip to Mexico had opened his eyes to poverty and he thought
that economics might be useful in alleviating it. He was also aware that there was
serious poverty in Texas. At that time, he accepted without question what he was
learning in economics – that growth was the way to eliminate poverty – but he

DOI: 10.4324/9781003094746-2

did not accept the view put about by the press and conservative politicians such as Republican President Eisenhower, that in the USA everything was fine. What with the Korean War, the Cold War and the threat of nuclear attack there was plenty to worry about – and worry he did. In the background was the thought that with economic growth everybody becomes wealthy and happy and the fighting ends. In retrospect Herman realized that this was a youthful miscalculation.

At about the time Herman was thinking of post-graduate studies he discovered a weekly newspaper called *The Texas Observer*, published at the University of Texas in Austin. It was the only liberal newspaper in Texas. His reaction to reading it was, 'Wow, could this possibly be true? All these things they say about the great Governor, silver haired Alan Shivers, is he really a crook?' Reading the *Texas Observer* gave Herman a new, critical perspective on politics and politicians which stayed with him throughout his life.

Another experience at Rice caused Herman to reflect later in life on the state of modern macroeconomics. In the summer of his senior year, he had a job with the Tennessee Gas Transmission Co. in Houston as a trainee accountant in the budget department. As far as he could tell, the department's job was to take last year's budget and increase each item by 3 per cent to get next year's budget. Herman asked his boss, who was a nice fellow and welcomed suggestions, if that was basically what the budget department was doing. He said it comes pretty close. Herman wondered if the procedure ever caused problems and his boss told him that one year it did because this way of calculating revenue, at the legally fixed price per cubic foot of natural gas, required a delivery of gas that exceeded the physical capacity of the pipelines at the time. This required a recalculation of the expected revenues based on this fundamental physical constraint. Herman realized and remembered that the financial side of the company only belatedly came into contact with the engineering side. This surprised him, but in retrospect seemed eerily consistent with mainstream economic thinking and the tendency of those who subscribe to it to assume that through technological change and substitution economic growth can overcome any and all physical limits.

Herman made the decision to remain at Rice University for a year of post-graduate study in economics after completing his B.A. in 1960. He was impressed by Edgar Edwards, his macroeconomics teacher who, along with Dwight Brothers and Gaston Rimlinger, was recruited from Princeton to start Rice's new Ph.D. programme in economics. There were four students in the first graduate cohort. In addition to Herman two were mathematically proficient Dutch students from Groningen and the third a woman who had graduated from Rice some years before and eventually became president of Berry College, a private liberal arts college in Georgia.

The Keynesian revolution in economics was well under way in the 1960s and Keynesian economics was featured in the graduate course on macroeconomics given by Edwards, who had his students study Keynes's *General Theory of Employment, Interest and Money* – one of the most influential books in the history of economics – from which much can be still be learned today. After a year at Rice, Daly was accepted at Vanderbilt University in Nashville, Tennessee in a programme in Latin-American economic development. It was the first time that Herman had lived away from home. Unlike Rice University, which admitted its first African-American student in 1965, four years after Herman had left, segregation at Vanderbilt had been breached in 1953 though there had been little progress towards integration by the time Herman arrived there in 1961. This soon changed as a result of the civil rights movement which had its origins in the Reconstruction era of the late nineteenth century but achieved its most important legislative gains in the 1960s.

Herman was very much in favour of ending segregation. He realized that it was a holdover from slavery, and he believed in the Golden Rule that he had heard so often in church: 'Do unto others as you would have them do unto you.' It was obvious to Herman that the Golden Rule did not permit segregation. As a freshman at Rice, he had written a paper on desegregation which earned him a high grade in philosophy. He thought that segregation was simply wrong – as did most of his college friends, though to varying degrees. Herman recalled his experience working in his father's store where all customers were welcome and he did not feel threatened by desegregation. Indeed, he welcomed it. He also saw and experienced the reality that bonds of individual friendship and respect frequently grew up between people across unjust racial barriers, like green shoots in rocky soil.

But desegregation was not easily won. One of Herman's friends at Rice, a math major named Ramon Mirreles from San Antonio, ended up in Los Angeles in the mid-1960s after working in New Mexico. Herman had lost touch with Ramon but a mutual friend from Rice reconnected them. Herman learned that one evening Ramon happened to be in Watts, the neighbourhood where there were race riots in August 1965. Ramon was Latino, not even white, but for some reason he was on the wrong side of something and got beaten up badly by a gang during the riots and it damaged his brain so badly that he had to give up mathematics. Herman talked with Ramon by phone sometimes. His language was not good, but he could make himself understood. Fortunately, he'd married a very wonderful woman, and she really took care of him. Herman made plans to see Ramon but they fell through at the last minute and they never saw each other again, which was unfortunate on many counts, not least because not living on campus, Herman made few close friends at Rice and Ramon had been one of them. He never forgot the tragedy that had befallen Ramon, who had shown so much promise as a mathematician.

When Herman went to Vanderbilt in 1961 to study for a doctorate in Latin American Development, he wanted to be supervised by Ray Carlson. Carlson had been Ambassador to Columbia and had worked with the Ford Foundation, but he had left Vanderbilt and Eric Baklanoff, who knew Latin America, took over Herman's supervision. At that time, perhaps less so now, a doctoral dissertation had to offer something original. It had to make a contribution to knowledge. Herman asked himself, 'What had not been done before that was still of interest regarding Latin America?' He was attracted to Uruguay, a small, relatively progressive country, considered by some to be the Switzerland of Latin America. Uruguay regarded international trade as an important part of its development policy. Herman was interested in international trade and proposed a research project looking at trade as a policy tool for Uruguay. The proposal was approved but then Baklanoff left Vanderbilt so Herman was once again left without a supervisor and there was no other senior professor with a knowledge of Latin America. This presented a problem for the department, as well as for Herman, which was resolved, fortuitously as it turned out, when Nicolas Georgescu-Roegen, the most outstanding economist at Vanderbilt at the time, was assigned by default to be Herman's supervisor.

Georgescu-Roegen was a Romanian mathematician and statistician who studied at universities in Bucharest, the Sorbonne and London, before going to the USA in 1934 to study with Joseph Schumpeter at Harvard. There he immersed himself in economics. Later, he returned to Romania but in 1948 came back to the USA to escape Communism. Georgescu-Roegen was given a permanent position as a professor in economics at Vanderbilt University, where he remained until his retirement in 1976.

Georgescu-Roegen – the first encounter

Herman knew of Georgescu-Roegen's reputation as an important theorist. At Vanderbilt Herman enjoyed Georgescu-Roegen's course on economic theory. Herman's mathematics was not strong, so it was fortunate for him, at least at the time, that Georgescu-Roegen did not emphasize higher mathematics in the course. Herman also took an excellent course in statistics from Georgescu-Roegen and, like other students in the department, was rather in awe of the man so when the Chairman of the Department of Economics said to Herman, 'Well, you know Georgescu's willing to take you,' Herman replied, 'Well I think I'm capable of writing an acceptable dissertation but I don't know if I'm capable of satisfying Georgescu, so I'm reluctant to do that.' The chairman sighed and said, 'I understand what you're saying but really there's nobody else.'

By this time, Herman was well into his topic, 'The Trade Control System of Uruguay as a Policy of Development'. He had developed an economic model for examining the effects of international trade on development in Uruguay and presented it to the Department.[1] Georgescu-Roegen, whom Herman considered brilliant, a genius even, had made a few comments which were supportive, so he thought the prospect of having him as a supervisor was not so bad. But supervision of a Ph.D. dissertation takes time and Georgescu-Roegen was busy with his own work, including his magnum opus to be published a few years later: *The Entropy Law and the Economic Process* (Georgescu-Roegen 1971). At one point, Herman waited in vain for six months for comments from Georgescu-Roegen on a draft of his dissertation. Meanwhile, he obtained a teaching position at Louisiana State University before his Ph.D. was finished. Finally, after three years, having moved to Baton Rouge, he received a letter from Georgescu-Roegen which read: 'Well Herman the news of your dissertation is not good. I think what you have here is basically something like three master's theses, but it doesn't add up to a doctor's thesis. So, I'm sorry to write you this letter and tell you this.' But he said nothing beyond that, no details of what was deficient and no suggestions for how to fix the problems alluded to but not described.

Herman was upset, angry and deeply disappointed. He wrote to the Chairman of the Department of Economics at Vanderbilt and explained the situation, saying it was unacceptable and he was not going to continue with Georgescu-Roegen. It was clear that nobody at Vanderbilt would stand up to Georgescu-Roegen and approve a dissertation that the big man, intellectually speaking, had rejected. The chairman, James McKie, understood the situation and resolved it unconventionally. He assigned a new supervisor, Rudolph Blitz, and set up a committee of the whole department for Herman's dissertation. Before completing his dissertation, Herman wrote five papers based on what he had written so far, four of which were published in very respectable journals (Daly 1965, 1966, 1967a, 1967b). Such a publication record by a doctoral student was quite unprecedented at Vanderbilt so nearly six years after arriving at Vanderbilt, the new examination committee passed Herman's dissertation. It was not going to go against the judgement of the journals.

Although dissertation approval was rocky Herman's earlier general qualifying written exams to be a Ph.D. candidate had gone very well. The last step in that process was an oral exam, usually a tense occasion. Herman was worried that day, but not about the oral exam, because it took place on the same day that the Cuban missile crisis reached its height. The USA had established a naval blockade to prevent ships from the Soviet Union from reaching Cuba where they were setting up missile sites. Herman thought he would pass the exam but probably would then be blown up along with everything else. As many feared at the time and now know for certain, it came very close.

After Herman completed his doctorate, Georgescu-Roegen, never an easy man, distanced himself from Herman. Later the two scholars reconciled, helped perhaps by the fact that Herman was now at Louisiana State University, some way from Georgescu-Roegen who remained at Vanderbilt. Georgescu-Roegen continued working on entropy and economics. Herman read and learned from him, writing papers citing him and supporting his path-breaking ideas on entropy and economics, first set out by Georgescu-Roegen in the long introduction to *Analytical Economics* (1966) and elaborated in *The Entropy Law and the Economic Process* (1971). Later, between 1973 and 1976, they served together on the Committee on Mineral Resources and Environment of the National Academy of Sciences. Herman was a strong supporter of what Georgescu-Roegen was saying about the dependency of economic growth on depleting sources of materials and energy, and the implications for the future of that dependency.

The Festschrift to mark Georgescu-Roegen's retirement in 1976 was attended by Professors Paul Samuelson, John Hicks, Kenneth Boulding and other economic luminaries. The papers they presented in Georgescu-Roegen's honour were published in a book. Georgescu-Roegen gave a copy to Herman with the telling inscription: 'To Herman Daly, my only follower whom I recognize with pride. With my warmest thoughts. Nicholas Georgescu-Roegen, October 27, 1976.' This was the high point of their relationship, but it was not to last. They had a dispute over the viability of a steady-state economy (see Chapter 9), which was quite civil followed later by a most unfortunate misunderstanding, quarrel even, over Herman's activities at the World Bank which we describe later in this chapter. With the passage of time the tensions between Herman and his much-admired teacher lessened but did not dissipate entirely, only to sour again later in Herman's career.

Marriage to Marcia

Soon after Herman went to Vanderbilt University, Herman met Marcia Demascano at a party in Nashville in 1963. Marcia, a charming, interesting and vibrant Brazilian, was a scholarship student at Scarritt Peabody College, a Methodist school in Nashville. At the party Herman listened to Marcia singing Brazilian songs, accompanying herself on the guitar, and he was smitten. He invited Marcia out for the best fried chicken in town and she gladly accepted but time was not on their side. Marcia had only a week to go before she was due to return to Brazil. They made the most of the one remaining week, seeing each every day, but Marica left as planned and with Nashville being so far from Rio, Herman thought it was a might have been and they would never see each other again. They wrote a few times and that was that, until Herman arrived unannounced at Marcia's home in Rio six months later.

Herman was travelling to Uruguay for his dissertation research. On his way down he stopped off at Rio to see a friend from graduate school. Since he had Marcia's address, he decided to take a chance just to show up at Colegio Bennett, a Methodist school where Marcia was teaching and living. She was amazed to see him. They rekindled their relationship but this time it was Herman who had to leave, to go to Uruguay for his dissertation research. Once there he realized that he could just as easily write up his research in Brazil as in Uruguay, so he returned to Rio and stayed in a hotel close to Marcia's college. During the day he would read while Marcia was working. Then they would spend the evenings together and very quickly fell in love. Herman was very attentive to Marcia who found him honest and genuinely interested in her. She was amazed to meet someone so mature though a few years her junior. Marcia soon realized that she had found the love of her life. Herman felt the same and after just five weeks they decided to get married.

Herman and Marcia wanted to have the ceremony in Rio in the presence of Marcia's family, but Brazilian regulations meant it would take too long so they decided to marry in an Anglican church in Montevideo where the laws were more flexible. Even in Uruguay, with its more relaxed approach to marriage, couples still had to 'post the banns' which meant waiting three weeks before they could be married. Herman went to Uruguay ahead of Marcia to make the arrangements. He had very little money, so he asked the Justice of the Peace, 'Can I sign for her now as well as for me so that when she gets here, we can get married immediately and I won't have to pay for her to stay in a hotel for all this time?' This was Herman the economist speaking. In those days sleeping together was something you only did after marriage so separate rooms were de rigueur. The Justice of the Peace replied, 'Well I don't know. That would sure save you a lot of money wouldn't it?' Herman said, 'yes', so then the Justice of the Peace repeated, 'that would sure save you …' the nickel dropped, and Herman thought to himself, 'Gee I've got pay this guy something. How much should I give him? I don't know.' Herman hadn't bribed a public official before and didn't know how it was done, but he really needed the man's help so he gave him 100 pesos. The Justice of the Peace took the money, put it in a drawer and said, 'Okay here you go. Sign here and sign her name underneath and try to make it look different.' Herman did just that. When it was time to go, the man said, 'By the way what are you doing here? Why are you in Uruguay?' So, Herman explained that he was writing a dissertation on controlling trade to help promote economic growth in Uruguay. The Justice of the Peace said, 'Oh that's interesting.' As Herman got up to leave, he heard the man say, 'Oh wait a minute, you forgot your change', and taking pity on a poor student gave Herman back 50 pesos.

Herman and Marcia married and stayed in Montevideo for another six months before returning to the USA. A few months later, back in Nashville, they had their first child, a daughter Theresa (Terri) Maria. This was a little sooner than Herman had anticipated, but like a typical new father, he thought the arrival of his baby daughter was wonderful. The burden of parenthood was assumed mostly by Marcia. This was part custom and part because of Herman's responsibilities at Vanderbilt, though he helped out as best he could. While writing his dissertation he worked as an Instructor in the Department of Economics. This brought in a very modest income on which they had to live but being young and in love made it bearable. More troubling to Herman than the lack of money was his inexperience as a teacher and the responsibilities of parenthood.

Parenting, then as now, is something people learn mainly from their own upbringing. Herman considered himself fortunate to have had good parents and he tried to follow their example, but only up to a point. From a very early age, Herman had been given considerable freedom to roam, much more so than had Marcia as a girl growing up in Rio – more dangerous than Beaumont and Houston. Between the two of them they found a balance when it came to raising Terri and later her younger sister Karen who was born in 1968. Herman tried to follow his parents' example and let his daughters make their own decisions, but times had changed, life was more difficult, and parenting norms were different, so he and Marcia played a more active role in guiding their children than his own parents had with him.

Louisiana State University

Newly married with little money, Herman realized that he ought to find a full-time position. Fortunately, in the early 1960s the academic job market was pretty good. Herman had two job interviews. One was at Lake Forest College near Chicago. Lake Forest was a very wealthy area and the small college catered to it. Coming from the warm, sunny south, Herman was not impressed by the snowfall that greeted him on the first day of spring in Lake Forest. He was also not impressed by the way the professors talked about one another. Every time he would leave one interview and go to somebody else, they would start gossiping about the person he had previously talked to. His interview at Louisiana State University in Baton Rouge, not too far from his parents in Houston, went much better. Herman preferred the southern climate over frigid Chicago, and he could relate much better to the people. After the interview, the Department chairman Bernie Sliger who later became President of Florida State University said, 'I think we can offer you …' 'I'll take it, I'll take it', said Herman, not waiting for Sliger to end the sentence. He had already made up his mind to take just about anything they offered. As it turned out, the offer wasn't too bad so Herman was well satisfied.

In 1964 when Herman began teaching courses on economics at Louisiana State University, the textbook he used was selected for him and other junior members of faculty by a senior professor. The choice of textbook largely determined the general approach to the subject that Herman was expected to convey to the students. He had to cover micro and macro principles in one semester in a course particularly designed for engineers and other nonbusiness students. Only later did he start bringing his own ideas into the classroom. Herman created a course called 'The Economics of Population and the Environment' and based it around the 1000+ page book *Ecoscience: Population, resources, environment* (Ehrlich, Ehrlich and Holdren 1970, 1972, 1977).

Religion had been important in both Herman and Marcia's childhoods, and so it was for their own children. Herman and Marcia joined the Methodist Church in Baton Rouge, Louisiana. Their children grew up in that church, regularly attending services and celebrating the festivals. The experience of parenting also changed Herman's views about being a parent. He came to appreciate what he had been given as a child and how difficult it is to be a good parent. Herman and Marcia made the long drive to Houston quite often to visit Herman's parents. They also discovered, somewhat belatedly, what a great place the Florida panhandle is. They started going to Pensacola every summer for a week of vacation and they bought a little motorboat to ride around in the bayous which gave Herman, if not Marcia, enormous pleasure.

In 1967–68 Herman had the opportunity to teach economics at the University of Ceará in northeast Brazil, funded by the Ford Foundation. This suited Herman and Marcia very well. Herman's task was to prepare a few students from a very poor part of Brazil for study abroad which appealed to Herman's sense of social justice. This was during the dictatorship in Brazil, which brought its own risks especially to people with 'radical' views such as one of Herman's students, 'a good and decent kid', who 'disappeared'.

In the 1960s northeastern Brazil was experiencing rapid population growth. Women in the lower class were having eight or nine children on average while in the much richer, but much smaller, upper class, the average fertility rate was about four children. Herman wondered how average incomes in the region were going to be raised, especially incomes of the poorest, under such conditions. He had been interested in the ideas of Thomas Malthus for some time. Malthus argued that human populations tended to grow faster than food production, making it very difficult – if not impossible – to sustain general living standards above subsistence levels for any length of time (Malthus 1798). A more sophisticated analysis, which earned W. Arthur Lewis the Nobel Memorial Prize in Economic Science in 1979, explained growth in a developing economy in terms two sectors: an agricultural sector with surplus unproductive labour; and a manufacturing sector where higher wages are paid and profits are reinvested in capital equipment expanding output in a self-reinforcing process (Lewis 1955). By this means surplus agricultural labour gradually moves into the expanding manufacturing sector. When he learned about this 'dual-sector model' it gave Herman a more optimistic perspective on economic development, though he continued to think that the population concerns of Malthus were relevant and should not be overlooked. He made this point on several occasions in later life, including in a friendly critique of Pope Francis' *Laudato Si'* (see Chapter 12).

A two-month strike at the University of Ceará against the Brazilian military dictatorship provided Herman with time to read everything he could find on development and demography. He also had time to read John Stuart Mill on the stationary state (Mill 1848) and Rachel Carson's *Silent Spring* (Carson 1962). Both books made a big impression on him. During this time, he wrote a paper on the population question in northeastern Brazil and the significance of high fertility for economic development. He noted the ideological conflict about this question between Marxists and the nationalists on the one hand and the militarists and the Catholic Church on the other. Herman realized that it was politically meddlesome and risky for a foreigner to get involved in population, a taboo subject, and he tried to be diplomatic in writing about it. When he presented his paper at the Fundação Getúlio Vargas in Rio (the Brazilian national research institute), it was favourably received to Herman's immense relief and surprise. The paper was published in the *Revista Brazilera de Economia* (1968b) and later, with revisions, in *Economic Development and Cultural Change* (1970a). This sequence of events is an early example of Herman's willingness, even compulsion, to examine controversial issues and to express unpopular points of view if that was where the evidence and his analysis led. Much later, when asked to explain this aspect of his character, Herman said,

> I sort of feel like it's a little bit Platonic. There's this realm of ideas, and ideas just reach down and grab me. Therefore, you have a duty to deal with them. As an academic or thinker, in this era of post-truth, I'm still interested in truth. I believe that some things are true, and some things are not. If you get a glimpse or you're grabbed by something which you think is true, well let's follow it, who knows where it will take you?

In addition to publishing his paper on population in northeastern Brazil Herman also published 'Economics as a Life Science' in the *Journal of Political Economy* (Daly

1968a). In this paper he explored some implications of thinking of economics as a life science more akin to biology and ecology than to physics (see Chapter 4).

After his year in Brazil Herman returned to Louisiana State University in 1968, where he was granted tenure and promoted to Associate Professor. Herman and Marcia bought a modest three-bedroom house which they enjoyed living in for the next twenty years with two sabbatical semesters in Brazil. Herman, in addition to a standard course on Economic Principles taught a course on Population, Resources and Environment and another on Comparative Economic Systems which included a large section on Marxist economics. What particularly resonated with him from Marx was the emphasis on social classes and the exploitation of one class by another. Then, in 1969–70, he went to Yale University as a research associate.

Driving almost 2,500 kilometers to Yale with Marcia and their two children, they passed through Nashville where Herman gave a summer course at Vanderbilt. This gave Herman the opportunity to see Georgescu-Roegen again and they re-established a more cordial relationship. Herman and Marcia visited Georgescu-Roegen at his home where they were received by him and his wife Otelia. In the house Herman noticed a photograph on the wall of a young man who resembled his former teacher and asked, 'Is that a picture of you when you were younger?' Georgescu-Roegen replied, 'No, that's my brother. My brother was what you call a likable fellow. I am not a likeable fellow'. Years later, economist John Gowdy who had also studied under Georgescu-Roegen, said in agreement with Herman, 'Well you know Georgescu, he really wanted to be a nice guy, but he just couldn't.' That summed it up for Herman. Georgescu-Roegen could be very kind, charming and generous, and just a total snake. He could be unreasonably critical. Something would set him off and he hated to climb down. As one example, Georgescu-Roegen had been lifelong friends with Wassily Leontief, another outstanding economist. Something Leontief said upset Georgescu who cut off relations with him. He was very mercurial, brilliant and difficult, an opinion that was widely shared among those he knew him well.

The year at Yale, with no teaching responsibilities, gave Herman the time to write. He wrote a paper on Marx and Malthus (Daly 1971a), and also an essay on the classical stationary state economy arguing for its renewed relevance (Daly 1971b). Herman gave a talk to the Yale chapter of Zero Population Growth in which he presented ideas that had begun to form the intellectual framework that would define his academic work for the rest of his life. He had come to understand the global economy as a sub-system of the biosphere subject to the first and second laws of thermodynamics. In brief, the first law teaches that the quantities of matter and energy remain unchanged in economic processes and the second law teaches that their quality declines. This perspective provided Herman with a rationale for limits on resource use and wastes, termed 'throughput' limits by Herman. It also gave him a reason for thinking about limits to growth. John Harte, a physicist who was present at the Yale talk, liked what Herman had said and asked if he was willing to write it up as a contribution to a book he was editing. Herman agreed. At the same time, the *Yale Alumni Magazine* asked him if they could publish a shorter version of his talk. The *Yale Alumni Magazine* is circulated to a lot of well-placed Yale graduates, one of whom happened to be Harrison Salisbury of the *New York Times*. Salisbury read Herman's article, and asked Herman to write an op-ed on the same topic which was printed under the title 'The Canary Has Fallen Silent' (Daly 1970b).

The op-ed was read by Dennis Meadows and colleagues, who at the time were working on *The Limits to Growth* (Meadows et al. 1972). Dennis thought, 'Oh wow, this is very close to the kind of thing we're doing.' Herman and Dennis agreed to meet at the upcoming AAAS meetings in New York to explore their mutual interests and concerns. Herman is mentioned twice in *The Limits to Growth*, including a lengthy quote in which he explains that the important issue for an economy not growing in physical terms will be distribution not production (ibid. p. 179). Dennis was the leader and spokesperson for the team that wrote *The Limits to Growth*, and a very effective speaker. Herman had a closer intellectual affinity with Donella Meadows whom he met at several conferences. Over the years he also visited them at their home in New Hampshire. Herman found them both kind and generous and he defended them against the onslaught of criticism of *The Limits to Growth*, especially from economists. Donella died at the age of 59. At a Quaker memorial service for her on Earth Day, 2001, Herman read a prayer he had written expressing his own, deeply felt religious beliefs. One sentence which could equally describe himself said: 'She was a faithful steward of Your Creation, an honest worker in Your vineyard' (Daly 2001).

The publication of *The Limits to Growth* in 1972 coincided with another seminal event: the UN Conference on the Human Environment. For the first time in history the leaders of the world's countries gathered to discuss the state of the global environment. The short book *Only One Earth: the Care and Maintenance of a Small Planet* (Ward and Dubos 1972) set out the main issues. Herman read the book which reinforced his interest in the relation between the economy and the environment and took him into the debate about nuclear power.

Not long before the Stockholm Conference Herman received a phone call from Dean Abrahamson, a health physicist interested in nuclear power and radiation, at the University of Minnesota. Abrahamson was part of a group mainly of physicists at the Hubert Humphrey School of Public Affairs whose concerns also included the costs of nuclear power, something that also interested Herman. They invited him to join their group, which he did. He had already been involved in 'a ragtag group', trying without success to oppose the River Bend nuclear power plant at Baton Rouge. Herman was one of only four Louisiana State University faculty opposed to the plant. He became more and more convinced that nuclear power was a bad investment that would never have gone ahead but for the regulated prices that guaranteed the utilities a 'fair' rate of return on their investment, and the Price Anderson Act that limited their liability if anything went wrong (Daly 1973c). He voiced similar concerns some years later at the World Bank when he reviewed a proposal for a nuclear power plant to be built in Brazil that did not go down well.

In the 1970s and 80s while still at Louisiana State University Herman worked with the World Council of Churches and their ecumenical work towards a just, equitable and sustainable society (Daly 1980d). This happened through friendship with philosopher-theologian John Cobb Jr with whom he later co-authored *For the Common Good* (1989, 1994) which has the revealing subtitle 'redirecting the economy toward the community, and the environment, and a sustainable future'. In 1972, while teaching at Louisiana State University, Herman received an invitation from John Cobb to contribute to a conference, 'Alternatives to Catastrophe', that he was organizing at Claremont in California. John was the leading figure in process theology (Alfred North Whitehead's philosophy plus theology). At the suggestion of his son, Clifford, John had read Paul Ehrlich's book, *The Population Bomb* (Ehrlich 1968) and that convinced him of the importance of ecology and its relation to economics. Clifford then gave him an article by Herman which led to an invitation to the conference. Herman was glad to see an emerging connection between Christianity and ecological problems and so he gladly accepted.

Later Herman invited John to contribute an essay to his book *Towards a Steady-State Economy* (Daly 1973a), which he did. They met at other conferences, including some of the World Council of Churches. Herman was very impressed with John's knowledge and intellect. His long-run project seemed to be to unify major disciplines under the umbrella of Whitehead's philosophy. His book *The Liberation of Life* (Birch and Cobb 1981) with Australian biologist Charles Birch, which Herman reviewed, was a step in that direction. Cobb's next step was to do something similar with economics, so he needed a partner. He did not have many sympathetic economists to choose from, so he chose Herman based on their past experience, growing friendship, and, according to Herman, the lack of an alternative – and Herman accepted.

In the 1970s Herman gave speeches at World Council of Churches conferences in Zurich, Boston, Rio and New York. He particularly liked their combination of environmental sustainability with economic justice as informed by Christianity, but in time he became disillusioned with the leadership of the Council. What disturbed him most was their attitude towards developing countries. Herman believed that it was disingenuous not to discuss population policy and family. He reacted negatively to what he viewed as an extreme Marxist position on these issues, seeing them as a subterfuge by rich countries to hold back economic growth among the poor. Any dissent from this position was resisted by the leadership of the World Council of Churches even if they did not agree with it. Herman thought this condescending, that open debate was more honest and respectful, and that 'nonsense is nonsense no matter who says it'. He was also suspicious of the amount of money that some 'third world representatives had'; too much, he thought, to truly represent poor countries. As happened a number of times in his career, Herman was not prepared to compromise his intellectual integrity and refrain from expressing disagreement simply because it made some people uncomfortable.

For the Common Good took Daly and Cobb several years to write. They wrote separate chapters and then read and rewrote each other's chapters. The book was published in 1989 not long after Daly had begun working at the World Bank. It is usual for authors to include a disclaimer relieving all those individuals and institutions that had helped the authors from any responsibility for the book's contents. Since *For the Common Good* was not the kind of book that might be expected from an employee of the Bank, to avoid any suggestion that it represented the views of the bank they added the following words: 'In Herman Daly's case this disclaimer should be explicitly extended to his current employer, the World Bank …' (ibid., p. viii). The senior management of the Bank might not have appreciated *For the Common Good*, or this farewell lecture, but others certainly did.[2] In 1992 it received the Grawemeyer Award for Ideas Improving World Order and has been frequently cited since.

Economics for a full world

It was not until the 1980s that Herman began calling explicitly for a different kind of economics suitable for a full world. In a paper written in 1983 but not published until 1987 he wrote, 'the economics of an empty world with hungry people is different from the economics of a full world, even when many do not yet have the full stomachs, full houses, and full garages of the "advanced" minority' (Daly 1987, p. 324). Over the years he developed the theme of the transition from an empty to a full world in a number of his writings, with the exposition of economics for a full world receiving its most complete statement in his textbook on ecological economics with Josh Farley (Chapter 7, Daly and Farley 2004, 2011), an article in *Scientific American* (Daly 2005) and an Essay for the Great Transition Initiative (Daly 2015a). Economics for a full world was the theme of his acceptance speech when he was awarded the Blue Planet Prize in 2014.

Herman's explicit call for a full-world economics was new but the ideas on which it was based had been in gestation for a number of years. In the 1970s Herman wrote some his most influential books and papers that pointed in this direction. His edited book, *Toward a Steady-State Economy* (Daly 1973a) was a collection of papers written by influential contributors including Georgescu-Roegen on entropy and economics, Paul Ehrlich and John Holdren on population growth, Kenneth Boulding on the economics of spaceship Earth, Garret Hardin on the tragedy of the commons, E.F. Schumacher on Buddhist economics, John Cobb Jr. on ecology, ethics and theology, as well as two papers by himself. This book was followed by the first of two editions of *Steady-State Economics* (Daly, 1977a, 1991a), the first comprehensive treatment of a steady-state (non-growing) economy bringing together and building on the ideas that had been percolating in Herman's mind for a decade or more. Then, in 1980, Daly published a revised version of his 1973 collection under the title *Economics, Ecology, Ethics: Essays Toward a Steady-State Economy* (Daly 1980a), followed in 1993 by a third edition, this time entitled *Valuing the Earth: Economics, Ecology, Ethics* (Daly and Townsend 1993). From one edition to the next the selection of essays changed somewhat but the basic thrust of the books remained the same, laying the foundations for an economics for a full world.

At his university Daly was rewarded with promotion in 1973 to the rank of Full Professor and in 1976 he was the recipient of Louisiana State University's Distinguished Research Master Award, which the Department of Economics chose not to celebrate. Despite his achievements – or perhaps because of them – and his growing reputation in the USA and abroad, Herman began to feel increasingly distant from others in the Department of Economics though he retained a few friends there. In contrast, his relations with the larger university were always good. He had many friends and colleagues in the departments of Geography, Chemistry and Coastal Studies. Bob Costanza was in the Coastal Studies Department and their friendship led from a special edition of *Ecological Modelling* devoted to ecological economics which they co-edited and introduced (Costanza and Daly 1987), to a partnership in founding the International Society for Ecological Economics and its journal *Ecological Economics* in 1989. In 1991 they co-authored the opening chapter in a collection of papers edited by Costanza that cast ecological economics as 'the science and management of sustainability.' Their chapter had a lasting influence on the goals, agenda and policy recommendations for ecological economics, which was the chapter's title (Costanza and Daly 1991). Costanza had already left for the University of Maryland and Herman would soon leave for the World Bank. The geographic proximity in the Washington DC area rekindled their partnership. After six years at the World Bank Herman went to the School of Public Policy at the University of Maryland so they were again colleagues at the same University until Robert moved to Oregon and then Australia.

While still at Louisiana State University, students sought him out, especially those disenchanted with mainstream neoclassical economics, but they were not well-received in the Department of Economics. This was grossly unjust to them and to Daly. Herman's first Ph.D. student was a former Jesuit with five Masters' degrees. Although he finally got his Ph.D. the arrogant and antagonistic attitude taken towards him taken by some members of the Department distressed Herman. Other good students of his included one with an undergraduate degree from Yale and another from Pacific Lutheran University with a double major in geology and biology and very high scores on the Graduate Record Exam. They were both highly capable students but were considered unsound by others in the Department because of what they wanted to study and why: the very topics Herman was interested in and for which he was becoming known. The first of the two students finally got through the doctoral programme but not with Herman as his supervisor. He had to bow out for the Department to pass him. The second one failed his exams and was not permitted to write a dissertation. Herman was furious because he thought the exam papers were certainly good enough to pass. The disappointed and disillusioned student left for the NGO world, where he has been doing excellent work ever since.

By the early 1980s it was increasingly apparent to Herman that the Department of Economics at Louisiana State University had shifted from the time that he first came there, when it was far more open and friendly, to becoming much more doctrinaire. It had been taken over by a right wing, strongly neoclassical group certain that they were right. At the same time, he had become more of a critic, and had begun to describe himself as an ecological economist. Naively, Herman thought that his criticism of neoclassical growth economics would be seen as an intellectual matter, based on logic and science, and that disagreements would be debated with due respect of other points of view. Instead, he was treated as a heretic who had rejected the fundamental faith in growth as the universal solution to economic, social and environmental problems, the very essence of progress. And as a heretic, if tenure protected him from being cast out, at least his influence on students should be kept to a minimum.

BOX 2.1

I deeply appreciated Herman's very fair and principled approach to inquiry and deliberation. It is so unfortunate that the mainstream dismissed his thinking so readily and that economics departments were uninterested in having him as a colleague – or to expose students to his work. Impressive that he persisted all the years.

Tom Green

The World Bank

From about 1983 Herman began looking for a way to leave academia. His writings had made him popular in Europe and other countries but not in his own country. Sometimes, when he was away giving talks or attending meeting, Marcia received phone calls deliberately intended to scare her and the children. On one occasion they left town to be with friends. Their religious convictions also gave them strength, but it was not a happy situation. In 1988 Herman was recruited by ecologist Robert Goodland to join the Environment Department at the World Bank as senior economist, an appointment that *Science Magazine* described as a surprising but hopeful move by the Bank. Goodland approached Herman about joining the Bank at the suggestion of Ragnar Overbeek, who had read Herman's 1973 book, *Toward a Steady-State Economy*, and thought that he could help them deal with the economists at the Bank.

BOX 2.2

Herman's time at the World Bank was very influential in framing the sustainable development case. He and Bob Goodland seemed to be having a good time disturbing feathers at the Bank.

Gus Speth

Herman's main responsibility in the Environment Department was to review projects that were being proposed for World Bank funding, and to identify their likely environmental consequences and suggest improvements. He greatly enjoyed working with Goodland, with whom he developed a close working relationship and strong friendship. Herman and Robert had much in common. They had both spent time in Brazil and they shared a willingness to speak up when the occasion called for it, despite the consequences. One difference was that Robert was a vegetarian. Herman agreed with him that reducing the human consumption of meat would significantly reduce the impact of humans on the biosphere but having grown up in Texas he found it too difficult to forego meat and so had to live feeling the guilt that comes when one's ideas and behaviour do not coincide.

Herman and Robert collaborated on a number of papers and activities. Most significant perhaps was their association with economist Salah El Serafy. They developed a series of conferences on 'Greening the UN System of National Accounts', which contributed to the development of the UN's System of Integrated Environmental and Economic Accounts. Daly wrote a brief paper about measuring sustainable social net product (Daly 1989) and the three collaborated in writing and editing *Environmentally Sustainable Economic Development: Building on Brundtland* (Goodland et al. 1991), which was inspired by their disappointment with the 1992 World Development Report, which they had seen in draft before it was published.

Georgescu-Roegen – the second encounter

By this time Herman and Georgescu-Roegen had enjoyed a long period of good relations. Herman was very pleased about this, but it was not to last. In addition to his work at the World Bank, Herman was very active in the newly formed International Society of Ecological Economics.

BOX 2.3

He is a good man, 'una buena persona'. He knows how to write clearly and also with extreme clarity. He has no sense of self-importance. He sees himself as part of a large school, ecological economics, that he helped to found. He never cared to be president, or in any way to be in command.

Joan Martinez-Alier

This included a volunteer position as one of several associate editors of the academic journal, *Ecological Economics* which, unbeknownst to Herman, published an article highly critical of Georgescu-Roegen, who went ballistic. He held Herman personally responsible because he was an associate editor of the journal and ignored Herman's remonstrations that he knew nothing about the article before it was published. Herman wrote to Georgescu-Roegen, saying he agreed the article was terrible and that he would use all his influence to make sure a reply by Georgescu-Roegen would be published. Georgescu-Roegen did not respond and Herman assumed he would calm down and decided not to press the matter. Meanwhile, Herman encouraged Gabriel Lozada, who had contacted him concerned about the article in question. Herman encouraged Lozado to write a response which he did and acknowledged Herman for encouraging him to do so (Lozado 1991). Despite Daly's best efforts to assuage Georgescu-Roegen's unjustified anger, Georgescu-Roegen remained bitter about this episode for the rest of his life (Letter from Georgescu-Roegen to Kozo Mayumi 1992 in Bonaiuti 2011, p. 232).

In 1992 something very strange happened. Herman's boss at the World Bank at the time was Mohan Munasinghe, whom Herman considered a good boss, given the constraints he faced from above. One day Munasinghe called Herman and said, 'Herman, have you been to Romania?' Herman said, 'no'. Munasinghe persisted, 'What you been doing in Romania?' Herman replied, 'I've never been to Romania.' Munasinghe said, 'Well there are reports in the Romanian press criticizing the World Bank and citing you.' Herman said, 'Well, I have no idea.' 'You sure you didn't go to Romania?' asked Munasinghe. Herman answered, 'The closest I've ever been to Romania was Hungary and I've never set foot in Romania in all my life.' 'Well, there's something really strange here,' said Munasinghe, who called Herman later to say: 'Maybe you can find out something about this by talking to Eugenio Lari at the IMF. He heard something.' Lari had been a student at Vanderbilt years before, but Herman had only met him briefly. Hoping to find an explanation for this unwelcome attention from the press, Herman called Lari and said, 'Maybe you can elucidate for me why there might be any reports about me in the Romanian press.' Lari said,

> Yes. The Romanian Embassy in DC gave an award to Georgescu. In his acceptance speech he criticized you for harming Romania by advocating sustainable development. He said you were the architect of sustainable development which was very bad for the Romanian economy. And he suggested you should be fired.

Herman reported what he heard from Lari back to Munasinghe at the World Bank, who shrugged his shoulders and said no more about it.

Herman wrote to Georgescu-Rogen, asking for some sort of explanation, but did not receive a reply and never heard from him again. Georgescu-Roegen died two years later, before the cantankerous teacher and devoted student could be reconciled yet again. In a letter Herman wrote to Mauro Bonaiuti in 1999, in his capacity as Associate Editor of the journal *Ecological Economics*, he said that Georgescu-Roegen

> could not tolerate anything less than total acceptance of his views among his friends. G-R's resentments increased with age, as did his bitterness over the fact that he did not receive the Nobel Prize, which in truth he certainly

deserved. He seemed, however, to blame this injustice on his friends for not praising him enough.

In my own work I have cited G-R more than I have cited anyone else, and more than anyone else has cited him. I have reprinted several of his key articles. I wrote a eulogy for him when he died. Later, I edited a special edition of *Ecol. Econ.* to honor his contribution. I hope others, like you, will build on his magnificent contribution. But as for his dementia, paranoia, personal resentments, and bitterness, I would prefer to let that part of him rest in peace and be forgotten.

Daly, 1999d

In the obituary essay Herman wrote for Georgescu-Roegen, he expressed similar sentiments about

the failure of the profession to give his work the recognition that it truly merited, and ... his own irascible and generally demanding personality. So great was his bitterness that he even cut relations with those who most valued his contribution. But none of that diminishes the great importance of his lifework, for which ecological economists must be especially grateful. He demanded a lot, but he gave more.

Daly, 1995d, p. 154

Herman and royalty

A much more pleasing incident happened to Herman than the unhappy conclusion to his relationship with Georgescu-Roegen, while he was working at the World Bank. It involved an invitation from Prince Charles for Herman to join a conference cruise on the Amazon. Herman had been friends with the Brazilian environmentalist José Lutzenberger. Lutzenberger was of German background and was fluent in German, Portuguese, English and Spanish. He was born and raised in Porto Alegre, in southern Brazil. Trained as an agronomist he worked for BASF, a chemical German company that sold pesticides and fertilizers all over the world. After returning to places where he had been a few years before selling pesticides, Lutzenberger saw the result of his work and said to himself, 'the net result of my economic activity is to lower the capacity of the Earth to support life. That's not a good thing to do'. He failed to convince BASF to change its ways and resigned.

Lutzenberger next worked as a landscape architect, designing natural systems for purifying water and sewage for municipalities around Porto Alegro. There was a tannery that was polluting the river and he designed a system where all the effluent from the tannery would go into a holding pond with plants which would purify the water by absorbing the contaminants. This work was more rewarding and much to his liking. Lutzenberger became very outspoken against the agrochemical industry in Brazil and developed quite a reputation as a firebrand. He became increasingly prominent as a leading environmentalist, winning 85 environmental awards over his lifetime. In 1990 the new, democratically elected Brazilian government appointed Lutzenberger to be Brazilian Secretary of Environment. This was a surprise to many people, not least Lutzenberger. It was largely thanks to him that Prince Charles, who has a strong environmental interest, hosted a conference in the mouth of the Amazon River on the Royal Yacht Britannia.

Lutzenberger wanted to invite Herman to the conference-cruise. They had met previously when Herman was teaching in Brazil and liked each other. But as Herman explained to him, 'I'm really low in the hierarchy of the World Bank. I'm not going to be able to represent the Bank, so thank you very much but it cannot be.' Lutzenberger said,

> No, no, no … Here's what we'll do. I'm not going to invite you directly. I'm going to first invite the President of the World Bank, but I'll wait until his calendar is likely to be full and then when he writes and says he can't come I'm going to say, "Oh, what a great disappointment. We understand. Could you possibly send Daly because we're interested in someone who knows Brazil."

And that's what he did, and it worked. Herman was sent by the Bank, not as an official representative but with the Bank's permission.

Herman knew that on the Britannia he would be outside his own social level and he wasn't really sure what to expect or what would be expected of him. He received notification that there would be a White tie dinner on board ship. Herman didn't know what that meant or what to wear. He assumed a tuxedo but in wintertime in the tropics is that a white tuxedo or a black tuxedo? Nobody he asked seemed to know until someone said, 'Well you should wear a white tuxedo because it's the tropics.' So, Herman made his way to Sears and Roebuck to get a white tuxedo coat and tie and all the accessories, which he packed neatly in his suitcase and off he went.

On board the Britannia there were a few discussions, but it was mainly a social occasion highlighted by the banquet. Herman gave himself plenty of time to dress for the occasion. His life experiences had not prepared him for dining with royalty. He really did not know what to expect and very much wanted to fit in. In his tiny cabin Herman, who stood over six feet tall, managed to dress himself one-handed in the unfamiliar clothing with the white jacket and all the accessories. Then he set off for the banquet. When he arrived, everybody was wearing black tuxedos … except for Herman and the servers. His awkwardness was only slightly relieved when later a few more white-jacketed men arrived. Herman overcame his embarrassment and thoroughly enjoyed the big occasion. Through Jonathan Porritt, who was to become leader of the Green Party in the UK and was close to Prince Charles, Herman was able to get a copy of one of his books into the hands of the Prince, suitably inscribed. Though he did not meet the Prince personally, he did fly back to Rio on the same private plane with Prince Charles and Princess Diana sitting up front, while he and a few other guests rode in the back.

Shortly after the conference cruise on the Britannia, Larry Summers, then chief economist and a vice-president of the World Bank, wrote a note that was leaked to the environmental community. In the note, Summers gave three 'economic' arguments for moving environmentally dirty industries to the least developed countries. He famously stated that 'under-populated countries in Africa are vastly "UNDER-polluted"' (emphasis in the original).[3] After the memo became public in February 1992, Lutzenburger wrote to Summers: 'Your reasoning is perfectly logical but totally insane ... If the World Bank keeps you as vice-president it will lose all credibility.' Lutzenburger was fired shortly after writing this letter. He died in 2002 and was buried as he wished: naked, without a coffin, close to a tree in a farm he restored in the state of Rio Grande du Sul. Summers, on the other hand, continued his highly successful career in government and academia.

As an employee of the World Bank, Herman was always very careful to make it clear that views he expressed in writing or in a speech were his own and not to be taken as those of the Bank. Nonetheless, the widening gap between the Bank's view of development with an emphasis on export-led growth, for example, and his own more favourable view of import substitution, inevitably led Herman to resign from the Bank in 1994. On his departure he invited some World Bank economists and the press to a farewell speech which is remarkable for its insights and candour. Herman began with a frank assessment of the Bank's internal management and its consequences. He wrote that 'top-down management, misguided by an unrealistic vision of development as the generalization of Northern over-consumption to the rapidly multiplying masses of the South has led to many external failures, both economic and ecological.' His advice was that the Bank should 'open up, listen up, and don't work weekends on anything you don't enjoy' (Daly 1994a). Looking externally, Herman offered four prescriptions derived from his scholarly work and experience at the Bank:

> 1) Stop counting the consumption of natural capital as income; 2) Tax labor and income less, and tax resource throughput more; 3) Maximize the productivity of natural capital in the short run and invest in increasing its supply in the long run. 4) Move away from the ideology of global economic integration by free trade, free capital mobility, and export led growth-and toward a more nationalist orientation that seeks to develop domestic production for internal markets as the first option, having recourse to international trade only when clearly much more efficient.
>
> *ibid.*

Herman's arguments against free trade – and against NAFTA in particular – were particularly unwelcome in the Bank. Whatever influence he had in the Bank was waning. It was time to return to academia. In retrospect, Herman's main influence in the Bank was that by occupying a position that within the Bank was considered environmentally extreme he had opened up a space for less "extreme" environmentalists to get a hearing as moderates by comparison. An indication that this made a difference can be seen in documents such as The World Bank Group's Environment Strategy 2012–2022 which, in its own words, 'lays out an ambitious agenda to support "green, clean, resilient" pods for developing countries, as they pursue poverty reduction and development in an increasingly fragile environment' (Awe 2012).

The University of Maryland

Herman's critique of free trade (see Chapter 13) had attracted the attention of Peter Brown, a political philosopher and environmentalist in the School of Public Policy at the University of Maryland. Herman might have preferred a position in an economics department, but none was available at the time. He was persuaded by Peter Brown to apply for a vacancy at the School of Public Policy. The School could not decide between Herman and another candidate and decided to appoint neither of them. Apparently, there was concern in some quarters that it would embarrass the University for Herman to be given a Professorship. His views were just too unorthodox. Not wanting to lose the opportunity of bringing Herman to Maryland, Brown raised some grant money and in 1994 the School hired Herman as a research scholar rather than as a tenured professor. It was the start of a period of enormous productivity for Herman. In the next two years he published numerous papers on many topics, collaborated again with John Cobb Jr. on a second edition of *For the Common Good*, and wrote a book on the economics of sustainable development under the revealing title of *Beyond Economic Growth* (Daly 1996).

As resourceful as Brown was, he was unable to persuade the School of Public Policy to hire Herman as a tenured professor despite this very creditable publication record. Then, in 1996, Herman was awarded two prestigious international prizes: the Heineken Prize in Environmental Sciences from the Royal Netherlands Academy of Arts and Sciences and the Honorary Right Livelihood Award, Stockholm from Sweden. Now it was a matter of the University avoiding embarrassment by having such a distinguished economist in its employ who was not a tenured professor. The President of the University stepped in and the appointment was made.

In his last years at the University of Maryland Herman's closest colleague in the School of Public Policy was the late economist Robert H. Nelson, author of *Economics as a Religion*, and many other articles and books elaborating that same theme. Among their colleagues Herman was the only one who was sympathetic to Robert's views. Robert had previously worked in the Bureau of Land Management, and while there was given the task of studying the growing environmental movement and defending economics against its criticisms. His starting thesis was that environmentalists' devotion to their cause was basically religious in nature, while economists' devotion to their cause was basically scientific. Of course, in the prevailing plausibility structure of the intelligentsia religious meant 'wrong' and scientific meant 'right'. At the end, however, he concluded that economists were in their attitudes at least as 'religious' in the sense of being dogmatically committed to their worldview and assumptions, as were environmentalists. Herman gladly wrote a Foreword to *God? Very Probably*, another of Nelson's books considering it 'clear, informed, and honest' (Daly 2015e).

From 1997 to 2010, Herman continued to teach and write. With others at the University of Maryland he developed a certificate programme in ecological economics following other universities in the USA and other countries. Herman's course was cross-listed as part of the University's programme in Conservation Biology which was a source of excellent students. But Conservation Biology was viewed by the Biology Department in much the same way as Ecological Economics was viewed by the Department of Economics. It too was eventually cancelled as being insufficiently scientific and too values oriented. Sadly, this modest attempt to provide students at Maryland training in ecological economics did not find favour with either the Economics or Biology Departments and the certificate programme failed to survive the departure from Maryland of Herman and Bob Costanza.

Until his retirement in 2010 and afterwards Herman received more prizes for his work, which is a tribute to the substantial influence he has had and continues to have around the world, if not yet in the classrooms of economics departments and the corridors of power. That is yet to come.

BOX 2.4

I admire Herman not only as a path breaking thinker and public scholar but also as a person. He is a generous and authentic human being. We need more public scholars like him especially at a time when we seem to have forgotten how to disagree respectfully and yet with conviction.

Sabine O'Hara

Retirement

After his retirement from the University of Maryland in 2010, Herman has carried on writing and inspiring others right down to the present day. He remains religious, perhaps even more so than before. This has always been key to his strong bond with Marcia. Herman is as attentive to her as when they first met over 60 years ago. They are a loving and caring couple. Herman and Marcia moved into a comfortable retirement home in Virginia in 2015. This was a return to Herman's southern roots, and it brought Herman and Marcia close to their daughter Karen who works for the Virginia Museum of Fine Arts and lives nearby in Richmond with her husband and two children. Karen's older sister Terri, an occupational therapist in a children's hospital, lives in Atlanta with her husband and Herman and Marcia's third grandchild. Life for Herman and Marcia in the retirement home is comfortable and quiet. He had given up playing tennis at 45 when he sought treatment for bursitis. The doctor looked at him and said, 'You know you've got more important things to do with that arm than play tennis', and he was right. Herman kept up his swimming well into his seventies. Now he spends much of his leisure time reading widely, listening to lectures on his iPad and to many kinds of music – from Cajun to classical. The radio, which he learned to love all those years ago in hospital when he was recovering from polio, remains a constant companion.

In his last years at the University of Maryland, Herman had been forced to reduce his teaching load because the post-polio syndrome *late sequelae* was causing him fatigue. The polio he suffered as a child had come back to haunt him, weakening his body but not his mind which remains strong and clear. In 2014 Herman published *From Uneconomic Growth to a Steady-State Economy* (2014a). It is a collection of his papers that sums up a lifetime of hard work, careful analysis, passion, professional risk and reward, all guided by a strong ethical compass and a religious commitment from which he has never wavered. In 2019 Herman wrote, 'Absence of the true and the good leaves a large vacuum to be too easily filled by the false and the evil. I have tried within my limits to serve the good and the true. In old age I hope and pray for a rebirth of and refocus on the transcendent power of truth and goodness. I cannot yet see that on the horizon. It will not come from the dominant worldview of Scientific Materialism, and unfortunately does not seem to be coming from the damaged and declining churches or universities. Optimism is foolish, yet hope is a virtue, and despair remains a sin' (Daly 2019a).

Notes

1 The non-linear programing model, which was quite a novelty at the time, sought to maximize public revenue subject to balance of payments constraints and subsidies to industry, while taxing traditional cattle and sheep taking account of the supply and demand conditions in each market.

2 An informal Bank employee publication, satirically titled *The Whirled Bank*, reprinted Daly's farewell address, indicating that there was at least some underground support for Daly's views among Bank staff, if not the official management (Daly 1994a).

3 Summers (1991).

3

PHILOSOPHY, ETHICS AND RELIGION[1]

> The world cannot stand another decade of narrow economists who have never thought about ultimate means or the Ultimate End, who are unable to define either entropy or a sacrament, yet behave as if there were no such thing as entropy and as if nothing were sacred except economic growth.
>
> *Daly*

Daly does not claim to be a philosopher, an ethicist or a theologian. However, his broad approach to economics has taken him into each of these areas. Unlike most economists of the modern era who have kept any thoughts about such matters they may have had to themselves, Daly has articulated views on philosophy, ethics and religion, especially as they pertain to economics. It is a demanding arena in which to work with numerous players in the Western tradition, ranging from Socrates in third century BCE to Whitehead and others in the twentieth century. Daly's gift is to distil the ideas most relevant to the theory and practice of economics as he sees them, and to express these ideas in clear and digestible form. In this chapter we consider the range of Daly's views on philosophy, ethics and religion, which underpin much of his work on economics. As we shall see, some of his views are simply adopted from other writers, often with a new twist in the particular way he applies them. Other views are more controversial in their substance as well as in their application. In all cases they are the thoughtful and considered ideas of a deeply original thinker.

The Ends–Means Spectrum

A popular definition of economics originating with Lionel Robbins is 'the science which studies human behaviour as a relationship between ends and scarce means which have alternative uses' (Robbins 1932, p. 15). Daly observed that the ends and

DOI: 10.4324/9781003094746-3

means referred to by Robbins, and by every other economist who has adopted this definition of economics, are 'intermediate' in the following sense. The scarce means referred to in Robbins' definition are typically labour and manufactured capital, which themselves are produced from the materials and energy provided by nature that Daly calls the 'Ultimate' means. Similarly, the ends referred to in the Robbins' definition of economics are intermediate in that they serve the higher objective of an 'Ultimate' end, which may be defined in religious or ethical terms but, says Daly, is needed to make choices among the intermediate ends (Daly and Farley 2011, pp. 38–43).

In Daly's mind, this broader classification of means and ends conjured up an Ends–Means Spectrum between Ultimate means and the Ultimate end, with Intermediate means and ends located between them. This expansion in scope from intermediate to ultimate in two directions is fundamental to Daly's approach to economics, so it is worthwhile exploring it in some detail. Two versions of the Ends–Means Spectrum are shown in Figure 3.1. Before publishing the Spectrum, Daly shared it with his students who found the graphic useful for unifying different subject areas and for placing economics within a wider context, just as Daly intended. The Ends–Means Spectrum first appeared in print as a triangle (Daly 1973a, p. 8). The simpler, linear version, and to my mind the more effective one, came in *Steady-State Economics* where he labelled it as the End–Means Continuum (1977a, p. 19). It was repeated almost identically in Daly (1980a, p. 9), except that Daly reverted to Spectrum in Daly and Townsend (1993, p. 20).[2] Both versions of the Spectrum show at the base the Ultimate means of low-entropy matter-energy, 'the useful stuff of the world … which we can only use up and cannot create or replenish' (ibid.).[3] As Daly emphasized, it is not the materials and energy that are used up. Their quantities are conserved, governed by the first law of thermodynamics. It is the low-entropy usefulness in colloquial terms, that is used up by human activities, dictated by the second law of thermodynamics. In the case of energy this means that the capacity of energy to do work is reduced when it is used. For example, steam from a boiling kettle cannot be used to boil the same quantity of water in a second kettle and then a third and so on. The total quantity of energy is conserved; it is its quality – its ability to do work – that is diminished.

On the left-hand side of the diagrams, we see that the study of the Ultimate means is the domain of physics. The Ultimate means are fashioned into Intermediate means, which are named in the linear version of the Spectrum, on the left-hand side of Figure 3.1, as stocks of artefacts and labour power. By comparing the two representations of the Ends–Means Spectrum in Figure 3.1, we can see that Daly initially named the discipline that deals with the production of the Intermediate means from the Ultimate means as 'technology' but switched to 'technics' in the later versions of the Spectrum. Technics means essentially the same as technique: the way or style of doing things. Technology is the application of scientific principles behind the working of gadgets and appliances. Different people can have different techniques for making use of the same technology. Both technics and technology are involved in the conversion of the Ultimate means into intermediate means.

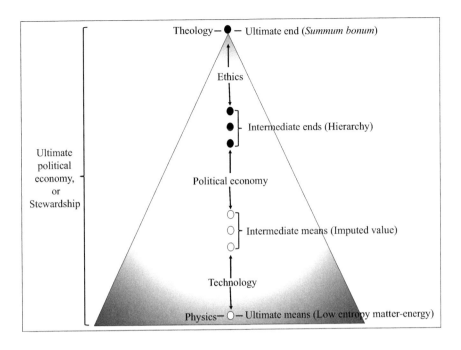

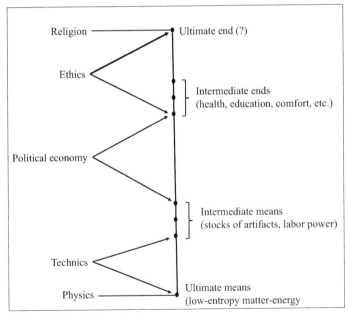

FIGURE 3.1 The Ends–Means Spectrum

Sources: Daly (1973, 1977).

Daly identified political economy as the academic discipline that studies the use of Intermediate means to produce Intermediate ends such as health, education, and comfort. It is interesting that Daly placed political economy rather than economics between Intermediate means and ends in both versions of the Spectrum in Figure 3.1. In Daly (1977a), where the Ends–Means Spectrum also appears, he used economics. Is this change significant? Political economy is usually treated by economists as an older and somewhat outdated term for economics, but that is only part of the story. Historically, political economy had a broader scope than modern mainstream economics. It placed more emphasis on economic history and the evolution of economic systems and also included considerations of political power, which today's political economists continue to do. Being aware of these differences, Daly's choice of political economy rather than economics in most versions of the Spectrum reflects the breadth of his intellectual outlook.

BOX 3.1

I regularly invoke Daly's Ends–Means spectrum when introducing my economics courses. It is important for students to know what an economy is for, and that is seldom outlined in economics courses.

Brett Dolter

'At the top of the Ends–Means Spectrum is the Ultimate End.' Daly tells us that the Ultimate end is 'that which is intrinsically good and does not derive its goodness from any instrumental relation to some higher good' (Daly and Townsend 1993, p. 20). It is similar to Aristotle's 'final cause'. This contrasts with Intermediate means derived from Ultimate means and Intermediate ends that serve the Ultimate end. Intermediate means and ends are similar to Aristotle's efficient and formal causes, which is the overwhelming focus of economics. Aristotle's 'material' cause corresponds to Daly's Ultimate means (Daly and Cobb 1994, pp. 41–42).

> Causation is both bottom–up and top–down: material cause from the bottom, and final cause from the top, as Aristotle might say. Economics, or as I prefer, 'political economy,' is in between, and serves to balance desirability (the lure of right purpose) with possibility (the constraints of finitude). We need an economics fit for purpose in a finite and entropic world.
>
> *Daly 2014b*

Figure 3.1 shows two changes in the Ultimate end between the original version of the Spectrum and the later versions. In the triangular version, after Ultimate end in parentheses is the Latin phrase *Summum bonum*. This means 'the highest good', the final ultimate end towards which all things strive, and was introduced by Aristotle. It was adopted with some reinterpretation by Christianity through the work of St.

Augustine and St. Thomas Aquinas. In later versions of the Spectrum, Daly replaces *Summum bonum* with a question mark. This change does not signify any change in the meaning that Daly gives to the Ultimate end or its role in determining priorities among the Intermediate ends, but it does make explicit his uncertainty about what in fact constitutes the Ultimate end. He writes, 'Ideally our starting point should be the Ultimate End, we can only see that end dimly and may find clues to its nature in our experience with ethical, economic, and even technical problems encountered on the way' (Daly and Townsend 1993, p. 24). Here Daly is referring to the links between Intermediate ends and the Ultimate end, which is the province of ethics as shown in Figure 3.1.

The second change between the two versions of the Spectrum shown in Figure 3.1 is the substitution of religion for theology. This seems inconsistent with Daly's stated goal of listing 'the traditional disciplines of study that correspond to each segment of the Spectrum' (ibid., p. 21). 'Religion is a specific system of belief and/or worship, often involving a code of ethics and philosophy whereas theology is the rational analysis of religious belief' (Aron 2011). It is quite possible that Daly's decision to substitute religion for theology in the Spectrum reflects his own personal belief that the Ultimate end, the Purpose of Life, can be found in Christianity or at least in religion. But where does that leave non-believers in an increasingly secular age? We will return to the question of the Ultimate end later in the chapter.

BOX 3.2

He is a wonderful human being. While I have been professionally less impacted by his religious/humanitarian views I think they are thoughtful, warm and wonderful.

Charles Hall

The Ends–Means Spectrum is not only an original taxonomic system of impressive scope, and it is not just a pedagogical device for teaching students trying to make sense of the world and the academic disciplines designed to understand it. It is also a framework within which the logic of Daly's critique of economics can be understood. He says it forces us to ask two questions:

> 1) What, precisely, are our ultimate means, and are they limited in ways that cannot be overcome by technology? 2) What is the nature of the Ultimate End, and is it such that, beyond a certain point, further accumulation of intermediate means (bodies and artifacts) not only fails to serve the Ultimate End but actually renders a disservice?
>
> *op cit., p. 22*

As we shall see in later chapters, Daly's answer to both questions is 'yes'.

Economics, says Daly, neglects the Ultimate end and simply assumes that ever increasing levels of consumption of goods and services improve human well-being. Because of this, Daly says that economics is too materialistic. Paradoxically, by also failing to consider where labour, and especially where capital, comes from – the Ultimate means – he says that economics is not materialistic enough. Labour and capital (in the economic sense of buildings and equipment) are made from the Ultimate means, which are provided free by nature. It is still commonplace for economists to assume that these natural sources are essentially abundant. Daly points out that focusing so much on Intermediate means and ends and so little on Ultimate means and the Ultimate end leads to mistaken choices. One in particular, the pursuit of economic growth – the ever-increasing production and consumption of goods and services – is causing irreparable and irreversible damage to the environment. Instead of asking what constitutes the good life, as philosophers, theologians and religious leaders have done for millennia, economists, with some notable exceptions, look no further than consumption. They simply assume that more consumption means a better life and since more consumption requires continual economic growth, economic growth becomes, to all intents and purposes, the Ultimate end by default.

One consequence of this exalted position of economic growth, says Daly, is the excuse it gives for not confronting inequality in the distribution of income and wealth both within and among countries. If economic growth makes everyone better off, meaning they are able to buy more commodities, which of course is not always the case, there is no need to bother with redistribution and the conflict it may generate. Even environmental pressures are dismissed as a transitory phenomenon best dealt with by even more economic growth. This is the message of the 'Environmental Kuznets Curve', which purports to show that as economies grow environmental pressures rise, but as growth continues environmental pressures peak and then decline. Unfortunately, this only applies in special cases where environmental damages are obvious and local, not when cause and effect are disputed and damages are regional or global, like climate change. Still, the Environmental Kuznets Curve provides a theoretical rationale, if one were needed, for most economists to pay little or no attention to the environmental consequences of economic growth. Such is the argument about economic growth that Daly has been making for half a century, the relevance of which becomes more obvious with each passing day.

Just as there are dangers in neglecting the relationship between Intermediate ends and the Ultimate end, there are parallel dangers in neglecting the relationship between Intermediate means and Ultimate means. This is what Daly means by economics not being materialistic enough. Most economists do not concern themselves with the sources of the raw materials used in the production and distribution of goods and services. They leave this to specialists in the economics of natural resources. Similarly, it is left to those who specialize in environmental economics to think about the problems of pollution stemming from the disposal of wastes onto land and into air and water. Add in assumptions about the ease with which substitutes can be found for anything in short supply, supported by a favourable view of technology's ability to solve any problem, and that prices will give the right signals, you end up with a world view in which the Ultimate means can be safely ignored.

BOX 3.3

Daly's greatest contribution was his vision that the economy should be governed in a way that serves moral and social ends while staying within planetary boundaries.

Richard Howarth

Ecological economics, to which Daly has contributed so much, arose in response to the narrow focus of mainstream economics on Intermediate means and ends. As we shall see in the next chapter, ecological economics views the economy as a sub-system of the biosphere, one that is totally dependent on Ultimate means for all the materials and energy used in the economy, without which there would be no commodities, no economic growth, no economy. In sharp contrast to mainstream economics, Daly attaches great significance to the conversion of Ultimate means – the materials of which the planet is made and the solar energy that falls freely upon it – into the Intermediate means made available in an economy. Following in the footsteps of his esteemed teacher Nicolas Georgescu-Roegen, Daly uses the second law of thermodynamics to show that economic activity necessarily depletes the sources of low-entropy matter-energy. He argues that to ignore this fact is a serious error with major consequences. It is one of the fundamental scarcities confronting economies and inevitably limiting economic growth. Another is the scarcity of assimilative capacity in the biosphere to absorb the unavoidable wastes from the human economy which have widespread, adverse consequences for all life on the planet. 'The absolute scarcity of ultimate means limits the possibility of growth …' (Daly 1980a, p. 10; Daly and Townsend 1993, p. 20).

Whitehead's 'lurking inconsistency'

Daly's Ends–Means Spectrum gives us an insight into his world view. Recall that when he was introduced to economics as a student, Daly mistakenly thought it was grounded in the natural sciences and the humanities. He discovered later that this was incorrect and set about trying to make it so. In approaching this monumental task, he became concerned about whether the determinism of the natural sciences could be reconciled with purpose. Without purpose the whole structure of economics collapses since it depends on people making rational choices and rational choices are choices made for a purpose. Purpose is also essential for public policy in which policy objectives are essential and policy objectives only make sense if they are derived from purpose. As Daly notes, there is an intimate connection between purpose and free will:

Determinists [i.e. people who consider free will to be an illusion] believe that there is only one possible future, rigidly determined by either atoms in motion, selfish genes, dialectical materialism, toilet training, or the puppet strings of a predestining deity. If there is only one possible future state of the world then there are no options, nothing to choose from, and therefore no need for policy'.

Daly 2014a, p. 157

Daly was not the first person to express concern about the apparent inconsistency between a scientific view of the world that is fundamentally deterministic, and purpose, although he may well be the first economist to do so. In several of his articles he quotes philosopher Alfred North Whitehead's description of the 'lurking inconsistency' in the Western worldview prevalent in the early twentieth century and possibly even more so today:

A scientific realism, based on mechanism, is conjoined with an unwavering belief in the world of men (sic) and of the higher animals as being composed of self-determining organisms. This radical inconsistency at the basis of modern thought accounts for much that is half-hearted and wavering in our civilization… It enfeebles [thought], by reason of the inconsistency lurking in the background… For instance, the enterprises produced by the individualistic energy of the European peoples presuppose physical actions directed to final causes. But the science which is employed in their development is based on a philosophy which asserts that physical causation is supreme, and which disjoins the physical cause from the final end. It is not popular to dwell on the absolute contradiction here involved.

Whitehead 1925, p. 76 quoted in Daly 2014a, p. 162

Or as biologist Charles Birch writes in his book *On Purpose* (Birch 1991) and quoted by Daly: '[Purpose] has become the central problem for contemporary thought because of the mismatch between how we think of ourselves and how we think and act in relation to the rest of the world' (op cit.).

BOX 3.4

Whitehead's thought has played no role in mainstream economic thinking. Indeed, the one major economist who has adopted it, Herman E. Daly, has been virtually ostracized from the guild of economists, at least in the United States. However, he has had a following in other circles, particularly ecologists and religious communities.

John Cobb Jr

Daly is most exercised by Whitehead's lurking inconsistency as it relates to Darwin's theory of evolution. He sees a fundamental contradiction between natural selection based on chance mutations and survival of the fittest on the one hand, and purposeful human actions based on the valuation of alternatives on the other. Daly rejects the idea that the widely shared sense of purpose among humans is just a useful illusion conferring on those that possess it a survival and reproductive advantage. And he questions, provocatively, whether this lurking inconsistency has 'lethal consequences for policy of any kind' (ibid., p. 163). Purpose, if real and not illusory, and causative, is fundamental to policies intended to effect change. If the world is governed by materialism and chance, if it is simply 'matter in motion', to use Daly's phrase, then where does Purpose with a capital P, meaning to seek the Ultimate end, come in?

To understand Daly's answer to these questions, we need an appreciation of Darwin's theory of evolution through natural selection since that is where Daly focuses his attention. At the simplest level Darwin's theory of evolution is built on four principles:

1. Variation – in all species, individuals differ in their genetic makeup, producing many variations in their physical features; individuals in a population vary from each other.
2. Inheritance – individuals pass some of their genetic material to their offspring; parents pass on their traits to their offspring.
3. Selection – some individuals have inherited character (genes) that allows them to better survive or produce more offspring. These offspring, in turn, are more likely to survive and create offspring of their own. As a result, their genes become more common in the entire population; some variants reproduce more than others.
4. Time – over time, selection results in changes in species. These changes may take days, or decades, or millions of years to occur; successful variations accumulate over time (KU Natural History Museum 2020).

Darwin used these principles to explain how life forms evolved. He realized that the simple mechanism embodying these principles 'had radical philosophical consequences … a causal theory stripped of such conventional comforts as a guarantee of progress, a principle of natural harmony, or any notion of an inherent goal or purpose' (Gould in Zimmer 2006, p. xxxv). Darwinists all accept that there is no purpose in evolution. It just happens. Humans, they say, evolved from a long series of random mutations, a few of which increased the chances of survival of the affected individuals who passed the mutations to their offspring. What happened to be successful mutations depended on the environment prevailing at the time.

Richard Dawkins, a biologist and a leading Neo-Darwinist,[4] says the absence of purpose in evolution is not just about the origin of species, which was Darwin's main interest, but of the whole universe:

> In a universe of blind forces and physical replication, some people are going to get hurt, others are going to get lucky, and you won't find any rhyme or reason in it, nor any justice. The universe we observe has precisely the properties we should expect if there is, at bottom, no design, no purpose, no evil and no good, nothing but blind, pitiless indifference … DNA neither cares nor knows. DNA just is. And we dance to its tune.
>
> *Dawkins 1995 p. 155, quoted in Midgely 2009*

Philosopher Mary Midgely says that 'this is Dawkinism, not Darwinism, but it should be looked at seriously' (ibid.), which is what Daly does and finds so unacceptable.

In terms of the Ends–Means Spectrum, Daly asks if there is no place for purpose in evolution; if science is fundamentally about mechanism, then where does the Ultimate end come from? Can there be an Ultimate end that is consistent with a deterministic view of evolution? What does it mean to have purposeful individuals if there is no Purpose in life? Daly's answer to these questions can be summarized as follows: Darwinian evolution is deterministic and purposeless, determinism and purpose are incompatible, purpose is an empirical fact supported by introspection and lived experience: therefore, determinism must be wrong or at least incomplete, the missing element being God who provides the Ultimate End.

In the language of philosophers, Daly is an 'incompatibilist', which is someone who holds that determinism and free will are incompatible. Put simply, if all actions, including ours, are determined by causes over which we have no control, then free will, which is the sense that we do have control, must be an illusion. Some incompatibilists argue that the absence of free will does not absolve people of moral responsibility for their actions. 'Hard' incompatibilists argue that determinism is incompatible with both free will and moral responsibility. Just as there are incompatibilists, there are also 'compatibilists' who, for different reasons, take the view that the kind of free will required for moral responsibility is not ruled out by determinism. One such view is based on the idea that the mind is separate from the brain and so conscious thought is not determined while everything else is.[5] This distinction between mind and brain has long been used by some philosophers and theologians to provide an entry point for God.

The following quote from Daly suggests that he is not just an incompatibilist, but a hard one:

> If the world we inhabit is an improbable ephemeral happenstance, long in coming and fated finally to dissolve, as naturalism teaches, and we as a part of it likewise are ephemeral chance happenings, then ethics is a sham. Ethics requires purposes, ordering of wants and actions relative to an objective value, final causation, teleology and a perception of ultimate value – all the things that the reigning naturalism and materialism deny. The prevailing view is that all is determined by the ancient Epicurean vision of atoms moving in their determined pathways through the void as reconstructed in modern scientific

materialism ... To refuse to recognize the devastating logical and moral consequences that result from the denial of purpose is anti-rational.

Daly, 2014a, pp. 117, 120

Faced with a choice between determinism and purpose, Daly chooses purpose. Mary Midgely's view on purposive behaviour is supportive of Daly's. She defines purposive behaviour as behaviour, 'striving to achieve goals'. She says that it is

something universal among earthly organisms... purpose, and values such as good or evil, aren't arbitrary colours painted onto the world by our vanity. They have grown up in that world and are intrinsic aspects of it – emergent natural properties, shapes that appear as soon as its inhabitants become complex enough to need them.

op cit.

Alan Holland, another philosopher, agrees that Daly is 'correct to point out that the human economy is rooted in the physical world and is bound by its constraints' (Holland 2002, p. 200) but finds Daly's account of evolution and purpose to be 'fundamentally flawed chiefly as regards the account Daly gives of neo-darwinism' (ibid., p. 203). In particular, Holland rejects the connection that Daly makes between 'Darwinism and the abandoning of criteria for making ethical judgements [and] between Darwinism and the embracing of philosophical determinism' (ibid.).

Charles Birch, a biologist with a deep interest in theology goes further than Midgley in looking for purpose within nature. In his book *On Purpose* (Birch 1991 which Daly quotes (Daly 2007, p. 245), Birch, following Whitehead, distinguishes 'internal' and 'external' relations. External relations are relations between separate things. When one thing collides with another, as when a car hits a person, that is an external relationship between the car and the person. Internal relations are part of the entity itself. Sense experiences, as when a person sees a pen for example, are an internal relation. Internal and external relations can coexist: 'Since the pen is not significantly affected [by being seen], we can say that the relation is external to the pen' (Birch 1991, p. xii).

In Birch's view, determinism applies more to external relations and not to internal relations. Free will and purpose relate to internal relations. Another way to put this and to partially reconcile free will and determinism, is to understand free will as action determined by conscious thoughts, an internal relation in Birch's sense. This is different from determinism in external relations where actions are determined by other actions. A problem with Birch's view is that he takes it to extremes. He suggests that all physical entities down to subatomic particles and up to the cosmos have internal relations of some sort. Intent on finding a place for God within this framework, Birch locates God in internal relations rather than in external relations as in the Judeo-Christian tradition. This is Birch's way of reconciling evolution, physics and cosmology with purpose.

There is a strong argument that there are many aspects of human conduct that seem to favour determinism. It makes sense for those asserting the existence of free will to acknowledge, as they do, that many human actions are involuntary or automatic. Blinking is an obvious example. Breathing is another. With so much going on in and around us it is advantageous not to have to consciously decide every one of our actions. To do so would overwhelm the brain's capacity for conscious thought. Automatic actions, the result of natural selection, leave us able to concentrate on decision-based actions (Kahneman 2011).

The role of the subconscious in determining human actions has been celebrated and studied since Freud, though much about the subconscious remains a mystery. Decisions made subconsciously are different from automatic, involuntary actions, but seem to be being made by forces over which we have no control, so this might be another part of human activity which cannot be said to be governed by free will. But when it comes to conscious, thoughtful decisions, the big decisions you might say, when people stop and think about what they are going to do, that is where those who believe in free will and those who believe in determinism part company.

Dawkins, who is far from alone,[6] goes much further than Midgely or Birch. In *Free Will* (Harris 2012) philosopher Sam Hill is equally uncompromising:

> Free will *is* an illusion. Our wills are simply not of our own making. Thoughts and intentions emerge from background causes of which we are unaware and over which we exert no conscious control. We do not have the freedom we think we have.
>
> *p. 5*

The question here is not so much about whether Daly is right or wrong about these complicated and far-reaching issues. It is about how the sharp differences in views about determinism, free will, purpose and values bear on the Ends–Means Spectrum that Daly uses to situate economics within a broader intellectual framework. Just trying to do this sets him apart from most economists. He urges us to consider the Ultimate means, that is, from where all economic resources come, and the Ultimate end, from which according to Daly values are derived for determining and assessing how these economic resources are used. He questions how we can in fact, undertake this quest if we subscribe to a view that the universe has no meaning, life has no purpose, determinism is rampant and free will is an illusion.

BOX 3.5

The spiritual core of his being is something that resonates strongly with me. But as humanity tends to marginalize all academic contributions when the spiritual aspect is hung out in front, it is not something I believe has been emphasized.

Stuart Scott

Given his characterization of science and evolution it is not surprising that Daly looks to religion to provide the Ultimate end from which purposes and values follow. His religious upbringing taught him about the Ultimate end from a Christian perspective. When he studied science, philosophy and history he found nothing better and when he immersed himself in economics, he found something not as good: purpose was reduced to preferences and tastes (Daly 2014a, p. 121). 'In denying objective value [i.e. the Ultimate end] we no longer have anything to appeal to in an effort to persuade. It is just my subjective preferences versus yours…' (ibid., p. 117).

Daly believes that the Ultimate end lies beyond what science has provided and can hope to provide, as does the Neo-Darwinist Stephen Jay Gould (1999). Richard Feynman, the brilliant American physicist and teacher, thought so as well.

Feynman's unanswered question

We do not know if Richard Feynman read Whitehead on the lurking inconsistency between a science based on mechanism and purposeful organisms. What we do know is that he was troubled by a similar issue: 'The great accumulation of understanding as to how the physical world behaves only convinces one that this behaviour has a kind of meaninglessness about it. The sciences do not directly teach good or bad' (Feynman 2005, quoted in Daly 2007a, p. 228). It is not a big leap from meaninglessness to purposelessness, so we come back to whether purpose is compatible with a deterministic science in which each effect has a cause, and each cause is an effect. Daly's answer is clearly 'no'. Feynman's concern was rather different. He thought that science had created doubt about the existence of God and that this had weakened the conviction that what one is doing has meaning. Feynman admitted that he did not know 'the answer to the problem of maintaining the real value of religion as a source of strength and courage to most men (sic) while at the same time not requiring an absolute faith in the metaphysical system' (ibid.).

Feynman posed this as a question in 1963 in a lecture published posthumously in 2005. Daly, writing about Feynman's unanswered question, is more pointed:

> Today the question is sharper and more obvious – how to maintain inspiration to provide the strength to do good in the face of, not a slight doubt as to the existence of God, but in the face of aggressive assertions by the high intelligentsia that the very idea of God is an infantile superstition.

Daly criticizes those who take this position, naming several eminent scientists and philosophers for failing to provide adequate answers to Feynman's question about the source of inspiration, Daly's Ultimate end, arising from his observation that 'the more we understand the behavior of nature, the more meaningless it seems, and consequently the harder it is to find inspiration to serve any goal at all' (ibid., p. 229).

Feynman's view that the sciences do not directly teach good or bad is reminiscent of Hume's distinction between positive statements of what is and normative statements of what ought to be and his argument that what ought to be cannot logically be derived from what is. Science deals with what is. What ought to be requires a distinction between good and bad, right and wrong, better and worse, which are all matters of ethics, and science is silent on these matters. This view is shared by Feynman and Daly and many others. In contrast to science, religion speaks to both is and ought questions. The challenge to religion from science has primarily focused on what religion has said, and sometimes still says, about what is with regard to the physical universe and evolution. Since religion has frequently been found to be mistaken about what is, there has been a loss of conviction in what it says about what ought to be, i.e., about purpose and values, even though the one does not follow logically from the other.

All this speaks to the Ultimate end that sits atop Daly's Ends–Means Spectrum. If Feynman is right, the Ultimate end cannot come from science. Science has nothing to say about Purpose in the grand sense of the term. This includes not saying that the Universe is purposeless as some scientists do, or at least, scientists should not suggest that such a statement is a scientific one. And if confidence in what religion says about the Ultimate end has been undermined because of errors in what it says about what is, then where else are we to look for the Ultimate end, if indeed one exists? Daly's assessment of the alternatives is not encouraging. He disparages the idea of a 'moral compass', something inside us that helps us distinguish between right and wrong.

> Calling for a moral compass within a … cosmology of scientific materialism, which considers the cosmos an absurd accident, and life within it to be no more than another accident ultimately reducible to matter in motion … is as absurd as calling for a magnetic compass in a world in which you proclaim that there is no such thing as magnetic north.
>
> *ibid., p. 230*

We should not be surprised that Feynman could not find Purpose in science. His science was physics and he is most famous for his work on quantum electrodynamics. Had he worked on cosmology, perhaps he might have found inspiration contemplating the universe as a whole rather than in sub-atomic particles. But science is more than physics and on the face of it, other branches such as biology (the study of life) hold out the possibility of finding Purpose (or at least purpose). Based on our own personal experience, we consider ourselves as purposeful and we have no problem assuming that others are purposeful as well. 'If I, the part of the universe I know best, experience freedom and purpose, then freedom and purpose are at least not absent from the part of the universe consisting of me' (ibid., p. 235). It only requires a small step to recognize purpose in other animals, though we might disagree about how far to take this. The point is that we are very familiar with purpose, with a goal towards which we direct our actions. This experience prompts us to ask whether there is Purpose in life itself. Is there an Ultimate end to which our lives should be directed, and our choices assessed? If there is, how do we determine what it is? Daly is keenly interested in these questions and while he allows that the answers may lie in the study of life, he does not think that Darwin's theory of evolution built on natural selection provides the answer. And he is not alone. He has Darwin for company: 'The term "Natural Selection" is in some respects a bad one, as it seems to imply conscious choice; but this will be disregarded after a little familiarity' (Darwin 1868, p. 6). As we saw earlier, Dawkins the Neo-Darwinist, goes further. In *The Selfish Gene* he wrote that people are 'robot vehicles blindly programmed to preserve the selfish molecules know as genes' (Dawkins quoted in George 2017, p. 55). Whereas Darwin was writing about natural selection, Dawkins is writing about life. To say that natural selection is not purpose driven, is not the same as saying that life has no purpose. As Daly says, humans are part of life and humans have purpose and so we might say that in our own way we are evidence for the evolution of life with purpose. All that can be properly concluded from evolutionary biology, the core of which is natural selection plus genetics, is that the purpose of one's life, and the purpose of life itself, are not to be found in natural selection.[7]

Daly understands this: 'One cannot rescue neo-Darwinism from the domain of purposeless and randomness by pointing to the role of natural selection' (Daly 2014b, p. 163). At the same time Daly recognizes that Neo-Darwinism is 'a good explanation of many facts' (ibid., p. 161). Natural selection provides 'factual insights into how the world works'. He just does not accept any more than Darwin did that the meaning of life, life's purpose if you will, can be found in natural selection: 'Selection may sound purposeful, but in the accepted theory of natural selection chance dominates' (ibid., p. 163).

Daly could have stopped there and simply said that if we are to find purpose in life we should look elsewhere, but he did not. The Neo-Darwinists would not have argued. They too see no purpose in evolution driven by natural selection (and other natural causes), no teleological end to which evolution is progressing. So why did Daly go beyond the conclusion that natural selection is not the key to the Ultimate end? Perhaps because of the 'evangelical atheism' (Daly 2007, p. 234) of some Neo-Darwinists, Daly felt an impulse to question the theory of evolution itself despite its acceptance by the vast majority of biologists. He dismisses the 'bible literalists' and those who 'deny the considerable evidence for microevolution (natural selection for differing characteristics within a species) … 'which has been confirmed by repeated observation' (Daly 2007, p. 231). He is more sympathetic to those who question 'macroevolution… an extrapolation of the same mechanism (random mutation and natural selection) to explain the development of all species from a presumed single ancestor over a very long period of time' (ibid.). Daly argues that macroevolution 'cannot be directly observed or repeated in a laboratory and is an extrapolation, a conjecture. Is it a reasonable conjecture? Certainly. Is there evidence for it? Yes. Are there gaps in the evidence and logical glitches in the theory? Yes' (ibid.).

Daly does not say which evolutionary biologists think there are logical glitches. As to gaps in the evidence, the researchers who work in this area agree that, at least for past evolutionary history, there are gaps simply because of the complexity of the fossilization process, but that is not really the issue. Any good scientific theory points to where new evidence might be found to support or refute it. The modern theory of evolution has been enormously successful in doing just that, and in virtually every case, the new evidence supports the theory. New evidence has also suggested ways in which the theory can be and has been improved, the integration of natural selection with genetics being the most notable example but also genetic drift, gene expression, epigenetics, group and sexual selection.

> We know how, over 3.8 billion years, evolution's blind, brutish and aimless methods filled a once-barren planet with the rich diversity of plants, animals, fungi and microbes that surround us. We understand how simple processes can produce astonishingly complex structures, from wings and eyes to biological computers and solar panels.
>
> *George 2017, p. xi*

It would be going too far to say that Daly believes in Intelligent Design, though he does write with some sympathy about criticisms made by creationists of Darwinian explanations of complex organs such as the eye and 'molecular machines at the subcellular level'. He is also sympathetic to their arguments that while natural selection can explain 'microevolution' like the different beaks on Darwin's famous finches, it cannot explain the emergence of entirely new species (see Daly 2007, pp. 231–232). These issues are much discussed by evolutionary biologists and answers with supporting evidence have been given, including the evolution of complexity and new species, but Daly seems to agree with the critics that the answers are not good enough, opening himself to criticism for doing so.

The most important point in all this is that even if Daly and the critics of Neo-Darwinism more generally were to be persuaded otherwise, we would be no further ahead in finding something that no one on either side of the debate thinks possible: the derivation of the Purpose in life, an Ultimate end, from the theory of evolution. On this matter evolutionary biologists and their critics agree.

If there is an Ultimate end, where is it to be found? Daly's answer is in religion. The problem this presents is that religions are unlikely to agree on a single Ultimate end or even if there is only one. For the time being at least, those who would look to religion for the Ultimate end are in much the same position as those who discuss ethical issues independently of any specific set of religious teachings. They all have to be content with something less ambitious, a set of ethical standards against which individual actions, group behaviour and public policy can be judged, which is almost as good. No doubt our lives would be simpler if we could agree on and accept a single Ultimate end, but it is not possible now and is unlikely ever to be. This does not mean we have to give up discussing and debating ethics, values and purpose. Perhaps this is what Daly meant when he said that we should look for clues to the Ultimate end through experience, but it is doubtful that he would be content to let it rest there: 'an enduring ethic must be more than a social convention. It must have some objective, transcendental authority, regardless of whether one calls that authority 'God', or 'the Force', or whatever' (Daly 2014a, p. 170).

BOX 3.6

I want to talk about Herman Daly's Christian religion influencing his writing for a moment because that's something that has stuck in particular with me. I realize how greatly heartened I am to see that the primary forefather of ecological economics left open this important place in our transdiscipline for people with different metaphysical beliefs.

Conrad Stanley

There remains the question of whether there is indeed one Ultimate end. Economist Tim Jackson, who sides with Daly on most issues, disagrees with him on this one: 'What makes Ends ultimate is not that they are singular, but that they are goods which are not held to be the means to some other end' (Jackson 2015). Jackson favours a pluralistic approach such as proposed by Sen and Nussbaum, who emphasize capabilities for flourishing, and further developed by himself in terms of 'bounded capabilities' taking account of the material conditions of a finite planet and the size of the human population (Jackson 2017, pp. 45–46). In practice, this leaves very little difference between Jackson who thinks there are a number of ultimate ends and Daly who thinks there is one but is unable to define it precisely and that it need be sought through experience.

Daly's religious beliefs lead him into eschatology, the part of theology that deals with last things. He looks forward to

> a future New Creation not made by man but a gift of a loving and just God to put right the evils suffered in this creation … It is not a doctrine of progressive historical improvement, but of hope for another miracle beyond the miracle of life as we know it.
>
> *personal communication*

Owing to the general lack of belief in a New Creation, Daly says we are

> irresistibly tempted to try to build a 'new creation' ourselves. That seems to mean a modern Tower of Babel and the economic growth that supports it… The net destructive consequence of the current scale and growth of the economy for the present creation that sustains it, is greatly downplayed, if not totally ignored… without faith in the New Creation, what final purpose is there beyond 'eat, drink, and copulate, for tomorrow we dissipate'? We may well improve the material and social conditions in which we carry out these

activities, and make it last a bit longer, but it all finally still dissipates unless there will be a radical renewal of the basic nature of creation. And who can renew creation other than the Creator?

Daly 2014a, p. 124

Conclusion

'Humanity's ultimate economic problem is to use ultimate means wisely in service of the Ultimate End' (Daly 1980a, p. 8; Daly and Townsend, 1993 p. 20). With the Ends–Means Spectrum Daly locates economics between the Ultimate means and Ultimate end. The Spectrum provides the larger philosophical framework for all his work on economics. It also helps make explicit the scope of his critique of economics when it neglects the dependence of Intermediate means on Ultimate means and places too much significance on Intermediate ends instead of asking what higher objectives, what Ultimate end in Daly's terms, they are intended to serve. As a result, economic growth, which is the expansion of Intermediate means to serve ever increasing Intermediate ends, has become a poor substitute for the Ultimate end.

Whether or not we agree that there are one or several ultimate ends has little practical bearing on Daly's economics. The same is true of his unorthodox views on evolution which run counter to the understanding of most contemporary evolutionary biologists. Whatever motivation people have for concerning themselves with the state of the world, as the following chapters will show there is something for all in Daly's economics for a full world.

Notes

1 In the previous chapters, describing how a young boy grew up to become a rather unusual economist, it was fitting to refer to Herman by his first name. In the remaining chapters, which are about his ideas, it is more appropriate to refer to him by his surname.
2 Since Daly did not attribute any significance to the change in wording, we will use spectrum throughout the book unless quoting him or others directly.
3 'By 'matter-energy' we mean just matter and energy, but with the recognition that they are convertible according to Einstein's famous formula, $E = mc^2$ (Daly and Farley 2011 p. 38). Strictly speaking, the second law of thermodynamics applies only to energy, not matter, though 'there is no dispute about matter being subject to entropy in the sense of a natural tendency to disorder' (ibid., p. 66).
4 Neo-Darwinism refers to the integration of natural selection and modern genetics.
5 A useful introduction to free will and determinism is provided in Griffith (2013).
6 In Daly 2014b (p. 118) Daly names Richard Dawkins, E. O. Wilson, Daniel Dennett, Christopher Hitchens, Alexander Rosenberg, and in Daly 2007 (p. 229) he mentions Stephen Jay Gould and Carl Sagan.
7 Daly distinguishes between a narrow version of Neo-Darwinism that characterizes mainstream biology, and a broader version in which the underlying philosophy is reductionist materialism. In reductionist materialism matter is considered the fundamental substance in nature, including mental states and consciousness, and objects or phenomena at one level are explicable in terms of objects or phenomena at a more reduced level. Daly's critique of Neo-Darwinism is directed at the broader version (Daly 2007, p. 226 fn 5).

4

ECONOMICS AS A LIFE SCIENCE

> Biologists, in studying the circularity system, have not forgotten the digestive tract. Economists, in focusing on the circular flow of exchange value, have entirely ignored the metabolic throughput.
>
> *Daly*

The origins of neoclassical economics lie in nineteenth-century classical physics with its emphasis on laws of motion and equilibrium. Yet, as Alfred Marshall, a major contributor to neoclassical economics, observed a century ago, 'The Mecca of the economist lies in economic biology rather than in economic dynamics' (Marshall 1920, p. xii). Daly takes this quote as his starting point in his insightful and provocative paper 'Economics as a Life Science' (Daly 1968a). By drawing out the parallels between economics and ecology, and suggesting heuristic and analytical ways they can and should be integrated, Daly anticipated not just ecological economics but also the integrated assessment models which have become widely used in studies of climate change.

BOX 4.1

I was delighted to come across Herman Daly's 1968 paper, 'Economics as a Life Science' around 1972. Daly's 1968 paper hooked me; I had thought I had a lot to say about how economics might benefit from the biophysical science behind ecology… but the more I read Daly, particularly pertaining to steady-state economics… the more I realized Herman Daly had arrived there first!

William Rees

DOI: 10.4324/9781003094746-4

Daly began working on 'Economics as a Life Science' at Louisiana State University in 1965 but it was unfinished when he went to Northeast Brazil in 1967 for 18 months as a Visiting Professor of Economics at the University of Ceará. He completed the hand-written paper but had a problem finding a way to get it typed. In 1967 secretaries did the typing for professors. The secretary in the University Economics Department did not know English but was game to try. She typed on poor quality paper, with smudges and hand-written corrections which was the condition it was in when Daly sent it to the prestigious *Journal of Political Economy*. He apologized for the form but not the content, hoping that at least he would get some useful criticism. Much to his surprise the paper was accepted for publication. The main reviewer was Professor Frank Knight, a founder of the Chicago School and teacher of several winners of the Nobel Memorial Prize in Economics. Knight commented that it was an 'interesting enough paper. I do not think that it would disgrace the JPE to publish it' (Daly, personal communication). While this was not the most enthusiastic acceptance, coming from Frank Knight Daly was most encouraged.

In this chapter we take a look back at what Daly wrote in the 1960s about economics and nature, and two rather different occasions when he critiqued prominent economists for not taking sufficient account of humans as living beings. We also examine three debates in which Daly became involved in the 1980s and 90s, and which are ongoing, regarding the relevance of the laws of thermodynamics to economics. Not only will we see how far sighted he has been throughout his academic career, but also how far economists still have to go to catch up with him.

Biology and economics

Daly's interest in economics as a life science stemmed from two sources. The first was the similarities between economics and biology that are 'rooted in the fact that the ultimate subject matter of biology and economics is *one*, viz., *the life process*' (Daly 1968a, p. 392, italics in original). He made this observation the starting point of his 1968 paper, noting several authors who had drawn connections between economics and ecology and who had influenced him: natural scientists Bates, Lotka, Schroedinger, Henderson, Teller, Carson, and economists Hobson, Spengler, Galbraith, Boulding, Georgescu-Roegen and Marx.

The second source of his interest in economics as a life science was that neo-classical economics, which reigned supreme in economics in the USA in the 1960s and is still hugely influential, is based largely on nineteenth-century physics:

> economists on the whole wanted economics to become increasingly scientific, and their idea of science was based on physics rather than on evolutionary biology. That meant that economics had to focus on formulating models and finding laws governing present economic behavior rather than seeking laws governing the changes of economic systems or asking about contingent historical patterns.
>
> *Daly and Cobb, 1989, 1994, p. 30*

To fully appreciate the significance of Daly's views on economics as a life science we need to understand the influence of physics 'on the methods (but not the content!) of physics [which] has been characteristic of the whole of modern economics' (ibid., p. 28). As Philip Mirowski in particular has made clear, this influence was no accident. It was the stated objective of the founders of neoclassical economics:

> What happened after roughly 1870 was … the influx of a cohort of scientists and engineers trained specifically in physics who conceived their project to be nothing less than becoming the guarantors of the scientific character of political economy … they had uniformly become impressed with a single mathematical metaphor that they were all familiar with, that of equilibrium in a field of force. They … copied the physical mathematics literally term for term and dubbed the result mathematical economics.
>
> *Mirowski 1991, p. 147*

Figure 4.1 shows a table from Irving Fisher where he made the direct correspondence between terms taken from mechanics and their counterparts in economics. Mirowski tells us that:

> Prior to the 1930s, the physical metaphor of the vector potential field adopted by neoclassical economics was the only coherent mathematical metaphor readily available to economists … [Following 1935 there] came an unprecedented wave of trained scientists and engineers into economics.
>
> *ibid., p. 148*

who discovered that the mathematics then being used in economics was that of the late nineteenth century. They set about modernizing economics by drawing on

§ 2.			
In Mechanics.			*In Economics.*
A particle	corresponds	to	An individual.
Space	"	"	Commodity.
Force	"	"	Marg. ut. or disutility.
Work	"	"	Disutility.
Energy	"	"	Utility.

Work or Energy = force × space.
Force is a vector (directed in space).
Forces are added by vector addition.
("parallelogram of forces.")
Work and Energy are scalars.

Disut. or Ut. = marg. ut. × commod.
Marg. ut. is a vector (directed in com.)
Marg. ut. are added by vector addition.
(parallelogram of marg. ut.)
Disut. and ut. are scalars.

FIGURE 4.1 Fisher's correspondence of terms in physics and economics
Source: Fisher (1892, p. 85).

new formalisms developed during the intervening evolution of physics – the increased stress on stochastic formalisms, the improved mathematics of vector fields and phase spaces, the increased familiarity with linear algebra and constrained optimization – which could be readily superimposed upon the older structure of discourse.

ibid., p. 153

The mathematization of economics taken largely from applications in physics was accompanied by the assumption, or assertion, that the entities studied in economics were essentially equivalent to those studied in physics. This is explicit in Irving Fisher's table shown in Figure 4.1, making it clear that it is more than a coincidence that economists conceive of people as utility maximizing 'atomistic individuals'. The mathematics used to describe the behaviour of people interacting in markets is drawn directly from physics which deals with atoms interacting in space.

Less well known than the adoption of analytical tools and principles in economics from physics are the analogies between biology and economics that have influenced the development of economics. Daly mentions 'the circular flow of blood and the circular flow of money, the many parallel phenomena of specialization, exchange interdependence homeostasis and evolution…' (op cit., p. 393).

When comparing biology and economics Daly uses the distinction due to biologist Marston Bates between 'within skin' and 'outside skin' life processes. Within skin is the domain of biology. 'Most of biology concentrates on the "within skin" life process, the exception being ecology, which focuses on the "outside skin" life process' (Bates, 1960, pp. 12–13, quoted in Daly op cit., p.392). Daly explains that 'Economics is the part of ecology which studies the outside-skin life process insofar as it is dominated by *commodities* and their interrelations' (ibid.).

Figure 4.2 is the first published version of a Daly diagram that, with various modifications, would become increasingly well-known and influential over succeeding decades. Daly literally draws a comparison between the within-skin life

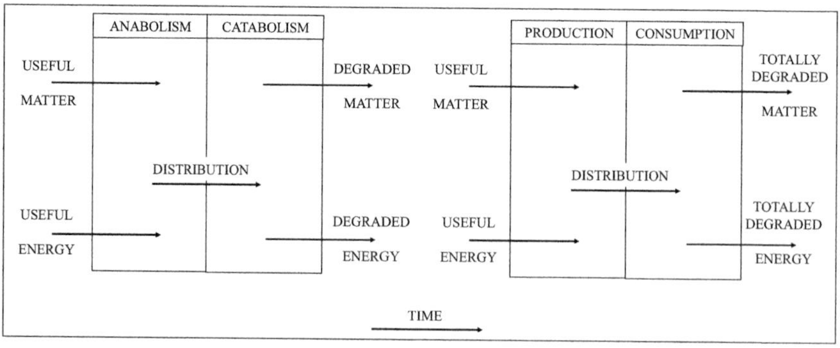

FIGURE 4.2 A comparison of biological and economic life processes
Source: Daly (1968, p. 395).

process of metabolism studied in biology and the outside-the-skin life process of economics.

On the left-hand side in Figure 4.2, under 'Metabolism', biological systems absorb useful matter and energy from the environment using anabolic processes to build molecules needed by living organisms. Catabolic processes break down the complex molecules releasing energy for the organisms to use.

On the right-hand side, under 'Production', Daly uses the same relationships as in Metabolism. The useful matter and energy absorbed by the economy is used to produce goods and services for consumption. In both processes, biological and economic, 'the only *material* output is *waste*' (ibid., p. 394). Daly compares the purposes of the metabolic and economic processes, the former being the maintenance of life, and the latter the enjoyment of [human] life.

Figure 4.3 is a more recent version of Figure 4.2. The upper image represents the world when the human economy was small in relation to the 'Ecosystem' in terms of the quantities of matter and energy it used. This represents Daly's 'empty' world. The lower image in Figure 4.3 shows a much-expanded economy using more matter and energy and possessing a much-increased stock of manmade capital. This represents Daly's 'full world'. In both images, the Economy is contained within the Ecosystem which is unchanged in size despite having to provide increased amounts of matter and energy to the economy and having to deal with increased quantities of wastes. Daly uses Economy and Ecosystem in the singular for simplicity. Within Economy some economies are much richer than others and use a disproportionate amount of matter and energy, and generate a disproportionate amount of waste. Types of ecosystems differ even more than economies, though all ecosystems, and Ecosystem writ large, depend on the fundamental principle of photosynthesis to support life and flourish.

BOX 4.2

Herman's drawing of Empty World Full World was the paradigm-changing diagram for me.

Kate Raworth

In the years between the creation of Figures 4.2 and 4.3 Daly published many versions of his empty world-full world picture, each one adding a new feature to more accurately represent how the Economy is embedded in the Ecosystem and fully dependent on it. It is an image that, in one version or another, has had a profound influence on many economists, especially those looking for a different conception of the relationship between the economy and the environment than offered by neoclassical economics.[1] During a webinar in 2020 with Daly and Kate Raworth, author of *Doughnut Economics* (Raworth 2017), Raworth made a point of saying how seeing the image opened her eyes to a new and refreshing way of

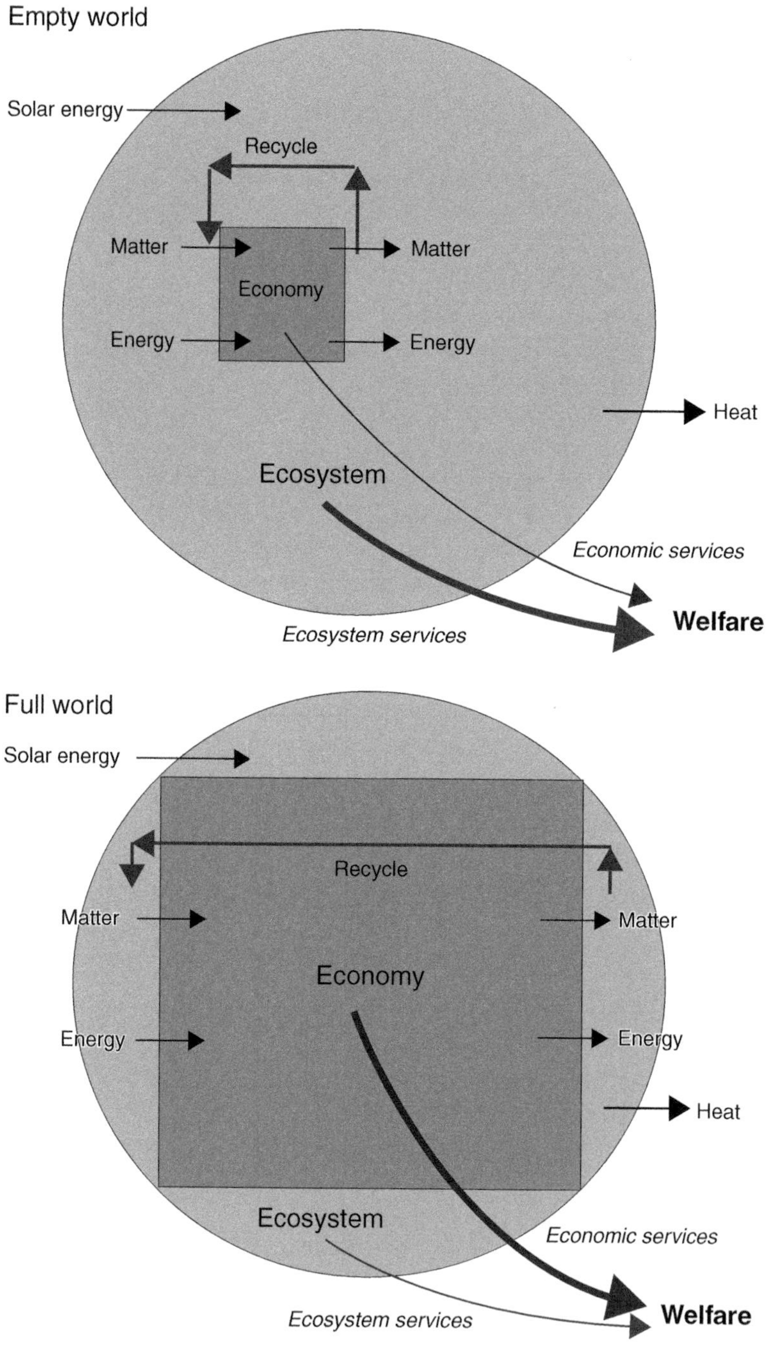

FIGURE 4.3 From empty world to full world
Source: Daly and Farley (2011, p. 18).

thinking about economics. Figure 4.3 is distinguished from some earlier versions by the inclusion of arrows representing the contribution of the services provided by the Economy to 'welfare' (i.e. human wellbeing) and the contribution to welfare of ecosystem services provided directly by the Ecosystem. The increased breadth of the Economic service arrows from the empty world to the full world signifies the increasing contribution to welfare of the Economy over time. The reduced breadth of the Ecosystem service arrows signifies the decreasing contribution to welfare of the Ecosystem. These opposite trends raise the question of whether human welfare has gained more from the growth of the Economy than it has lost from the decline in the Ecosystem's capacity to provide ecosystem services. Daly has addressed this question in numerous books and papers, never losing sight of the very different circumstances faced by people living in different places, and we return to it in later chapters.

Life, thermodynamics and economics

The most fundamental question that the life sciences has yet to answer completely is 'what is life'? One of the most famous answers to this question was provided by physicist Erwin Schrödinger in a short book published in 1944. Daly summarizes Schrödinger's answer in this way:

> Erwin Schroedinger… has described life as a system in steady-state thermo-dynamic disequilibrium which maintains its constant distance from equilibrium (death) by feeding on low entropy from its environment – that is, by exchanging high-entropy outputs for low-entropy inputs. *The same statement would hold verbatim as a physical description of the economic process.*'
>
> *Daly 1968, p. 396, italics in the original*

Schrödinger was interested in the material basis of life and he looked to the second law of thermodynamics to provide the key. In 2005, scientists Eric Schneider and Dorian Sagan explained:

> The second law in its original formulation foretold things inexorably losing their ability to do work, burning out and fading away until all states are in or near equilibrium with no energy left to run organisms or machines. *But life demonstrates an opposite tendency, of complexity increasing with time…* [Equally] The second law states… entropy (atomic or molecular randomness) will inevitably increase in any sealed system. *Yet living beings preserve and even elaborate exquisite atomic and molecular patterns over eons.*
>
> *Schneider and Sagan 2005 p. 7, emphasis added*

Schneider and Sagan called 'life's apparent defiance of the second law of thermodynamics' Schrödinger's paradox. Solving this paradox was no mean task, especially in light of what Sir Arthur Eddington, the leading British physicist of his day, said:

> The law that entropy always increases – the Second Law of Thermodynamics – holds, I think, the supreme position among the laws of Nature. If someone points out to you that your pet theory of the universe is in disagreement with Maxwell's equations – then so much the worse for Maxwell's equations. If it is found to be contradicted by observation – well, these experimentalists do bungle things sometimes. But if your theory is found to be against the second law of thermodynamics I can give you no hope; there is nothing for it but to collapse in deepest humiliation.
>
> *Eddington 1928, p. 81*

In 1943 in a speech to dignitaries, diplomats, government and church leaders, artists, socialites and students at Dublin's Trinity College, Schrödinger explained his resolution of the paradox later named after him. His answer was simple. Living organisms exist and grow by using high-quality energy obtained from the environment outside their bodies. They increase the organization of their own bodies by increasing the disorganization (i.e. entropy) of their surroundings. Taken together, the overall increase in entropy increases in accordance with the second law. As Schneider and Sagan say, 'In the case of ecosystems and the biosphere, increasing organization and evolution on Earth requires disorganization elsewhere. You don't get something for nothing' (op cit., pp. 15, 16).

It is a small step from understanding the entropic nature of living organisms and whole ecosystems to an equivalent understanding of economic activities and entire economies. Nobel prize-winning chemist Frederic Soddy had said as much in 1926 but it was only when Nicolas Georgescu-Roegen published his magnum opus *The Entropy Law and The Economic Process* in 1971 that the step was finally and forcefully taken. As one of his students, and later a friend and colleague, Daly knew the lines along which Georgescu-Roegen's was thinking well before the book was published. In his *Analytical Economics* (1966) Georgescu-Roegen had already written about economics and thermodynamics but it would be another five years before he fully elaborated his path-breaking ideas in a book that demands a great deal from the reader. Georgescu-Rogen did write more accessible versions of his main arguments and their implications for economic growth (Georgescu-Roegen 1975) but the main responsibility for explaining their relevance to economics to a wider audience was taken up with enthusiasm by Daly.

The second law of thermodynamics features prominently in much of Daly's work and it has become one of the fundamental principles of ecological economics along with the first law of thermodynamics which relates to the conservation of matter-energy.

> The first and second laws of thermodynamics should also be called the first and second laws of economics. Why? Because without them there would be

no scarcity and without scarcity, no economics... if we could create useful energy and matter as we needed it, as well as destroy waste matter and energy as it got in our way, we would have superabundant sources and sinks, no depletion, no pollution, more of everything we want without having to find a place for stuff we do not want. The first law [of the conservation of matter-energy] rules out this direct abolition of scarcity. [Even with limited amounts of matter and energy we could still abolish scarcity if] '... we could use the same matter-energy over and over again for the same purposes – perfect recycling. But the second law rules that out.

ibid.[2]

Energy always becomes less available to do work when it is used, and materials can never be recycled 100 per cent. While the validity of the first and second laws of thermodynamics is not in question, their relevance to economics certainly is. Following the lead of Georgescu-Roegen, Daly has been making the case throughout his career that economics should be consistent with these laws. He contends that the failure to do so can lead to serious misunderstandings of the nature of economies with adverse consequences for economic and environmental policy and for society.

Daly's efforts to impress upon economists that economies are sub-systems of the biosphere and the relevance of thermodynamics to economics have met with only limited success. The sub-discipline of ecological economics which he helped found, and biophysical economics with which it has much in common, have adopted these principles but they have been largely ignored by mainstream economists – although there have been exceptions. For example, Stuart Burness and colleagues questioned the relevance of thermodynamics to resource-use policies. Their position was that 'the notion of *value* may lie at the heart of interface between thermodynamic and economic concepts' (Burness et al. 1980, p. 7). Commenting on the paper Daly pointed out that, contrary to Burgess et al., Georgescu-Roegen did not claim an entropy-theory of value. 'Their misreading ... underlies their misconception of how the entropy law is relevant to economics' (Daly 1986, p. 319). Then he explained how: 'In the light of the entropy law a previously neglected aggregate constraint on the physical scale of the economy relative to the ecosystem is seen to exist' (ibid., p. 320). If, Daly continued, this constraint is to reflect the collective value of sustainability, it cannot be derived from market prices based on individual values (see Chapter 5 for more details). Daly recognised that aggregate constraints are ecological and ethical but insisted that the physical law of entropy does not go away just because people ignore it. In their subsequent response to Daly, Burness and Cummings failed to acknowledge their misreading of Georgescu-Roegen, concluding incorrectly that the difference between them 'is with ethical issues rather than concern with the allocative efficiency of markets' (Burness and Cummings 1986, p. 324). Daly, in contrast, was quite clear 'entropic considerations (depletion and pollution) are objective facts evident in the present, not value judgements, and not speculation about future millennia' (op cit., p. 321).

In the 1990s Daly became engaged in two more substantive debates on the relevance of the laws of thermodynamics to economics, the first with Jeffrey Young and the second with Robert Solow and Joseph Stiglitz.

The Young/Daly debate

In 1991 Jeffrey Young, a professor of economics at St. Lawrence University in New York state, asked 'Is the Entropy Law Relevant to the Economics of Natural Resources Scarcity?' (Young 1991). He was

> 'concerned only with the claim that entropy imposes a long-run, absolute scarcity which technological change, resource substitution, and exploration cannot reverse... particularly... to stress the extent to which such a claim requires a concept of materials entropy and the validity of such a concept'
>
> *ibid., p. 170*

Young describes the position of Georgescu-Roegen and Daly on this question as 'fundamentally flawed' (ibid., p. 169). He acknowledged the second law of thermodynamics in relation to energy but denied its relevancy to resource scarcity on Earth 'since the earth is an open system with respect to energy any inevitable entropic decay or dissipation in the earth which sets a long run physical limit on economic activity must occur for matter not for energy' (ibid.).

Following Young we will confine this discussion to the scarcity of materials.

Both Daly and Young agree that economics has a lot to say about scarcity. The difference between them is that Young thinks 'the economist's concept of *relative* scarcity' (ibid., p. 171) is sufficient. He sees no value in '*absolute*' scarcity, which he says, is 'difficult if not impossible to quantify... impossible to define'... [and] ... can only be quantified if we are talking about a single material resource and a single unchanging technology...' (ibid., p. 178). In contrast, Daly regards low entropy matter-energy as the ultimate means from which all intermediate means are produced: 'nature really does impose "an inescapable general scarcity," and it is a serious delusion to believe otherwise' (Daly 1976, p. 69) Young believes otherwise. He thinks that resources can be scarce in relation to each other but there is no general, irreducible scarcity that cannot be overcome with new, yet to be discovered technologies. As an example, he takes a resource that has been dissipated through use. A new technology becomes available for collecting the dissipated material and recycling it and in this way, Young says it is possible for entropy, in the sense of 'disorderliness or unavailability, to be decreasing even though the system is closed' (ibid., p. 178). He is not entirely clear if he means closed to materials and energy, though it seems both, because in the next sentence he says, 'It [the system] is, however, open with respect to knowledge which has increased exogenously.' In other words, technological change based on knowledge which is conserved (i.e. one person's use of knowledge does not diminish its availability to others) can compensate, Young says, for increasing entropy.

Continuing his critique of absolute scarcity based on the second law of thermodynamics Young says there is 'a classic aggregation problem in applying entropy to matter which does not exist for energy (ibid., p. 178). He contends that unlike energy for which there are unambiguous, single units of measurement, there is no common unit of measurement for assessing changes in entropy for the many different material resources. Young concludes that:

> either the entropic analogy is not relevant for material resources or the system boundaries must be drawn in very peculiar ways. In either case the entropy law is not particularly relevant to the economics of long-run resource scarcity. If it is confined to energy resources only, as it should be, the relevant system is not closed, although the continual flow of solar energy, being beyond human control, may impose a long-run, ultimate constraint on energy use. This, however, is so far into the future that it is hardly relevant to current resource allocation issues. Attempts to extend it to matter fail because of the problems associated with defining available matter in a technically progressive world with numerous kinds of material resources.
>
> *ibid., pp. 178–179*

Daly acknowledges that Young's simple model of an economy obeys the laws of thermodynamics and is 'faithful to [his and Georgescu's] conception of the economic process' (Daly 1992a, p. 91). This is significant because it shows that they share some common ground. Then Daly says, contrary to Young, that 'Physicists routinely apply entropy to matter, and, although this extension may involve some difficulties, it is far more than an analogy' (Daly, p. 91).

Daly dismisses new discoveries of resources as bookkeeping, not to be confused, as Young does, with a decrease in entropy that can compensate for increases in entropy due to economic activities. He acknowledges that new technologies can provide access to previously unusable, concentrated, deposits of available matter. The introduction of hydraulic fracturing (fracking) of shale oil and gas is a good example. But Daly insists that this is quite different from and not to be confused, as Young does, with 'finding a new use for the dissipated high-entropy remains of old resources' (op cit., p. 93). Daly also points out that new knowledge can reduce the availability of resources as well as increase them, and gives the example of the capacity of the atmosphere to absorb carbon dioxide being found less than previously thought.

Daly distinguishes between materials that are unavailable because they are dispersed and materials that are concentrated but as yet there is no knowledge for how to use them. He asks rhetorically, 'Does the fact that we discovered uses for aluminum imply that we can invent a technology to recycle all the particles of rubber scraped from tire on curbs and inter-state highways?' (ibid., p. 93). He then defends his position that absolute scarcity exists with respect to materials and that it does not depend on the veracity of a fourth law of thermodynamics applied to matter as proposed by Georgescu-Roegen. This law essentially states, 'that matter will dissipate in a closed system (i.e., a system closed with respect to matter but with a plentiful throughput of energy), or, alternatively, that complete recycling is impossible' (ibid., p. 92). Instead, Daly adopts a pragmatic position based on 'common-sense evidence that for all practical purposes complete recycling is impossible... I am not competent to assess the claim that it is physically impossible, so I leave that to the physicists'.[3] Daly insists that 'for present purposes it does not matter whether Georgescu-Roegen is right or wrong about his fourth law' (ibid., p. 92).

In the case of recycling the rubber worn from tires, Daly says we do have a technology: people on their knees using tweezers, but 'Since disordered matter requires more energy for processing, and since that extra energy will at some point make the recycling of dispersed matter uneconomic, we need no rigorous law of material entropy with a physical (as opposed to an economic) limit on recycling matter. This economic limit stems from the physical fact that enormous amounts of energy, as well as of other materials, are required to recycle highly dispersed matter' (ibid., p. 93).

The two protagonists are furthest apart on the question of whether 'the model of entropic decay is relevant for modeling open systems' (ibid., p. 93). Daly refers to Schrödinger's account of living organisms that are open systems dependent on converting low entropy matter-energy obtained from their environment into high entropy matter-energy. The parallel he draws with economies which are also open systems, and which behave in much the same way, seems beyond question. Daly's position is that all living organisms are constrained by the availability of low entropy matter-energy and this includes humans and our economies which rely on a continual flow of low entropy from the environment. Young and Daly agree that improvements in technology can reduce the rate of increase in entropy in economies, but continued to differ on whether technology, discoveries of new deposits, and the substitution of one resource for another can negate the increase in material entropy entirely.

Economist Ken Townsend also disputed some of Young's claims (Townsend 1992). Two years after Daly's and Townsend's responses to Young, Young published a reply to Daly and Townsend. In his reply Young conceded that 'the entropy law applies to matter and that materials entropy is well-defined' but maintained his position that 'entropic considerations are either already incorporated into relative price information or represent such a long-run constraint as to be irrelevant to foreseeable human welfare' (Young 1994, p. 210). The assertion that relative prices already incorporate entropic considerations was a new argument and a surprising one since the only reference to prices in Young's original paper was a statement to the contrary, referring to economists David Pearce and R. Kerry Turner, 'who consider the entropy law to be a constraint which is not reflected in market prices...' (ibid., p. 171). The point is a simple one. Some entropic considerations may well be included in market prices. The concentration of mineral ores in the Earth's crust is an entropic consideration, as is the release of carbon dioxide when fossil fuels are combusted. Depending on the market structure, the richness of concentrations of ores will be reflected in relative prices. But by definition, externalities are not included in market prices. These externalities include damage to health and the environment from releases of air and water pollutants including unpriced and under-priced emissions of carbon dioxide from the combustion of fossil fuels which are causing climate change. Since prices are affected disproportionately by this omission, relative prices are disturbed which undermines the significance that market prices might have for economic efficiency. The pervasiveness of externalities means that, contrary to Young, it is mistaken to rely on relative prices to incorporate entropic considerations of the kind that concern Daly and Georgescu-Roegen.

The second reason Young gives for the irrelevancy of entropic considerations is that they are such a long-term consideration that they can be safely ignored. Young was writing before the concept of planetary boundaries was introduced into the literature (Rockström et al. 2009; Steffen et al. 2015). A group of earth scientists identified nine environmental issues of planetary scope, each of which can be understood as the result of economic activities in which low entropy matter energy is converted to high entropy matter energy.[4] Of these issues, Rockström et al. and Steffen et al. presented evidence to suggest that for four of them humanity was already exceeding safe levels. Daly did not continue the debate with Young but if he had he would probably have argued that entropic considerations are ever present and that the inescapable general scarcity of which he first wrote in 1968 is more evident than ever.

BOX 4.3

An eminent economist, Daly is a rare expert who understood the notion of limits and the role that nature plays in the economy. Limits to growth and incorporation of nature into any economic endeavour are critical for any discussion of sustainability.

David Suzuki

The Solow Stiglitz/Georgescu-Roegen Daly Debate

A few years after his debate with Jeffrey Young, Daly become engaged in a second and potentially more significant debate with Robert Solow and Joe Stiglitz, both recipients of the Nobel Memorial Prize in Economic Sciences. Daly had been complaining for 20 years or more that Solow and Stiglitz had never answered Georgescu-Roegen's critique of neoclassical economics, and in particular of the theory of production.

> I have insisted that in our haste to mathematize economics we have often been carried away by mathematical formalism to the point of disregarding a basic requirement of science; namely, to have as clear idea as possible about what corresponds in actuality to every piece of our symbolism.
>
> *Georgescu-Roegen 1970, p. 1*

Solow and Stiglitz each wrote papers in the 1970s in which they argued against the idea that a scarcity of natural resources represented a limit to economic growth (Solow 1974, 1978; Stiglitz 1974). Their papers were written in response to the highly popular and controversial book, *The Limits to Growth* (Meadows et al. 1972). The book was distinguished by its use of a computer model – the 'World model' – for projecting inter-related, long-term global trends for industrial output per capita, food per capita, population, resource use and pollution. This was before personal computers and the vast majority of people had no direct contact with the large 'mainframe computers' used by governments, the military, large corporations and universities. Watching Dennis Meadows, one of the authors of *The Limits to Growth*, on television standing beside a printer the size of a small stove as it printed out line by line a forecast of the future made an impression on the public psyche that is hard to appreciate today.

In *The Limits to Growth* the authors described several global scenarios out to the year 2100 based on different assumptions. The scenarios had very different outcomes, ranging from expansion and collapse to relative stability. Meadows et al. emphasized that the scenarios described 'modes of behaviour' of the world system and were not predictions of the future. Nevertheless, they sometimes blurred the distinction between prediction and modes of behaviour. For example, '*We can thus say with some confidence that under the assumption of no major change in the present system population and industrial growth will certainly stop within the next century at the latest*' (ibid., p.126, italics in the original). This certainly sounds like a prediction, albeit one that is contingent on an assumption of no change in the present system.

Immediately upon publication, *The Limits to Growth* was roundly criticized, especially by economists who found fault with the systems dynamics methodology used in the World model, and the assumptions about technology and resource stocks. They decried the lack of the price mechanism in the model, which they said would provide the necessary signals to avoid the threat of resource scarcity. Solow and Stiglitz each took up the challenge laid down by *The Limits to Growth* to show that the outlook for sustained economic growth was not all that bleak, basing their analysis on the analytical workhorse of the day, and still in widespread use in economics, the Cobb-Douglas production function. To appreciate Daly's disagreement with these distinguished economists it is necessary to have some appreciation of this (in)famous equation (though readers with a distaste for equations may want to skip the next section).

The Cobb-Douglas production function

The Cobb-Douglas production function is an algebraic equation in which capital (i.e. buildings and equipment) are multiplied by employed labour to determine the aggregate output of an economy.[5] The equation was invented by Kurt Wicksell, a Swedish economist who used it in his doctoral dissertation in 1895 (Sandelin 1976).

It is named after mathematician Charles Cobb and economist Paul Douglas, who undertook the first empirical test of the equation using data for manufacturing in the USA (Cobb and Douglas 1928). One reason for the popularity of the Cobb-Douglas production function in neoclassical economics is that it can be interpreted as consistent with the theory that labour and capital are paid the value of their marginal products (i.e. the value of the change in output attributable to a small change in each input holding the other input constant). This is the favoured theory of income distribution in neoclassical economics first introduced by American economist John Bates Clark (Clark 1899).

In the Cobb-Doulas equation, labour and capital are raised to the powers α and β respectively. These are interpreted as the output elasticities of labour and capital, (i.e. the proportional change in output divided by the proportional change in each input holding the other input constant). If it is assumed, as is commonly the practice, that $\alpha + \beta = 1$ then it can be shown that α and β equal the shares of labour and capital in national income providing they are paid according to the value of their marginal products. The Cobb-Douglas production function also exhibits constant returns to scale, meaning that a proportional increase in the inputs of labour and capital results in the same proportional increase in output.

There are many well-known problems in describing the aggregate production of an economy with a Cobb-Douglas production. They can be grouped into two categories: i) inadequate conceptual foundations and ii) aggregation problems. (See the annex to this chapter.) Daly describes it as 'a recipe without ingredients' since nothing can be made solely from capital and labour. Energy and materials are also required. Daly is very fond of this cooking analogy and used it often to make his point. Here is one example:

> Since the production function is often explained as a technical recipe, we might say that Solow's recipe calls for making a cake with only the cook and his kitchen. We do not need flour, eggs, sugar, etc. nor electricity or natural gas, not even firewood. If we want a bigger cake, the cook simply stirs faster in a bigger bowl and cooks the empty bowl in a bigger oven that somehow heats itself.
>
> *op cit., p. 261*

Most, if not all, of these problems with the Cobb-Douglas production function were well-known to Solow and Stiglitz when they challenged *The Limits to Growth* in the 1970s on whether resource scarcity was a significant threat to economic growth. Yet they based their challenge on the Cobb-Douglas production function, with the addition of natural resources as a third input, giving it an exponent and including it in the multiplication. Then they each proceeded in slightly different ways to analyse the implications of resource scarcity for economic growth using this version of a Cobb-Douglas production function.

Substitution and technological change

Solow (1978) emphasized the potential for capital to substitute for resources in response to resource scarcity, whereas Stiglitz emphasized the potential for technological change to overcome any scarcity of resources. In a widely quoted statement and one that Daly comments on, Solow said 'If it is very easy to substitute other factors for natural resources, then there is in principle no "problem." The world can, in effect, get along without natural resources, so exhaustion is just an event, not a catastrophe' (Solow 1974, p. 11). Or in more technical terms:

> If the elasticity of substitution between exhaustible resources and other inputs is unity or bigger, and if the elasticity of output with respect to reproducible capital exceeds the elasticity of output with respect to natural resources, then a constant population can maintain a positive constant level of consumption per head forever.
>
> *ibid.*[6]

This result follows directly from Solow's use of a Cobb-Douglas production function to describe what is possible in an economy in the face of increasing resource scarcity. He obtained the result assuming a constant population. This is at least one step towards a steady-state economy that Daly would support (see Chapter 8). Also of interest, and sounding very much like Daly, is that in his 1974 paper, Solow said that 'the laws of thermodynamics and life guarantee that we will never recover a whole pound of secondary copper from a pound of primary copper in use, or a whole pound of tertiary copper from a pound of secondary copper in use. There is leakage at every round' (ibid., p. 2). And later in the same paper, Solow said that to answer the question of whether the world can get along without natural resources 'it takes economics as well as the entropy law' (ibid., p. 11). Unfortunately for economics, whenever Solow returned to the question of limits to growth as he did several times in later years (e.g. Solow 1992), he never again mentioned the laws of thermodynamics.

Stiglitz used essentially the same theoretical model as Solow to reach very similar conclusions but relying more on the potential for technology rather than substitution to alleviate resource scarcity. 'With technical change, at any positive rate, we can easily find paths along which aggregate output does not decline' (op cit., pp. 130, 131). The difference between substitution and technical change is not exactly clear and neither Solow nor Stiglitz defined them precisely. One view of the difference is that the substitution of capital for resources is based on existing technologies, whereas over time the technologies change and become more efficient in terms of resource use.

Daly contested this line of analysis (Daly 1997a) in what might be called the battle of the production functions. Thirty years previously Solow had written,

> I have never thought of the macroeconomic production function as a rigorously justifiable concept. In my mind it is either an illuminating parable,

or else a mere device for handling data, to be used so long as it gives good empirical results, and to be abandoned as soon as it doesn't, or as soon as something better comes along.

Solow 1966, pp. 1259–1260

Daly's contention was that the Cobb–Douglas production functions used by Solow and Stiglitz in their analyses of resource scarcity and economic growth were not an illuminating parable, did not give good empirical results and that something better had come along. The something better was the production function proposed by Georgescu-Roegen (1970, 1971).

Georgescu-Roegen's better approach to production

Georgescu-Roegen's proposal for characterizing economic production at any scale, from the individual firm to the macro economy, is based on two key propositions. The first is that production functions should be consistent with physical laws, specifically the first and second laws of thermodynamics. The second is that a distinction should be made between stocks and funds on the one hand and flows and services on the other. As Daly explains, 'any material process consists in the transformation of some materials into others (the flow elements) by some agents (the fund elements)' (Daly 1997a, p. 262). The flow of materials are obtained from stocks such as copper which is mined, concentrated, smelted, refined and then used in multiple products. The capital equipment used in each of these stages of production represents funds that provides particular services. Unlike the copper, the

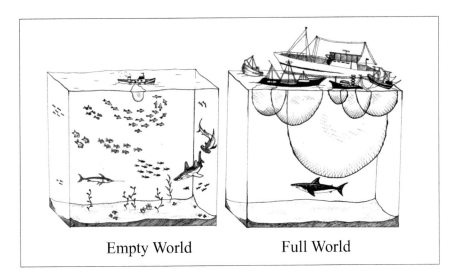

Empty World   Full World

FIGURE 4.4 Fish and fishing fleets: complements not substitutes

Source: Adapted from Daly (2015).

capital equipment does not become part of the product. As Daly points out, funds and flows are complements not substitutes. Both are required in any production process. His example of fishing vessels and fish is apropos. No matter how powerful and technologically advanced fishing vessels become, without fish their holds will remain empty.

Daly faults Solow and Stiglitz for assuming that manufactured capital, which is a fund that provides services, can be readily substituted for a stock of resources that provides a flow. This is a category error, built into the Cobb-Douglas production function which fails to distinguish stocks from funds and flows from services. In Daly's words, in which he adopts Aristotle's distinction between efficient and material causes:

> Georgescu's fund-flow model of the production process is superior to the neoclassical production function. It emphasizes that physically what we call 'production' is really transformation – of resources into useful products and waste products. Labor and capital are agents of transformation (efficient causes), while resources, low-entropy matter:energy, are 'that which is being transformed' (material causes). We can often substitute one efficient cause for another, or one material cause for another, but the relation between efficient and material cause is fundamentally one of complementarity, not substitutability.
>
> *ibid., p. 265*

Requiring that production functions in economics should not contradict the laws of physics would seem to be incontestable. Neither Solow nor Stiglitz contest it, they just do it. In his response to Daly's question: 'Why is it that neoclassical production functions do not satisfy the condition of mass balance?' Solow dismisses the relevance of the first law of thermodynamics in the form of the mass balance principle used in engineering: 'Because up until now, and at the level of aggregation, geographic scope and temporal extent considered, mass balance has not been a controlling factor in the growth of industrial economies' (Solow 1997, p. 268). In Daly's view this simply goes to show that Solow thinks materials are unimportant, as indeed they are if the economy is viewed through the lens of a Cobb-Douglas production function. Economists do not usually include resources as an input in production functions but even when they do, as in the papers by Solow and Stiglitz, the assumption is made that they can be substituted virtually without limit.

The reason that Solow gives for his lack of interest in the second law of thermodynamics is that it 'is of no immediate practical importance for modeling what is, after all, a brief instant of time in a small corner of the universe' (ibid., p. 268). Solow is thinking cosmologically. Daly and Georgescu-Roegen's interests are more prosaic. In his response to Solow, Daly uses the example of burning a lump of coal (Daly 1997b, pp. 272–273). The second law explains why it cannot be burned twice: total energy remains the same but available energy declines, which is why new coal (or some other source of energy) must be found if the same level of energy service is to be maintained. The first law, the conservation law, explains why the combustion of coal results in ash that must be disposed of and carbon dioxide and other emissions are discharged into the atmosphere bringing health risks and climate change. Through this simple example, one among countless others available to Daly, it should not have been so difficult for Solow to see the relevance of the first and second laws of thermodynamics to the economics of production.

BOX 4.4

It is odd, then, that alone among disciplines with any pretence to analytic rigor, economics has steadfastly resisted the thermodynamic revolution that swept physical and life sciences in the nineteenth and early twentieth centuries. Physics, biology, chemistry, geology, even the study of history were transformed, but not economics.

Eric Zanecy

Schroedinger explained, and others have amplified, that all living systems extract low entropy from their environment and return high entropic wastes. This is true of individual plants and entire ecosystems. It is also true of economies. All are complex systems that maintain themselves with material and energy 'throughput', a favourite term of Daly's which links the materials and energy that enters an economy with the wastes that emerge from it. Throughput is a natural consequence of the first law of thermodynamics. Much of the materials and all of the energy that enter an economy in a year become wastes discarded back to the environment. Fossil fuels, uranium and biomass, which are used to supply energy, also enter the economy as materials. Only renewable energy from sources such as solar, wind, hydro, geothermal and wave energy, enter the economy directly and, apart from a small amount kept in storage, leave immediately upon use. Daly's contention is that the combination of throughput and mass balance, together with the second law of thermodynamics which accounts for the degradation of energy and matter, provides an indispensable framework for understanding and analysing the dependence of economies on the environment. This has been Daly's consistent message since the 1960s.

One consequence of ignoring this message has been the creation of two largely separate sub-disciplines in neoclassical economics: the economics of natural resources and environmental economics. Solow's claim that 'mass balance has not been a controlling factor in the growth of industrial economies' partly stems from a mindset which fails to see the inescapable link between what comes into an economy from the environment and what goes back out. Rather than generate 'illuminating parables', the omission of resources and wastes from production functions, or the inclusion of resources but not the wastes they eventually become, has resulted in generations of economists with little or no appreciation of the fundamental dependency of economic output and growth on the environment. It has also limited the scope and meaning of the empirical results obtained from economic analyses based upon such an inadequate conceptualization of production. Daly's final word on his debate with Solow and Stiglitz in 1997 was, 'In any event Georgescu-Roegen's criticisms remain unanswered' (ibid., p. 273).

The Solow Stiglitz/Georgescu-Roegen Daly debate revisited

Little attention was paid to Daly's disagreement with Solow and Stiglitz in the years that followed their debate. Then in 2019 two substantial papers were published that examined the debate afresh (Couix 2019; Germain 2019). Quentin Couix's paper arose from his doctoral dissertation. When writing it, he corresponded with Daly. After reviewing the debate in detail Couix concluded that, 'neither side has been able to provide definitive proof of the validity of its own claim because both face important conceptual issues' (Cuoix 2019, p. 1370). What Cuoix would consider 'definitive proof' is unclear but he is correct that neither Georgescu-Roegen nor Daly give a sufficiently clear interpretation of the thermodynamic limits confronting an economy to know what should be measured for discerning their practical significance. But contrary to Cuoix, this is not so much a conceptual issue that calls into question the relevance of the laws of thermodynamics to economics. Rather, it is an issue requiring further research. At the conceptual level is the unexplained relationship between entropy and value, a relationship that Georgescu-Roegen and Daly assert but 'a quantitative relationship between the two… remains mostly unexplored' (ibid., p. 1371).

It should not be surprising that there are larger conceptual difficulties on the Solow/Stiglitz side of the debate simply because the Cobb-Douglas production functions on which they rely are not based on a concept of production in which inputs are transformed into outputs through physical processes. They are convenient mathematical equations that can be estimated using data from the national accounts, though the real meaning of what is obtained from the estimates remains controversial even without considering natural resources. In 1979 Herbert Simon explained that estimates of Cobb-Douglas production functions were simply capturing a national income accounting identity (Simon 1979). For some time, Simon corresponded with Solow about the problems of modelling an economy using a Cobb-Douglas production function. In a letter to Solow in May 1971 (Carter 2011, p. 263) Simon specified the production function in physical terms but unfortunately did not develop this idea in his 1979 paper (Felipe and McCombie 2011–12). Had he done so, the debate between Daly and Solow and Stiglitz might have taken on a different character or not taken place at all.

Solow and Stiglitz add more conceptual difficulties when they include resources in their Cobb-Douglas production functions with 'the assumption of unbounded resource productivity [which] rests on the concepts of "substitution" and "technical progress"' (Cuoix 2019). They provide ex post explanations of what this means in terms of changes in the type of resources, capital or produced goods, but as Cuoix notes, none of these possibilities are represented in their models.

Couix's final conclusion about Daly's debate with Solow and Stiglitz is that:

> The debate of 1997 underlines that the notion of production itself needs to be questioned in order to improve this integration [between economic and thermodynamic concepts]. In particular, the relationships between natural resources, produced goods, and the non-material services they provide seem to be at the heart of the issue.
>
> *ibid., p. 1371*

Though he refrains from saying so, this appears to be a definite nod in Daly's direction and the critique of the neoclassical theory of production adopted from Georgescu-Roegen.

The second paper published in 2019 revisiting the Solow/Stiglitz–Georgescu-Roegen/Daly debate took a decidedly different approach from that of Couix. Economist Marc Germain sought to answer

> the question of whether a production function that ignores the constraints of physics on the production process (such as the Cobb-Douglas) can generate a good medium-term approximation of the trajectory of the economy obtained with a "true" production function, which takes these constraints into account.
>
> *Germain 2019, p. 168*

In his reply to Daly, Stiglitz defended the kind of analytical model built around a Cobb-Douglas production function that he and others have used, saying: 'They are intended to help us answer questions like, *for the intermediate run—for the next 50–60 years*, is it possible that growth can be sustained?' (Stiglitz 1997, p. 269 italics in the original). This statement can be taken at face value, but then one wonders why there is no mention of this time limit on the usefulness of the model in Stiglitz's 1974 paper? There is nothing in his analysis in the paper that suggests that it is time limited. Had it been otherwise it might have led others, if not Stiglitz himself, to think about possible physical constraints on endless economic growth and how to represent them in the models. This might have led them to Georgescu-Roegen's alternative conceptualization of production. But that was not to be.

Instead, Stiglitz was content to describe Daly's critique of his and Solow's work as a 'tirade' and claim that Daly lacked an 'understanding of the role of the kind of analytic models that we (and others) have formulated... we write down models as if they extend out to infinity but no one takes these limits seriously... ' (Stiglitz 1997, p. 269). It is unclear what limits Stiglitz is referring to, but it seems just as likely they are the limits of mathematical functions as the kind of physical limits that Daly is concerned about. In any case, having said that he was only talking about the next

50–60 years, Stiglitz restates his position that the substitution of capital for resources and technical change guided by prices where markets work well with corrections for market failures where they do not, will be sufficient to keep resource scarcity at bay. Stiglitz says that: 'No one, to our knowledge, is proposing repealing the laws of thermodynamics! (ibid., p. 270).' True enough, but it is a fine line between that and ignoring them.

Germain compared consumption over time in an economy described by a conventional Cobb-Douglas production function with capital and labour as inputs, with an economy described by two alternative 'true' production functions consistent with the first and second laws of thermodynamics. The first of these 'true' production functions is a CES function[7] that satisfies two physical conditions not met by the Stiglitz-Solow Cobb-Douglas production functions: 1) output must be less than infinite; 2) the productivity of resources has an upper bound which constrains the ability of capital to substitute for resources. Germain developed a second production function for comparison with the unconstrained Cobb-Douglas production which he called the Alternative 'True Function' or ATF. This function is closer to the Cobb-Douglas function than the CES.

Germain examined a variety of scenarios with different values of key parameters.

> The different simulations show that if the substitution elasticity between capital and resource is significantly less than 1, the approximation generated by the Cobb-Douglas of the path obtained with the CES is generally rough… and this, even for the first few periods.
>
> *ibid. p. 17*

A main reason lies in the different effects of technology in the two functions. This difference is neutralized in the ATF function but even so, trajectories of the two functions diverge early on and 'generate totally different long-term trajectories following the fact that one takes into account the limits of physics and the other not' (ibid., p. 169).

Germain concludes that the Cobb-Douglas production function is acceptable 'only when the constraints of physics act sufficiently weakly. Unfortunately, various empirical studies suggest that it is not the case, especially in the case of energy' (ibid., p. 182). He does allow that Cobb-Douglas production functions generate paths with a similar shape to those of the "true" functions in the medium term, but not if some precision is required such as 'errors must be less than 10% for all variables' (ibid., p. 180). This is a partial vindication of Stiglitz's ex post rationalization of the medium-term relevance of his analysis.

Many interesting questions remain arising from this debate among economists. Why, since 1997, has it not been taken up in mainstream economics journals? Why are students still drilled in the 'nice properties' of 'well-behaved' production functions that lack a physical foundation and are inconsistent with the laws of physics? In particular, can the environmental and resource problems facing the globe be addressed adequately with such tools? Is there anything in the Solow-Stiglitz models to indicate that the sustained growth they thought was possible in the 1970s for the next 50 to 60 years has run its course, or do we need a different framework and a different set of tools to answer that question, one example of which is Daly's steady-state economics (see Chapter 8)?

The asexual revolution

Daly was not and is not opposed to the use of mathematics in economics, though he has not made much use of it himself. This was due in part to his lack of adequate mathematical training dating back to high school. It is also because, in his view, the mathematization of economics has led to 'elaborate and beautiful logical structures [that] heighten the tendency to prize theory over fact and to reinterpret fact to fit theory' (Daly and Cobb 1994, p. 38). We saw examples of this in his debate with Solow and Stiglitz. A more egregious example on which Daly commented (Daly 1982) was offered by economists Gary Becker and Nigel Tomas in a different context (Becker and Tomes 1979).

Becker was a prominent member of the Chicago School of Economics, known for its commitment to neoclassical economics with a particularly strong emphasis on individualism. Becker and Tomes addressed the important question of inequality and intergenerational mobility with a mathematical model in which they make the extraordinary assumption that 'children have the same utility function as their parents' (Becker and Tomes 1979, p. 1161). Anyone who has been a parent or child would know the falsity of assuming that parents and children have the same preferences, but it makes the mathematics easier. An even more extraordinary assumption, and one that Daly chastises the authors for, is that children 'are produced without mating, or asexually'. No one who has any inkling that economics is, or should be, a life science could be content with this absurdity. As Daly says, 'it reveals just how far some members of the Chicago School will go in amputating those limbs of human society that do not fit the Procrustean bed of individualistic utility maximization' (Daly 1982, p. 308).

In his critique of the Becker and Tomes paper Daly makes an intriguing argument about providing for the future once bisexual reproduction is brought back into the frame. He shows how a combination of biology and economic theory can yield interesting conclusions about community and shared interests. Daly starts from the biological truth that:

> One's great-great grandchildren will also be the great-great grandchildren of fifteen other people in one's own generation, people whose identities cannot be determined before the fact. They could be almost any fifteen people. More generally… a given person in a given generation will normally have 2^n ancestors in the nth previous generation.
>
> *ibid., p. 308*

Daly reasons that:

> Even if everyone felt a strong moral obligation to the distant future, he or
> she would be foolish to try to fulfill that obligation by individual action, for
> the reason just given – that anonymity and multiplication of ancestors make
> provision for the distant future a public good.
>
> *ibid.*

Public goods are goods and services that if provided to one person are provided
to all. Common examples include national defence and street lighting. Defining
characteristics of public goods are that, once provided, people cannot be excluded
from using them, and one person's use does not reduce the availability to others.
The non-exclusion feature of public goods makes it infeasible for profit-seeking
enterprises to provide them since few people would pay for something that is freely
available. Public goods must be provided collectively, financed through taxation
rather than sales revenue.

Having shown that provision for the future beyond a generation or two is essentially a public good, Daly argues that since we do not know the other 14 people
(excluding spouse or partner) that are alive today and with whom we will share
descendants, we have an interest in caring for everyone.

> … concern for the future is often seen as blunting or diverting our eth-
> ical concern away from the more pressing problems of present injustice.
> Considered logically, however, moral concern for the distant future should
> strengthen rather than weaken the bonds of brotherhood in the present,
> because we are all potential co-progenitors of each other's descendants.
>
> *ibid., p. 310*

Clearly, the assumption of atomistic individuals who procreate asexually may be
convenient for a mathematical analysis of intergenerational mobility, incomes and
equity, but bisexual reproduction, which would not be overlooked in economics as
a life science, makes all the difference in the world to the conclusions we reach on
intergenerational equity.

Environmentally extended input-output analysis

Daly seldom uses mathematics or mathematical symbols to make his case. One
exception is in his 'Economics as a Life Science' paper, where he constructed what
today would be called an 'environmentally extended input–output table'. To understand this term, we will start with a conventional input–output table in which the
environment does not appear and then show how Daly proposed adding it.

Input–output analysis was developed by Russian born economist Wassily
Leontief, whose work was recognized by the award of the Nobel Memorial Prize
in Economics in 1973. It is based on the idea that an economy can be understood

as the inter-connections among different sectors. This idea dates back to Francois Quésnay, a French physician turned economist who published his Tableau Economique in 1758. According to Quésnay the only real production in an economy is from agriculture, the rest is processing. This seemed obvious to Quesnay in mid-eighteenth-century France, before industrialization began to obscure the dependence of economies on the Earth and on what nature provides.

Leontief moved to the US in 1931. His objective was to '… construct, on the basis of available statistical data, a Tableau Economique for the US economy' (Leontief 1936, p. 105). Leontief designed a table displaying the 'intermediate demand' transactions among economic sectors – the inputs bought from each other to produce outputs – and 'final demand' sales to households, businesses, government and net exports (exports minus imports). Figure 4.5 shows a simplified input-output table for a national economy. It includes all purchases and sales in an economy in a specified time period, usually a single year. Companies are grouped into sectors and all are included in the input-output table. In Figure 4.5 the grouping is based on seven sectors. Real input-output tables are much more detailed and can have hundreds of sectors. Regardless of the number of sectors the input-output tables cover the entire economy.

The rows of the table are divided between producers and value added, and the columns are divided between intermediate demand and final demand. Reading across a row for each producer shows the sales in dollars of the companies in that sector to companies in each sector, including itself. Continuing along each row beyond the shaded cells for intermediate demand are entries showing the sales to the different categories of final demand. The row total (T) shows the total sales of the sector for the year.

Reading down a column for each producer shows the purchases from companies in each sector, including itself. In addition to these inter-industry purchases, companies pay taxes and receive subsidies and pay their employees. What remains from revenues, after allowing for the depreciation of capital equipment, is profit for the business owners. The calculation of profit as a residual ensures that the row and column totals for each sector are always equal. The gross domestic product (GDP) of the economy can be calculated by summing final demand or value added as shown in the bottom right-hand corner of the table.

So far this is just accounting. Leontief's genius was to use his input-output table to create a model of how an economy functions. His model was based on the simple assumption that each dollar of output in a sector requires a constant amount of purchases from each sector. He derived these 'input-output coefficients' by dividing the value of purchases from each sector by the sector's total sales. For example, if the annual output of the automobile sector is $1 billion and the automobile sector purchased $50 million dollars of steel, the input-output coefficient for steel required to make automobiles is .05 (i.e. $50m/$1b).

Leontief used these coefficients to estimate the total output of each sector required to satisfy any specified level of final demand. The logic is simple. If, for example, households spend $500 million on automobiles in a year the automobile

		INTERMEDIATE DEMAND								FINAL DEMAND				
		Agric	Mining	Constr	Manuf	Trade	Transp	Services	Other	Personal Consumption Expenditure	Gross Domestic Investment	Govt Purchases of Goods & Services	Net Exports of Goods and Services	Total
PRODUCERS	Agriculture													T
	Mining													
	Construction													
	Manufacturing													
	Trade													
	Transportation													
	Services													
	Other													
VALUE ADDED	Government		Indirect business taxes less subsidies							GROSS DOMESTIC PRODUCT				
	Employees		Employee compensation											
	Business owners and capital		Profit and capital consumption allowances											
Total		T												

FIGURE 4.5 A simplified input-output table

Source: Adapted from Miller and Blair (2009, p. 3).

sector would have to buy $25 million (.05 x $500m) of steel. To produce this steel the steel sector would have to buy certain amounts of inputs from other sectors (and itself) as determined by the input-output coefficients. This could include purchases from the automobile sector which would then need more steel from the steel sector and so on, round after round. These inter-industry purchases and sales occur right across the economy, with all sectors potentially buying from all other sectors. Leontief showed with some fairly simple mathematics, that this process of inter-industry purchases and sales comes to an end, since each round of purchases gets smaller and smaller and so the production required of each sector to satisfy the initial $500 million demand for automobiles can be determined.

Although the math was comparatively simple, the computational requirements in the 1930s to do the calculations for all sectors simultaneously were considerable. It was only when electronic computers became available after World War II that input-output models became really practical. It is an interesting historical foot-note that for a time, during the Cold War, the construction of input-output tables was suspended in the USA because 'it smacked of communist "central planning"' (Milland Blair 2009, p. 732). Meanwhile, in the Soviet Union, where input-output analysis had been independently developed, it was used for that very purpose. Western countries other than the US were less troubled by these considerations and work on input-output analysis continued, including the publication of detailed input-output tables by national statistical agencies.

In the 1960s when environmental issues became a matter of public concern, a few economists realized that input-output analysis could be extended to include interactions between the economy and the environment. Daly (1968) was one of these economists.[8] Daly's objective was to bring purely economic interactions, purely environmental interactions, and interactions between the economy and the environment into one comprehensive framework. The simplest version of this framework is shown in Figure 4.6.

Daly described the table in the following way:

Cell or quadrant (2) is the domain of traditional economics, that is, the study of inputs and outputs to and from various subsectors within the human-to-human box. Cell (4) represents the traditional area of concern of ecology, the

From	To	
	Human	Non-Human
Human	(2)	(1)
Non-Human	(3)	(4)

FIGURE 4.6 An input-output representation of the total economy
Source: Daly (1968, p. 401).

inputs to and outputs from subsectors in the non-human- to-non-human box. Cells (1) and (3), respectively, contain the flows of inputs from human subsectors to non-human subsectors and from non- human subsectors to human subsectors. All of the items exchanged in (2) are economic commodities, by which we mean that they have positive prices. All items of exchange in cells (1), (3), and (4) may by contrast be labeled ecological commodities, which consist of free goods (zero price) and 'bads' (negative price). The negative price on bads is not generally observed, since there usually exists the alternative of exporting the bad to the non-human economy, which cannot pay the negative price (that is, charge us a positive price for the service of taking the 'bad' off our hands, as would be the case if it were transferred to another sector of the human economy). Ecological commodities that are bads are bad in relation to man, not necessarily to the non-human world. The difficulty, however, is that these more than gratuitous exports from the human economy in cell (1) are simultaneously inputs to the non-human economy and as such strongly influence the outputs from the non-human back to the human sector-that is, cell (1) is connected to cell (3) via cell (4), and cell (3) directly influences human welfare… the basic vision is still a 'world of commodities,' although a bigger world that now includes both economic commodities (…in quadrant [2]) and ecological commodities (… in quadrants [1], [3], and [4]). … quadrants (1), (3), and (4) are the 'biophysical foundations of economics.'

ibid., pp. 400–401

A more detailed version of Figure 4.6 is shown in Figure 4.7. Quadrant 2 is a three-sector, conventional input-output table. It is in the other three quadrants that we see the kind of detail contemplated by Daly for keeping account of economy-environment (quadrants 1 and 3) and purely ecological interactions (quadrant 4). Daly's description of row 10 anticipates his debates years later with Young and Solow and Stiglitz about the relevance of entropy to economics: 'In row 10 we have a primary-service sector providing the ultimate source of low-entropy matter-energy, the sun, and, in column (10), the great thermodynamic sink into which finally consumed high-entropy matter-energy goes, forever degraded as devil's dust.' (Daly, 1968, p. 403)

Daly's tables are very useful for bringing to light the interdependencies between the human world of production and exchange and its natural counterpart in the biological and physical world. It is a vastly expanded accounting framework of the one that Leontief designed, which was confined to quadrant 2. But Leontief's table has the practical advantage that all the entries are in terms of money. This allowed Leontief to build his input-output model from unitless coefficients obtained by dividing the entry in each cell by the total output of the relevant sector. Daly suggests in his paper that the same method could be used for calculating coefficients from his environmentally extended accounting framework and he writes an equation in physical terms to show how this could be done. The problem is that all the entries

OUTPUT FROM	Agriculture (1)	Industry (2)	Households (Final Consumption) (3)	Animal (4)	Plant (5)	Bacteria (6)	Atmosphere (7)	Hydrosphere (8)	Lithosphere (9)	Sink (Final Consumption) (10)	Total
	Quadrant (2)						Quadrant (1)				
1. Agriculture	...	q_{12}	...	...	...	...	q_{17}	...	...	...	Q_1
2. Industry	q_{21}	(q_{22})	q_{23}	...	...	...	q_{27}	...	...	...	Q_2
3. Households (primary services)	...	q_{32}	...	...	...	...	q_{37}	...	...	...	Q_3
	Quadrant (3)						Quadrant (4)				
4. Animal	...	...	...	...	...	...	q_{47}	...	...	...	
5. Plant	...	...	...	...	...	...	q_{57}	...	...	...	
6. Bacteria	...	...	...	...	...	...	q_{67}	...	...	...	
7. Atmosphere	q_{71}	q_{72}	q_{73}	q_{74}	q_{75}	q_{76}	(q_{77})	q_{78}	q_{79}	$q_{7,10}$	Q_7
8. Hydrosphere	...	...	...	...	...	...	q_{87}	...	...	...	
9. Lithosphere	...	...	...	...	...	...	q_{97}	...	...	...	
10. Sun (primary services)	...	...	...	...	...	...	$q_{10,7}$	...	...	...	

INPUT TO

FIGURE 4.7 Daly's environmentally extended input-output table

Source: Daly (1968, p. 402).

in his expanded input-output table would have to be in terms of a common physical unit to be summed to get row totals. In an example, Daly suggests weight in pounds could be used (ibid., p. 403) but this hardly seems satisfactory for both the purely economic transactions in quadrant 2 and for the environmental entries in the other quadrants as well.

Daly did not attempt to fill his environmentally extended input-output table with data from a real economy or to build and implement an input-output model; but many others have, starting with Victor (1972). The approach most have taken is to limit the scope of the environmental extensions to quadrants 3 and 1 and to calculate hybrid coefficients expressed in physical units per dollar, e.g. kilograms of greenhouse gasses released per $1 million dollars of output by each sector. Then when sector outputs are estimated in the normal way from the outputs required to satisfy final demand these hybrid coefficients are used to estimate the material and energy inputs from the environment and outputs (wastes), both in physical terms. This procedure avoids having to convert financial transactions into physical terms or environmental flows into dollars.[9]

Conclusion

Daly's scholarly contributions have almost all been conceptual. He has challenged many of the assumptions of neoclassical economics, and with his Ends Means Spectrum he put economics in its place. By the standards of most writings in economics his books and papers are literary gems. They are clear, accessible, often entertaining and they point out important directions for further work.

Daly's consideration of economics as a life science was a bold step for a young academic to take in the 1960s. His landmark paper has been read widely and cited many hundreds of times in the past 50-plus years as economists have struggled to understand the dependence of the economy on the environment. In his debates with Young and with Solow and Stiglitz, Daly made the case that economics itself has to change if economists are to properly engage in what is necessarily an inter-disciplinary endeavour. Following the lead of Georgescu-Roegen, he argued strenuously that economics should take fully into account the first and second laws of thermodynamics. Life and the economy cannot be understood without them.

Annex to Chapter 4: A summary of inadequacies of the Cobb-Douglas production function

Inadequate conceptual foundations

- The Cobb-Douglas production function is not based on an engineering description of production but is favoured by economists because of its mathematical characteristics. A multiplicative function presupposes substitutability of the inputs as an artefact of the math rather than as a representation of the process of production.

- Substitution between the inputs is assumed to be smooth, instantaneous and reversible. There is no path dependency in which decisions about the choice of inputs at a point in time constrain opportunities for substitution in the future.
- It is inconsistent with the laws of physics. Output cannot be produced solely from capital and labour. It requires materials and energy which are conserved in quantity but degraded in quality.

Aggregation problems

- The aggregation of diverse outputs and types of capital measured in terms of money is problematic. It was the basis of the long-running, unresolved Cambridge Capital Controversy in which economists from Cambridge University (UK) argued with economists from MIT (Cambridge, MA) about the aggregation of capital. One concern of the British economists is that the distinction between the price and quantity of capital, which is critical in neo-classical theory, is blurred when a rate of interest (a price) is used to calculate the quantity (value) of capital (Cohen and Harcourt 2003).
- The Cobb-Douglas function has been used to describe production in individual firms, specific economic sectors and entire economies, each one being an aggregation of the former. And yet, Cobb-Douglas functions cannot be summed to get another Cobb-Douglas function unless they are proportional to each other and $\alpha + \beta = 1$ (Bao Hong 2008). This means that if a Cobb-Douglas production function describes production at one level of activity in an economy, it cannot also describe production at another level such as a region or a firm.
- The Cobb-Douglas production function has been shown not to be a production function at all, but an income distribution function based on an accounting identity. It simply reflects the distribution of national income between labour and capital which has been fairly stable over time (Carter 2011, p. 258).

Notes

1 Daly uses the labels neoclassical, mainstream, orthodox, conventional, and standard to signify an approach to economics that prioritizes economic growth with an unduly strong emphasis on the role of markets. The labels are used as synonyms in this book.

2 This is not entirely correct because it neglects the time required for recycling. If all the materials and energy used in an economy in a year were available for use again in the following year the output of the economy could be maintained year after year, but in any year the output of the economy remains limited, scarcity remains, and choices have to be made. Increases in the rate at which materials are used in the economy through changes in technology or in the composition of output could ameliorate the situation but not without limit.

3 According to Swedish physicists Tomas Kåberger and Bengt Månsson (2001), 'the division into material and energy entropy, which Georgescu-Roegen used extensively and in particular in his 'fourth law' suggestion… is fallacious. There is only one kind of entropy…'

(p. 167). However, they side with Georgescu-Roegen and Daly on the main point in dispute, 'Physicists have discovered a large number of immutable laws of nature. Some of these are particularly important for the social sciences. Foremost among these are the first and second laws of thermodynamics' (ibid., p. 166).

4 The issues are: climate change, novel entities, stratospheric ozone depletion, atmospheric aerosol loading, ocean acidification, biochemical flows (phosphorous and nitrogen) freshwater use, land-system change, biosphere integrity (functional diversity and genetic diversity) (Steffen et al., 2015).

5 The simplest form of the Cobb-Douglas production function is:

$$Y = AL^{\alpha}K^{\beta}$$

where:

Y = total production of goods and services in constant dollars produced in a period of time

L = labour input in person hours

K = capital input in constant dollars

A = total factor productivity. It measures changes in output not attributable to the inputs

α and β are the output elasticities of labour and capital

A time variable t is often included to account for changes in the inputs over time

(Note – dollars are used for convenience. Any currency could be used in its place.)

6 The elasticity of substitution is the proportional change in the ratio of two inputs in production divided by the proportional change in their marginal products.

7 The constant elasticity production function (CES) is another widely used production function in which the elasticity of substitution among inputs is constant but can take on any value. When it has the value 1 it is equivalent to a Cobb-Douglas production function.

8 Others include Cumberland (1966), Isard (1969), Leontief (1970) and Victor (1972). Only Daly and Victor included material balances in their extended tables. Victor applied his model empirically to the Canadian economy.

9 Environmentally extended input-output tables and models have now been developed for multiple regions right up to the level of the globe. Kitzes (2013) provides a useful introduction to environmentally extended input-output analysis though he overlooks thermodynamics, materials balance and entropy that were features of Daly (1968) and other early contributions. The UN has also made progress in expanding the system of national accounts framework to include an environmental dimension but not to the full extent originally proposed by Daly (UN 2014).

5

SCALE, DISTRIBUTION, ALLOCATION

We all recognize that 'you can't kill two birds with one stone', at least not if the birds are flying independently. If they are flying in tandem or sitting on the same fence, then one might manage to do it. In economic theory today we are trying to kill three birds with two stones.

Daly

There is a strong focus on policy throughout Daly's writings. Daly's approach to policy is to base his proposals on clearly articulated principles. Among his most influential principles is the distinction he makes among scale, distribution and allocation. Scale refers to the material size of the economy in relation to the environment in which it is embedded. Distribution refers to the claims on economic output among people according to income and wealth, within and across generations. Once scale and distribution have been established, allocation can be determined. Daly argues that there is a logical policy sequence starting with sustainable scale, followed by just distribution and then efficient allocation. He also says there should be policies for each according to Tinbergen's principle of one instrument for each policy objective (Tinbergen 1952). Mainstream economists, Daly says, concentrate too much on efficient allocation to the neglect of sustainable scale and just distribution, neither of which can be adequately determined within the market but must be seen as market determining.

In this chapter we examine the background to Daly's tri-partite distinction of scale, distribution and allocation, and the assumptions and theory underlying it. More attention is given to scale than to distribution and allocation because, of the three aspects, it is the one that is most neglected in mainstream economics and it is the one that Daly has emphasized in his work. As we shall see, his views on scale have not gone unchallenged. Never one to shy away from debate, it is in his disagreements with eminent scholars such as philosopher Mark Sagoff and

DOI: 10.4324/9781003094746-5

economist Kenneth Arrow, that provide the greatest insights into the issues that are in dispute.

BOX 5.1

Aspects of economies of scale, distribution and allocation are aspects that have greatly helped me in trying to understand the relationship between the exploitation of forest resources in Indonesia and the sustainability of economic development.

Mamat Rahmat

From an empty world to a full world

Throughout virtually all of human history, human societies and their economies have been small in relation to the biosphere, that is, small in terms of the human use of materials and energy and the transformation of land. For some 200,000 years since *homo sapiens* first appeared, the kinds of societies that could survive and thrive were shaped by what nature provided locally and regionally, quite independently of any global limits. Nature was rich in extent and diversity and the human population was small. There were less than 270 million humans worldwide as recently as year 1000 (Maddison 2006, table B-10) – about the same as live in Indonesia today – buildings were modest, and possessions were few. Daly refers to this long expanse of history as an 'empty world', empty that is of humans. It was still empty in the sense of human impacts on the biosphere as recently as 1800 when the Industrial Revolution was just getting underway in the United Kingdom. All this changed in the matter of 200 years. By the turn of the millennium, 'world population rose 22-fold… per capita income increased 13-fold, world GDP nearly 300-fold' (ibid., p. 19). What Maddison could have added and which is fundamental to Daly's conception of a full world, is that during the twentieth century the global use of resources consisting of biomass, fossil fuels, industrial ores and construction materials increased eight-fold and at an increasing rate which has continued into this century (Krausmann et al. 2009). A study by Klee and Graedel (2004) found that by the end of the twentieth century humans were moving greater quantities of a majority of the elements than nature. No wonder Earth scientists are troubled by environmental degradation at the global level and are urging humanity to live within planetary boundaries (Steffen et al. 2015). With nearly 8 billion people, many consuming at levels far surpassing those of all but the very richest a century or so ago, the sixth extinction underway and barely anywhere on the planet untouched by human presence, Daly has more than enough reason to claim that we are living in a 'full world'.[1]

A different world, a full world – from the point of view of human beings that is — calls for a different economics. Daly argues that in an empty world the limiting factor on economic growth was capital. The physical capital stock of an economy, comprised of infrastructure, buildings and equipment is built by investment. Investment in any year is made possible when consumption is less than output. What is not consumed today can be invested to make more consumption possible tomorrow. Investment is not easy and may not be possible when most people in a society are living at subsistence level with nothing to spare. It was investment and the buildup of the capital stock, accompanied by political and cultural changes, that made the remarkable economic growth of the past two centuries possible, though the process has been nationally and regionally uneven, sometimes massively so, with the benefits often accruing mainly to a small proportion of the population, bringing increasing inequality in its wake (Piketty 2014).

Ever since his first trip to Mexico as a teenager Daly has been well aware of this pattern of inequitable economic development. For reasons that will become clear in this and later chapters, Daly has resisted the idea that continual expansion of the economy is the answer to inequality. His contention is that we are living in a full world and that the limiting factor has shifted from human-made capital to nature. To make his point, he often gives the example of fishing being limited not by too few fishing boats but by too few fish. Daly argues that nature – or 'natural capital' as it is sometimes called in this context – has replaced manufactured capital as the limiting factor on economic growth, not just because the biosphere which contains the expanding economies, is not growing. It is also because economic growth has depleted both the regenerative capacity of natural systems and the deposits of minerals that lie within the Earth's crust and on which economic growth depends.[2]

Daly's view that the limiting factor on economic growth has shifted from manufactured to natural capital is not easily accepted by economists whose fundamental assumption is that all inputs into production are substitutes for each other. This is different from a view of production in which capital and labour funds process material and energy flows. Funds can be substituted for one another, and flows can be substituted for one another, but in general funds and flows cannot be substituted for each other (Daly and Farley 2011, pp. 156–157). Funds and flows are essentially complements; they must be used together to produce output. A shortage of one cannot be overcome by a greater use of the other. For example, ample forestry equipment and a large forest will produce a considerable supply of lumber. But if equipment is unavailable the supply of lumber will be small even if the forest is large as it will be if ample equipment is available, but the forest is small. The difference between the two situations is that in the first, manufactured capital is the limiting factor and in the second natural capital is the limiting factor. Daly's contention is that the transformation of the world from empty to full has resulted in natural capital displacing manufactured capital as the limiting factor on economic output.[3]

When Daly was writing about economics as a life science in the 1960s it seemed obvious to him that this change in conditions merits attention to the *scale* of the economy. He was influenced in this by an essay entitled 'On Being the Right Size' by biologist John B.S. Haldane (Haldane 1926), which he read early on and remembered later when the idea of scale in economics became important to him. Haldane explained that, 'For every type of animal there is a most convenient size, and a large change in size inevitably carries with it a change of form' (ibid., p. 1). Daly wondered if similar considerations apply to economies. Accordingly, he began to ask three related questions: How big *is* the economy; second, how big *can* it be without collapsing its ecological foundation; and third, how big *ought* it be? The questions alone caused quite a stir. The answers, many of which are still forthcoming, promise to be even more disturbing.

A hierarchy of objectives

The first time that Daly presented his idea of the scale, distribution, allocation hierarchy was in an invited lecture at the Dutch Ministry of the Environment on 28 November 1991. He was suffering from jet lag having flown from the USA, and unable to sleep, intensely awake, he completely rewrote his prepared speech for the next day around the idea of a logical hierarchy and policy sequence among the three goals. The speech and discussion went well. Perhaps jet lag can stimulate creative thinking. Daly re-wrote it again for publication (Daly 1992b). In his seminal article, which has been cited nearly 1000 times, Daly offers the following definitions:

> *Allocation*: … the resource flow among alternative product uses - how much goes to production of cars, to shoes to plows, to teapots, etc.
>
> *Distribution*: … the relative division of the resource flow, as embodied in final goods and services, among alternative people. How much goes to you, to me, to others, to future generations?
>
> *Scale*: … the physical volume of the throughput, the flow of matter-energy from the environment as low-entropy raw materials, and back to the environment as high-entropy wastes … relative to the natural capacities of the ecosystem to regenerate the inputs and absorb the waste outputs on a sustainable basis … [the economy's] scale is significant relative to the fixed size of the ecosystem.
>
> *ibid., pp. 186–187*

Following each of these definitions, Daly offers a normative criterion for setting an appropriate policy objective:

> *Allocation*: A good allocation is one that is efficient, i.e. that allocates resources among product end-uses in conformity with individual preferences as weighted by the ability of the individual to pay.
>
> *Distribution*: A good distribution is one that is just or fair, or at least one in which the degree of inequality is limited within some acceptable range.

Scale: A good scale is one that is at least sustainable, that does not erode environmental carrying capacity over time.

To complete his taxonomy, Daly provides a policy instrument for achieving the allocation and distribution policy objectives:

Allocation: The policy instrument that brings about an efficient allocation is relative prices determined by supply and demand in competitive markets.

Distribution: The policy instrument for bringing about a more just distribution is transfers – taxes and welfare payments.

It is only when Daly offers tradeable pollution permits as a 'beautiful example of the independence and proper relationship among allocation, distribution, and scale' (ibid., p. 188) that he becomes more explicit about a policy instrument for achieving the desired scale. He identifies three essential components to any emissions trading scheme such as the ones introduced in 1990 in the USA to deal with acid rain, and in the European Union in 2005 to reduce greenhouse gas emissions. Scale is the first component of any emissions trading scheme, set by government in terms of total allowable emissions from specified sources in a given region. Distribution is the second component. Emissions allowances or permits are assigned to the sources. They may be given away, auctioned or some combination of the two. Daly prefers the auction option. The third component, allocation, allows the sources to trade their permits. Sources that can reduce their emissions most cheaply have an incentive to do so since they can profit by selling surplus allowances to other sources with higher emissions reduction costs. These higher cost sources save money as long as the price of the additional emissions allowances are less than their avoided emissions reduction costs. The trades have no impact on the total allowable emissions, but they result in a lower overall cost of emissions reduction, and allocation efficiency is improved.[4]

From this example Daly draws two main conclusions that he holds are generally applicable in a full world. The first is that scale, distribution and allocation form a logical sequence. The desired scale of the economy must be chosen first, taking into account the benefits that come from a high level of economic output and the associated environmental and social costs. Also relevant are the implications for future generations, especially from depleted mines and wells, loss of biodiversity, and the accumulation of pollutants such as greenhouse gases. Once scale has been determined, the next consideration is a just distribution of the output of the economy. In a modern economy this comes down to the distribution of incomes and wealth which can be adjusted through taxes and transfers. Then Daly says, competitive markets can be used for allocating resources to alternative ends through prices determined by demand and supply with the important caveat that markets do not work well for natural monopolies (e.g. centralized electricity generation) and non-rival commodities (i.e. goods and services that if consumed by one person do not reduce opportunities for consumption by others, such as street lighting and information).

> **BOX 5.2**
>
> As an Associate Professor of Economics I use the concepts of scale, distribution and allocation as the overarching framework for my Principles courses.
>
> *Gerda Kits*

Tinbergen's principle of an equal number of policy instruments and objectives

From the logical sequence of scale, distribution and allocation Daly moves to his second main conclusion concerning the need for a policy instrument for each of them. The principle of equality between the number of policy instruments and policy objectives was proposed by Jan Tinbergen, another winner of the Nobel Memorial Prize in Economics (Tinbergen 1952). Tinbergen observed that, in general, any policy instrument, be it a regulation, a tax, or a subsidy which is intended to meet a specific policy objective will also affect other policy objectives. Take, for example, an income tax and a carbon tax. Both taxes are instruments that raise revenue for government, which is one objective; and both reduce carbon emissions, which is another objective. A carbon tax, such as that introduced by the Canadian government in 2019, makes it financially attractive for sources subject to the tax to reduce their carbon emissions. The lower their emissions, the less they have to pay in tax. An income tax which reduces what people can spend on consumption, reduces production and so reduces carbon emissions. Both taxes also generate revenue for government. Because both taxes affect both objectives, to rely on only one *exclusively* to meet a revenue objective and also an emissions reduction objective is almost bound to fail. For example, the tax rate for a carbon tax which meets an emission reduction objective is most unlikely to also generate the revenue required to meet the revenue objective. Tinbergen showed that in a situation like this with two objectives it is necessary to employ two instruments, both of which affect both objectives. More generally he showed that any number of policy objectives requires an equal number of policy instruments if all objectives are to be met simultaneously.

Having established the requirement for at least as many policy instruments as there are objectives, Tinbergen showed that the matching of instruments and objectives was also very important. Building on the example of an income tax and a carbon tax, consider a situation with two government ministries: the Ministry of Finance which is responsible for raising revenue with the income tax and the Ministry of Environment which is responsible for reducing carbon emissions with the carbon tax. Even though both taxes affect both objectives, Tinbergen showed that both objectives can be met even if the Ministries set their rates quite independently of each other. Through a process of learning and adjustment, the Ministries will adjust the tax rates they each control until the objective for which they are responsible is met. The key is to match instruments and objectives so that each Ministry controls the instrument to which the objective they are responsible for is most responsive.

The Prakash and Gupta/Daly and Stewen/Daly Debates

Daly's application of Tinbergen's principle of equality between the number of instruments and objectives would seem to be uncontroversial and yet he was criticized for it on more than one occasion:

> The neat scheme proposed above is beset with problems. What if a single policy instrument affects more than one goal? … this Tinbergian one-to-one correspondence among goals and instruments is highly contestable.
>
> *Prakash and Gupta 1994, p. 89*

> The interpretation of 'independence' as the possibility of dealing with scale issues isolated from other goals is fatal… All goals and instruments are interactive and must be worked together.
>
> *Stewen 1998, pp. 122, 124*

Daly responded to both critics in his usual respectful way, showing his willingness, even desire, to engage in discussion and debate. He concedes a small point about terminology: 'Perhaps I should have said three *separable* problems instead of three *independent* problems' (Daly 1994b, p. 90) but holds to his position about numbers of policy goals and instruments. In response to Stewen, Daly says, 'What I meant to convey, following Tinbergen, is independence in the sense of simultaneous equations' (Daly 1999a, p. 1). With sets of simultaneous equations, basic algebra establishes that you need one equation for each unknown precisely because of the inter-relationships among the variables. It follows, says Daly, that three policy objectives cannot be simultaneously met with less than three instruments. 'From a policy perspective once cannot expect prices (one instrument) to serve all three goals. One needs a separate instrument for distribution, and also a separate one for scale' (ibid.). This conclusion follows from the interdependence of instruments and objectives, which Daly fully appreciated. What Daly did not explain is how the simultaneous interdependency assumed in Tinbergen's analysis applies when the policy objectives are hierarchical, as Daly assumes in the case of scale, distribution and allocation.

A key insight from Tinbergen which Daly failed to stress clearly enough in his 1992 paper and in his responses to Prakash and Gupta and Stewen, is the matching of policy instruments with policy goals over which they have the most influence, as explained above. What he does do, and with considerable force, is to argue that just distribution cannot be derived from competitive prices and that optimal scale cannot be derived from just distribution and allocation. Daly's claim is that it makes logical sense to proceed in the opposite direction, from scale to distribution and from distribution to allocation. He gave the best account of this argument in a keynote address to the Canadian Society for Ecological Economics in Jasper, Alberta in October 2003. He reminded the audience that 'Economists have long accepted that an optimal allocation of resources (Pareto optimum), with its resulting set of prices, requires a given distribution of income' (Daly 2007a, p. 99). A maximally efficient economy is one that has achieved a Pareto optimum, i.e. when the allocation of resources cannot be changed without making at least one person worse off. This happens when all mutually agreed trades have been exhausted. Daly then points out that '… there is a different Pareto optimal allocation or each possible distribution of income' (ibid.) which is also a standard result in mainstream welfare economics. It follows, therefore, that prices cannot be used to determine independently a just distribution of income and wealth since prices reflect the prevailing distribution. A different distribution would yield different prices.

Daly extends this argument to scale. Prices depend on scale as well as on income and wealth distribution.

> We cannot know what new prices would correspond to optimal scale unless we already know the optimal scale… It is circular to calculate the optimal scale on the basis of equating marginal costs and benefits measured by prices, which assume that we are already at the optimal scale to begin with.
>
> *ibid., p. 101*

From this, Daly concludes that while optimal scale can be conceived in terms of not pushing the scale of an economy beyond the point where marginal benefits equal marginal costs, 'We need some metric of benefit and cost other than prices, other than exchange value, other than the ratio of marginal utilities' (ibid.). Daly does not say what this other metric should be, but he does provide a principle on which it should be based, 'ecological sustainability, including intergenerational and interspecies justice' (ibid., pp. 101–102).

The size of the economy in relation to the biosphere in which it is embedded, i.e. its scale, matters because the size of the biosphere is limited. More than that, its ability to provide the goods and services on which the economy depends can be reduced as the physical size of the economy increases. Daly has reiterated these points many times in many places. In his speech in Jasper he added two more reasons why scale matters. Just as it is impossible, as Haldane noted, to increase all dimensions of a physical object in the same proportion, so it is equally impossible to increase all dimensions of an economy in the same proportion. This has implications for prices. Even if the quantities of all commodities were to increase by the same proportion, their marginal utilities would not and so neither would their prices. Prices, Daly concludes, should be determined by scale and distribution, not the other way around.

To argue that there is a logical sequence moving from scale to distribution to allocation but not in reverse, is not the same as saying that the objectives for each of them can and should be set in isolation of each other. 'Daly's paper could be read as suggesting that a strict hierarchy of goals is normatively desirable, that a target planning of "optima" is independently possible and that trade-offs between the goals are of minor significance' (Stewen 1998, p. 119). Daly's principal concern is that scale is largely ignored by policy makers and by most economists who implicitly or explicitly assume that economic growth can continue without end. Having identified the need for a suitable policy objective for scale, Daly explains that it cannot be derived from allocation and distribution but should be determined by considerations specific to itself, i.e. in relation to the biosphere within which the economy is embedded. Contrary to Stewen, this does not mean that when deliberating about an objective for scale that implications for distribution and allocation are irrelevant and can be ignored, only that the scale objective cannot be arrived at based on only those considerations.

In his critical appraisal of Daly's hierarchy of scale, distribution and allocation Stewen claims that Daly retains too much of the neoclassical framework. He quotes economist Fay Duchin in support for just adding the problem of scale to the 'scope of neoclassical economics… while leaving its analytical core untouched' (Stewen 1998, p. 122). It is the separation of scale, distribution and allocation that Stewen objects to, arguing instead for a 'co-evolutionary' view. 'Daly argues in terms of comparative statics [comparing different states of an economy without explaining the time path from one state to another], but his models and metaphors could be misleading in a dynamic, evolutionary world' (ibid., p. 120).

Stewen says that 'The great danger is to overlook the complexity of the problem…' (ibid., p. 121), to which Daly replies:

> Fair enough, but I am more worried about the opposite danger of failing to seek and take advantage of simplicity whenever we are fortunate enough to find it. To quote William of Ockham, 'Entities should not be multiplied beyond necessity…' Neither should *co-evolutionary interdependencies* lest we be reduced to inaction by the perceived necessity of having to do everything before we can do anything.
>
> *Daly 1999a, p. 2*

Stewen politely but clearly tells Daly that he has failed to wean himself away from neoclassical economics by neglecting complexity and politics, a charge that others have made against Daly (see Chapter 10). Daly's response is to stress the importance of simple rules where they can be found and to avoid the impotence of thinking that change is all or nothing. He sees value in using taxes and transfers to reduce inequality and he has proposed a maximum as well as a minimum income as a matter of public policy. And he believes that markets have a useful role to play in allocating economic output, providing they are reasonably competitive and subject to regulation to avoid abuse of market power, employees and the environment (Daly and Farley 2011, Chapter 10). To that extent Daly's views are not so different from neoclassical economists on the virtues of markets for allocation, but when it comes to scale especially, and to distribution, they could hardly be further apart.

The Sagoff/Daly debate

When Daly left the World Bank in 1994 and joined the School of Public Affairs at the University of Maryland, one of the people he met there was Mark Sagoff. The two professors had a lot in common. Sagoff was a well-established philosopher with a keen interest in the economy and nature. He was known as a critic of environmental economics with its instrumental approach to the environment and the valuation of its services to humans in terms of money. Daly, of course, was also a critic of environmental economics but for different reasons, and he had an interest in philosophy. This could have led to a meeting of minds, a productive collaboration even, but that was not to be.

Their disagreements became public in 1995 when Sagoff turned his sights on ecological economics in an article that took aim at Daly in particular, mentioning him 29 times in the text and citing 10 of his books and papers (Sagoff 1995). At the root of their disagreement was and still is their fundamentally different philosophical positions. Sagoff is a Kantian ethicist whose approach to environmentalism is close to that of Henry David Thoreau and John Muir. They 'looked to the intrinsic properties of nature, rather than its economic benefits as reasons to preserve it' (ibid., p. 610). For them, and for Sagoff, environmentalism should be grounded in spirituality, not in the uses that humans seek to make of nature, especially when choices among uses are evaluated in terms of money using benefit–cost analysis.

In his 1995 paper Sagoff criticized ecological economists who seek 'to vindicate environmental protection on instrumental grounds' (ibid.). It is not clear to what extent he thinks Daly is guilty of this, but since he referred to his work so much in his article and knowing that Daly is one of the founders of ecological economics, it is only reasonable to assume that Sagoff includes Daly in his charge that ecological economists are hardly different in their consequentialist ethics and instrumental values from mainstream economists.

It is not spirituality that separates Daly from Sagoff. We saw in his Ends-Means Spectrum, Daly's belief in a higher, spiritual End that all intermediate ends – the focus of mainstream economics – should serve. For Daly, the Ultimate end resides in 'God's creation and its evolutionary potential', which has intrinsic value (Daly 1995b, p. 624). To protect and enhance this intrinsic value Daly looks to 'prudential reasoning in terms of costs and benefits arising from the consequences of our actions' (ibid.). In this context, costs and benefits are to be understood widely and not simply as those included in a conventional benefit–cost analysis with all its attendant conceptual and methodological problems.

Daly posed several rhetorical questions designed to reveal weaknesses in Sagoff's philosophical position: 'Fine. Now in the light of that philosophy tell me how large the human population should be, what is the proper level or range of per capita resource consumption, and how much of the habitat of other species we are justified in preempting for human use (ibid., p. 621). According to Daly, Sagoff's view on what is to be done is to do what is inherently right in light of the intrinsic value of nature and not on any rational assessment of the consequences of human actions. What makes this work for Sagoff, says Daly, is the technological optimism that he displays throughout the article:

> technological optimism mixed with Kantian deontology is an alchemist's elixir. It means that we do not have to be seriously concerned with consequentialist ethics because technology can always neutralize any unfortunate consequences. No criterion is left but the inherent rightness of an act (based on the authority of the philosopher's intuition), because all offsetting negative consequences can be erased by technology.
>
> *ibid., p. 621*

It is Sagoff's technological optimism, which Daly does not share, that he says allows Sagoff to essentially ignore consequences, because if there are bad ones, technology can be relied upon to find a solution.

The particular issue that Sagoff chose to build his critique of ecological economics and Daly on is carrying capacity, the very issue for which Daly, in his rhetorical questions, suggests that Sagoff has no answer. The idea that the capacity of the Earth to support life, especially human life, is limited and is being exceeded is fundamental to Daly's interest in the scale of the economy.

Sagoff identified five theses concerning the Earth's carrying capacity which he criticizes. First is the assertion of ecological economists that entropy limits economic growth. Because the amount of solar radiation received by the Earth greatly exceeds the energy obtained from the combustion of fossil fuels and because of potential improvements in energy efficiency, Sagoff says: 'It is not obvious how the second law of thermodynamics limits economic growth' (op cit., p. 613). Sagoff continued this line of attack with his second thesis by supporting the view of mainstream economists that knowledge, ingenuity or invention 'are likely always to alleviate resource shortages' (ibid., p. 611) by increasing reserves, substitution with less scarce resources and using resources more efficiently.

For Sagoff, says Daly, 'Knowledge is the key, and resources are of minor importance…', as if knowledge is immaterial. But Daly says that 'for knowledge to mean anything for the economy, it must be imprinted on the physical world'. This can be in books, brains, bytes and all kinds of physical structures, none of which are immutable. 'Entropy melts those structures, giving rise to the need for a continuous input of low entropy [energy/matter] from the environment for maintenance…' (op cit., p. 622). Where Sagoff sees knowledge as a substitute for resources and produced capital, Daly sees complements.

> Even though there are many possibilities for substitution of one source of low entropy for another, there is no substitute for low entropy… itself. Intelligent substitutions and technical adaptations should not blind us to the existence of the fundamental constraint to which they are still only adaptation.
>
> *ibid.*

As Daly says, it is a mistake to think that a recipe (i.e. knowledge) is a substitute for ingredients (i.e. resources). Both are required to make a cake. Daly's keen awareness of what it actually takes to make things may well have come from the many years he spent working in his father's hardware store.

Sagoff's third critical thesis is the redefinition of economic growth by ecological economists in terms of material and energy throughput rather than GDP, the money value of final output. Daly, more than anyone, has insisted on the distinction between quantitative growth and qualitative development. Just as growth is possible without development, development is possible without growth. Growth in physical terms is what Daly means by an increase in scale. Sagoff objected to this distinction between growth and development for two main reasons. First, if ecological economists measure improvements in the quality of life with metrics used by conventional economists, then there really is no difference between what ecological economists mean by development and mainstream economists mean by growth. Second, while acknowledging in passing that 'growth is not a scientific term in mainstream economics...' (op cit., p. 614), Sagoff says that it generally refers to the rate of increase in gross domestic product.

Daly was unpersuaded by Sagoff's arguments. The distinction between growth (quantitative) and development (qualitative), says Daly, 'is straight from the dictionary, so it is not idiosyncratic as Sagoff claimed' (ibid., p. 623). At the same Daly said that gross domestic product is a mixture of growth and development and that conflating them in a single measure is confusing. For one thing, it conceals the material component of economic output, which Daly contends cannot grow without limit, allowing others to argue that growth has no limit because they are thinking about the qualitative component as represented by market prices. Even then, as Daly has commented in other writings (e.g. Daly and Cobb 1994, Chapter 3), national income statisticians hold prices constant when calculating growth in GDP. This is so that they can use changes in inflation adjusted 'real' GDP as a measure of quantitative growth. It is an error, says Daly, to treat real GDP as a measure of value independent of its material dimension since it leads to the mistaken conclusion that the economy can grow forever.[5]

The most direct challenge by Sagoff to the idea that carrying capacity is a constraint on economic growth centred on a paper published several years before Sagoff's 1995 critique of ecological economics written by a group of biologists led by Peter Vitousek. They estimated that

> nearly 40% of potential terrestrial net primary productivity [NPP] is used directly, co-opted, or foregone because of human activities... NPP provides the basis for maintenance, growth, and reproduction of all hecterotrophs (consumers and decomposers); it is the total food resource on Earth.
>
> *Vitousek et al., 1986, p. 368.*

Ecological economists, Daly included, interpreted this estimate as a measure of the scale of the human economy and drew from it the implication that economic growth would eventually be limited as humans appropriated an increasing percentage of the net products of photosynthesis, and in the process deprive other species of habitat and food pushing them to extinction.

Sagoff objected to this interpretation of the data and the prognosis suggested by Vitousek, noting that increases in global food production between 1950 and 1989 had kept pace with the human population. This was the result of improved yields and not the use of more land in agriculture. Sagoff also questioned the premise that economic growth entails co-option of more and more organic matter. He described the service sector, information, communication, medical technology, education and finance as the 'great engines of economic growth' (op cit., p. 616) and said they do not depend on net primary production. Consequently, there is no reason why net primary production should limit economic growth. Daly on the other hand argued that Vitousek and his colleagues' calculation 'is a reasonable attempt to put some quantitative dimension on the scale of the human economy relative to the total ecosystem' (op cit., p. 623). In this instance, Daly did not point out – as he has elsewhere – that the increase in the output of food that Sagoff highlights was due largely to the greater use made of petroleum fuels, which for reasons of limited supply and climate change is not something that can be relied upon for much longer (Daly 1977a, p. 10).

Sagoff's fifth and final thesis concerned support of ecological economists for use of the precautionary principle in the face of environmental uncertainty. The only reason he gives for his disdain for this sensible sounding approach is that they offer 'little instruction as to what this means' (op cit., p. 618). Daly's response was to invite a comparison between the precautionary principle and Sagoff's technological optimism; and to use the opportunity to promote precautionary limits on throughput: 'If one is a technological optimist and believes that resources are unimportant for the economic process, then one should not object to a policy of limiting the resource throughput, thereby raising its price. Such a policy would induce exactly the technological advances that use resources more efficiently – the very technology in which the optimists have so much faith.' In this way, Daly attempted to find a bridge between their two positions. Take precaution, limit throughput which will raise the prices of natural resources and encourage the kinds of technological changes that Sagoff is confident will see us through. Sadly, it did not work. Sagoff remained unpersuaded by Daly and continued for many years to attack ecological economics using many of the same arguments as in 1995 (Sagoff 2012). Despite their enduring disagreements, Daly found it impossible not to like Sagoff, 'a philosopher who loved to argue, was fair in debates and willing to hear opposing arguments, and he was smart enough to win an argument even when he was wrong' (Personal communication).

The Arrow et al./Daly et al. debate

A similarly interesting, constructive and for the most part, polite exchange of views took place in 2007 between two groups, each comprising economists and ecologists.

It is one of the few times that neoclassical economists have taken Daly seriously enough to contest his arguments in writing. One group was led by Kenneth Arrow, among the most accomplished neoclassical economists of the twentieth century, and the other by Daly. Of Arrow's many contributions that earned him the Nobel Memorial Prize in Economics, two bear his name. The first from his doctoral dissertation in 1950 is known as the Arrow Impossibility Theorem (Arrow 1951). Arrow was interested in the conditions under which a community-wide ranking of alternatives could be obtained from the rankings of individual members of the community. He proved that no such 'social' ordering of alternatives could be obtained from any voting system that satisfied a set of very reasonable conditions associated with democracy such as that all rankings of alternatives by all voters be allowed and that here be no dictator.

The second contribution named after Arrow is the Arrow–Debreu model of general equilibrium (Arrow and Debreu 1954). Arrow and fellow economist Gérard Debreu proved that under certain conditions there is a set of prices such that supply equals demand for every commodity simultaneously in the economy. This proof remains the theoretical bedrock of the neoclassical analysis of competitive markets.

Arrow and ten other prominent economists and ecologists met in the early 2000s to discuss whether 'humanity's use of Earth's resources [is] endangering the economic possibilities open to our descendants' (Arrow et al. 2004, p. 147). One of their objectives was to reconcile the 'conflicting intuitions' of the pessimistic ecologists and optimistic economists. The pessimism of the ecologists was based on the same sort of information used by Daly in his description of a full world. Following the lines of argument developed by Solow and Stiglitz (see Chapter 4), the optimistic economists emphasized the potential for manufactured capital, human capital and technological change, to compensate for diminishing natural resources.

Arrow's group offered two different criteria for evaluating consumption: i) maximization of the present value of utility; and ii) a sustainability criterion. In the first criterion 'utility' (i.e. well-being or welfare) is obtained from consumption. They defined consumption 'broadly to encompass all goods and services, (including biodiversity) whether marketed or not'. They went on to say that it 'does not capture everything that contributes to well-being – elements such as companionship and community' (Arrow 2007, p. 1365). Having defined consumption, they 'discounted' it so that consumption enjoyed over time by existing and future generations is given less weight today when comparing alternative consumption paths. This is accomplished mathematically by dividing utility obtained from consumption whenever and by whomever experienced by a discount factor that increases exponentially with time. Discounting converts a stream of utility obtained from consumption from the present into an infinite future, which is then summed to get its equivalent present value.

Arrow's group sought a consumption path that maximizes the present value of utility from consumption to determine whether consumption today exceeds the consumption level compared to the optimal path. 'To put it another way, current consumption is excessive if lowering it and increasing investment (or reducing dis-investment) in capital assets could raise future utility enough to more than compensate (even after discounting) for the loss in current utility' (op cit., p. 149). The answer that Arrow et al. give to the question of 'are we consuming too much?' based on the maximum present value of utility criterion is that 'several factors – the inability to pool risks perfectly, the taxation of capital income and the underpricing of natural resources – contribute toward excessive consumption… Among these imperfections, the underpricing of natural resources strikes us as the most transparent … Although the evidence is far from conclusive, we find some support for the view that consumption's share of output is likely to be higher than that which is prescribed by the maximize present value criterion.' (Ibid., pp. 159, 167)

The empirical evidence cited by Arrow in support of this conclusion is rather indirect since 'no one can seriously claim to pinpoint the optimal level of current consumption for an actual economy' (ibid., p. 155). They considered elements in their theoretical model such as the discount rate, the absence of a complete set of forward markets, and taxation and their likely effects on consumption and investment over time. However, they overlooked the efforts of the marketing industry and other social pressures to consume, which place an additional wedge between the assumptions underlying Arrow et al.'s theory of consumption and real life.

The largely theoretical answer to the question 'Are we consuming too much based on utility?' contrasts with the way in which Arrow et al. use empirical evidence and their sustainability criterion to answer the question. In this case, they specify that the welfare of each generation from consumption, also defined broadly, must be equal to or greater than the welfare of the preceding one. They reasoned that this condition requires that 'the productive base [of the economy] be maintained, but this does not necessarily entail maintaining any *particular* set of resources at any given time' (ibid., p. 151).

The requirement to maintain the combined value of productive resources is known as 'weak sustainability,' based on the assumption that manufactured capital can be readily substituted for whatever nature provides. Weak sustainability is contrasted with 'strong sustainability' in which a distinction is made among components of the productive base according to whether they can be substituted or not. This relates directly to Daly and Georgescu-Roegen's position that in general funds (e.g. manufactured capital) and flows (e.g. resources) are complements rather than substitutes, which is consistent with strong rather than weak sustainability.

To test for weak sustainability, Arrow et al. used World Bank estimates of *genuine wealth* which had only recently become available from the World Bank. Genuine wealth which is 'the accounting value of all capital assets,' (ibid., p. 152), i.e. monetary value where 'shadow' prices are estimated and used for all non-priced goods and services from nature. Unusually, Arrow et al. included population in the computation of genuine wealth by attaching a shadow price to people as a way of accounting for changes in the size of the population over time. The assumptions required to do this (i.e. people are substitutable 'assets' and part of the productive base of the economy) and the data used to estimate this 'human capital' (i.e. expenditure on education) do not encourage confidence in the results.

In terms of their sustainability criterion of non- decreasing value of productive assets, Arrow et al. stated that,

> We… find evidence that several nations of the globe are failing to meet a sustainability criterion: their investments in human and manufactured capital are not sufficient to offset the depletion of natural capital. This investment problem seems most acute in some of the poorest countries of the world.
>
> *ibid., p. 167*

They also say that 'high levels of consumption in rich countries may promote excessive resource degradation in poor countries, which imperils well-being in the poorer countries' (ibid., p. 166). This means that sustainability of rich nations according to their sustainability criterion of non-declining productive assets may be a cause of non-sustainability in poor countries. Other than commenting that 'This negative by-product of rich nations' consumption is not captured in existing measures of changes in per capita wealth' (ibid., p. 166), this potent observation is pursued no further.

The *Journal of Economic Perspectives* in which the Arrow et al. paper was published does not publish comments, so Daly et al. sought an opportunity elsewhere to make their observations and present counter-arguments. They found it in *Conservation Biology* (Daly et al., 2007) where they made a number of criticisms of Arrow et al. – the main one concerning the question of scale. Daly et al. began by asking:

> Are we consuming too much for the rest of the planet? In other words, is the scale of the human economy so large relative to the containing biosphere that it displaces biospheric functions that are at the margin more important than the extra production and consumption?
>
> *ibid., p. 1359*

Daly had been pondering this question for years, decades even. It made him think that Arrow et al. were asking the right question but in the wrong way:

> To ask if we are consuming too much within their analytical framework means only, 'Is consumption too large relative to investment' – is the total product, whatever it may be, allocated properly among alternative uses – especially between consumption and investment? The consumption-investment decision is allocation within existing scale and does not affect existing scale since investment is just as resource-intensive as consumption. But it does affect growth. It seems that for neoclassical economists the economy can never be too big, but it can grow too fast if investment is too large relative to consumption?
>
> *Daly personal communication*

Daly et al. continued their comments on Arrow et al. with an insistence that the question of scale is distinct from allocation, the preoccupation of neoclassical economics, which 'does not recognize any problem of optimal scale... Whether the scale of the human economy is optimal, or even sustainable, is one question. Whether the allocation of the amount of transformed stuff between current consumption and investment is optimal, is a second, entirely different, question' (ibid.). Even more pointedly, Daly et al. wrote:

> to them excessive consumption does not mean reduction in biodiversity or overuse of resources resulting in observable consequences such as the loss of ecological services or increased pollution or even the widespread purchasing of meretricious junk. They mean only that current consumption is too large relative to investment if one wants to allocate resources so as to maximize or at least maintain, the unobservable, subjective utility of mostly yet-to-exist individuals theoretically aggregated into society from now to infinity.
>
> *ibid., p. 1361*

Arrow et al. responded to this critical comment by agreeing that 'the most fundamental criticism raised by Daly et al. is that our framework fails to consider what they call the scale of the economy' (Arrow et al. 2007, pp. 1363–1364). They argued that to satisfy their 'sustainability criterion the economy must not allow this intertemporal index [of social well-being] to fall, which in turn requires that its productive base – its capacity to generate material and nonmaterial goods and services into the indefinite future – be maintained' (ibid., p. 1364). They advanced

> the idea that society's productive base is its genuine wealth – the social worth of the economy's entire set of assets, comprising not only reproducible capital (roads, buildings, machines) and human capital (knowledge, skills, health) but also natural capital (various nonrenewable resources such as fuels and minerals and renewable resources such as forests and biodiversity).
>
> *ibid., p. 1363*

This is quintessentially a weak sustainability perspective and is the main point of dispute in this exchange with Daly et al. However, Arrow et al. do allow that 'there can come a point where no amount of feasible investment in reproducible capital can offset further declines in natural capital' (ibid., p. 1364). In saying so they allow substitution at the margin but not necessarily in total, which is a step in the direction of strong sustainability. However, they offer no clear opinion on whether that situation has been reached, but they imply it has not been by pointing to the possibility that future changes in the shadow prices of different types of capital will reveal a 'point at which no further substitution is possible' (ibid., p. 1364).

This exchange of views shows the neoclassical reliance on actual or estimated 'shadow' prices as the best, if not the only, source of information about the relationship between the economy and the biosphere. This is in marked contrast to the much greater use of biophysical data expressed in natural units (not money), in ecological economics and advocated tirelessly by Daly. In unpublished notes for a further rejoinder Daly says, 'It is disingenuous for them [Arrow et al.] to claim in reply that their original paper had been motivated by a concern for scale, as well as composition, when they did not even discuss the scale of the economy.'

In response to Daly et al.'s criticism that the maximation of the present value of utility criterion gives 'no clue as to what the "intertemporal social welfare" criterion implies for the larger question of optimal scale relative to the biosphere,' Arrow et al. maintain their neoclassical position. They agree with Daly et al. that 'intergenerational equity is one of the cornerstones of sustainability' but then claim that, 'because a judgment of intergenerational equity is a comparison of subjective utilities across yet-to-exist individuals, we are at a loss to understand how their methodological position differs from ours' (ibid.).

One reason they may have failed to see the difference in positions is that they overlook in their response the distinction that Daly et al. make between real, material natural resources and utility. Unlike living natural resources that have growth rates which provide a rationale for discounting, utility does not. Utility cannot be set aside today in anticipation that it will be larger tomorrow. Since materials can be bequeathed but not utility, Daly et al. contend that it makes more sense to think about intergenerational equity in terms of real, tangible resources and not utility which is, 'a psychic phenomenon' (Daly et al. 2007, p. 1360).

To close this discussion of the Arrow et al. and Daly et al. debate, it is to the credit of both Arrow and Daly, the intellectual leaders of the respective groups, that although the differences between them was considerable they took each other's position seriously in their published exchanges and in their personal correspondence. Arrow ended a letter in 2005 to Daly saying, 'with all appreciation for your stimulation' and Daly, in his reply wrote 'all good wishes for the New Year, and thanks for your patience.' Given the general antipathy between neoclassical economists and ecological economists, perhaps this represents something of a high-water mark.

Conclusion

Daly's distinction among scale, distribution and allocation has come to be considered foundational in ecological economics – though it is not without its critics. For example, John Gowdy, another of Georgescu-Roegen's distinguished students, finds Daly's distinction too 'mechanical'. He faults Daly for providing 'no notion of how inequality, growth, accumulation, and the dynamism of a capitalist economy are intertwined' and suggests that 'the focus on scale of the economy alone [which Daly does not do] as something that could be frozen with some tinkering, keeping the basic nature of the system intact, led ecological economics in a very conservative, even neoclassical direction' (Gowdy 2016, p. 81).

From the commentaries by Prakash and Gupta and by Stewen on Daly's 1992 paper, we can see that the distinction Daly was making among scale, distribution and allocation was not as well understood as he would have hoped. They are separate in the sense that they each require their own policy instrument, but inter-connected because each policy instrument affects all three dimensions, just as Tinbergen had said in 1952. The difference, which may explain some of the confusion, was that Tinbergen's analysis concerned equally significant policy objectives whereas Daly argues for a hierarchy and sequence of scale, distribution and allocation.

Daly's insistence on the importance of scale follows from his observation that the world has gone from empty to full. Humans and our artifacts continue to expand as does the energy and materials required to keep them functioning, while the Earth remains the same size. If we do not attend to scale, or if we continue to think that it is satisfactorily dealt with through market prices, then we should expect to encounter increasingly severe consequences delivered *gratis* by Mother Nature. Such has been Daly's warning for 50 years and for which the evidence continues to grow.

If scale matters, then so does our capacity to measure it. What we choose to measure and how we measure it depends on our understanding of what it is that we want to measure. On the one hand there is the view of neoclassical economists, like Arrow in this chapter and Solow and Stiglitz in the previous chapter, who focus on GDP, a value measure. On the other hand, there is the view of Daly, an ecological economist, who sees scale first and foremost in physical terms. These different perspectives will surface again in the next chapter, which looks at Daly's contributions to measuring the economy and the inspiration he has given to others to build on his ideas.

Notes

1 Daly was not the first to observe that the world has become 'full' in this sense. John Bellamy Foster quotes E. Ray Lancaster, a zoologist, evolutionary biologist, and pioneer ecologist who wrote in *The Kingdom of Man* (1911): 'The world, the earth's surface, is practically full, that is to say, fully occupied' (Foster 2020, p. 63).

2 The designation of nature as natural capital is not without problems, especially when valued in terms of money using 'shadow' prices (see Victor 2020).

3 The principle of limiting factors is very common in biology. It is 'the principle that the factor (such as a particular nutrient, water, or sunlight) that is in shortest supply (the limiting factor) will limit the growth and development of an organism or a community' (Oxford Reference 2017).

4 There are many other features of emissions trading schemes. The OECD is a good source of information on emissions trading.

5 GDP can also be measured as the sum of value added in an economy. Daly says, 'We must focus more on "that to which value is added," by which he means the '… flow of fresh natural resources [which has become] … the limiting factor in production' (Daly 1995c, p. 451).

6
MEASURING THE ECONOMY

If GNP were a cigarette, then the ISEW would be a better cigarette with a charcoal filter. If you are addicted to cigarettes it's better to smoke one with a charcoal filter; if you are addicted to numerical measures of welfare, it's better to use the ISEW.

Daly

This chapter is devoted to Daly's ideas about measuring the scale of an economy and the work it has inspired in others. The scale of an economy is fundamentally a quantitative issue; it is about size, usually the size of one thing in relation to another, for example the scale of a map or the scale of a model airplane. When Daly refers to the scale of an economy, he means the size of the economy in relation to the biosphere or some part of it; and by size he is usually referencing either the throughput of materials and energy or the stocks of material embodied in humans themselves and in their accumulated artifacts. Sometimes he means both. Throughput and stocks are, of course, intimately related. Throughput is a flow and is measured in terms of materials and energy per unit of time. Stocks are built from the accumulation of materials over time and are measured at a point in time.

The measurement of scale leads naturally to questions about whether there is an optimal scale of an economy, be this a local or national economy or for the global economy? If there is and an economy has grown beyond it, then Daly says that such growth is 'uneconomic' in the sense that at the margin the benefits the economy is providing are likely to be less than the costs it is imposing.[1] In practice this can be difficult to determine, especially when the benefits and costs accrue to different people in time and space. What is optimal for a national economy in the short term may be sub-optimal at the local or global level or when a longer time horizon is considered. To paraphrase Georgescu-Roegen, more Cadillacs today means fewer people tomorrow (Daly 1996, p. 196).

DOI: 10.4324/9781003094746-6

Daly argues that since GDP does not distinguish systematically between benefits and costs, or between quantitative growth and qualitative development, it cannot be used to determine if growth has become uneconomic. Daly is one of the originators of the Index of Sustainable Economic Welfare (ISEW), which attempts to fix some of these problems of using GDP as a normative measure of an economy. We will assess how successful this has been adding in some comments about distribution, the second of Daly's triumvirate: scale, distribution and allocation.

Measuring scale

At the global level, the scale of the economy is its size in relation to the whole planet. This is the sense in which Daly uses scale in his discussion of the transformation from an empty world to a full world. At the national or local level, the scale of an economy can be its size in relation to the national or local environment or as a proportion of the global environment though Daly is not especially clear on this.

In neoclassical economics discussions of scale generally refer to specific activities. 'Increasing returns to scale' prevail when an increase in inputs gives a proportionately larger increase in output such as when a larger truck reduces lower costs per unit transported. But neoclassical economists seldom concern themselves with the overall scale of an economy. Daly gives two reasons for this (Daly 2017, p. 90). One is the assumption of a virtually infinite environment: abundant fish in the oceans, seemingly endless forests, ample capacity of the environment to absorb human wastes, an immense flow of solar energy vastly greater than the energy obtained from the combustion of fossil fuels, and in earlier times, new lands with unexplored and unexploited resources. This may have been a reasonable world view from a European perspective in the late eighteenth and early nineteenth centuries when the foundations of classical economics were laid by Smith and Ricardo. It is not reasonable today and has not been for decades, though neoclassical economics have yet to wake up to these changing circumstances.

The second assumption that has allowed most economists to ignore or pay little attention to scale is the belief that any shortage in materials, energy or absorptive capacity of the environment that may arise can and will be dealt with by new scientific discoveries and technologies. While it was becoming increasingly obvious in the nineteenth century that the environment was anything but infinite, the astonishing developments in science and technology impressed on the minds of many, economists included, that through invention and ingenuity any and all environmental constraints were short term and would be overcome. These two assumptions, an infinite environment and unlimited possibilities of science and technology, though not necessarily made explicit, have made it possible for most economists to continue their work largely ignoring the issue of scale. This, at any rate, has been Daly's contention throughout his career. Add to this the assumption that humans wants are infinite, limited for individual commodities by the law of diminishing marginal utility, but not for commodities in general, and you get Daly's conclusion: 'Give the economist infinite wants and infinite environments and he will give you eternal growth' (Daly 1980b, p. 83).

Gross Domestic Product (GDP)

A lack of interest in scale is not the same as a lack of interest in size, quite the opposite in fact. Daly has observed that economists who disagree on very fundamental issues, say most Marxist economists and neoclassical economists, are of the same mind when it comes to growth (Daly 1991a, p. 196). They both think that economies can and should grow without limit. The most common measure of such growth is the increase in real, inflation adjusted, Gross Domestic Product. GDP is the summation of price times quantity across all final transactions, final in the sense that they are not made in order to make something for resale. (These 'intermediate' transactions are excluded from GDP to avoid double counting.) GDP includes purchases of goods and services by households and government, and by businesses and government when they invest in new construction, equipment and software. GDP also includes exports, since they are produced in the country for which GDP is being calculated, but not imports which are produced abroad. This is what makes GDP 'domestic'. It is 'gross' because the depreciation of capital is not subtracted even though it is a cost of production. The main point here is that the multiplication of quantities, which are measured in all sorts of different units – kilos of apples, litres of gasoline – by their prices, makes it possible to add them all up to get a measure of the size of an economy in terms of money.

An equivalent measure of GDP and therefore the size of the economy is obtained by summing the value added, i.e. payments of labour and capital at every stage of the production process.[2] Daly finds this definition of GDP especially convenient because it makes us ask value added to what? Daly's answer is to resource throughput: 'low-entropy matter/energy is our ultimate means without which we cannot satisfy any of our ends, including that of staying alive' (Daly 2007, p. 89). By attributing all value in the economy to capital and labour, this accounting convention overlooks any contribution from nature. 'To omit this necessary contribution from nature both from our theory of production and from our accounting of value is a monumental error (ibid.). And as Daly makes clear, the error is magnified the further we move from an empty world, where capital is the limiting factor, to a full world where nature becomes the limiting factor.

To avoid the appearance of growth in economic output being due to an increase in prices rather than quantities, prices are held constant when the GDP for one year is compared with the GDP of another. These comparisons are generally made with GDP calculated as the sum of final expenditures. The calculation of 'real' inflation adjusted GDP is not as simple as it may sound but the basic principles as stated here should be sufficient to understand why GDP is not a satisfactory measure of the size of an economy in purely physical terms. A small quantity of a high-priced commodity and a large quantity of a low-priced commodity appear as equal in GDP. This is one reason why mainstream economists think that GDP can grow without limit while the physical quantity of physical objects that are produced, traded and consumed, declines. They contend that changes in the composition of GDP can reconcile economic growth with constant or declining use of resources and the generation of wastes, though for how long is unclear.

In Daly's view this conclusion is not well-founded and is not borne out by the historical record. The total quantity of materials and energy used as GDP has grown has risen enormously since industrialization began and especially since 1900 (Krausmann et al. 2018). Furthermore, the materials embodied in human-made capital have increased as well, a 23-fold in the twentieth century according to one estimate (Krausmann et al. 2017). When it comes to energy, between 1930 and 2017 at the global level, energy use and GDP rose at virtually identical rates (Keen 2020, Figure 5).

Even the best minds can go astray

Daly shows how reliance on GDP as a measure of the size of an economy can lead even the best minds astray. Before the 'marginal revolution' in economics in the late nineteenth century it was considered a paradox that water was life giving and valuable but cheap, and diamonds were simply decorative but expensive. Marginal analysis, in which attention is focused on the value of incremental changes in utility and costs rather than total utility and costs, provides a good explanation of the diamonds–water paradox since price is determined by marginal utility, not total utility. According to the law of diminishing marginal utility, the marginal utility of each incremental unit a person consumes in a given period of time declines. Ice cream on a hot day is a good example. The first one is wonderful. The second might be pleasant, but the third, fourth and fifth will bring diminishing pleasure, none at all at some point and then negative. If something, like water, is comparatively abundant then both its marginal utility and price will be low. Diamonds in comparison are rare and desirable so their marginal utility, like their price, is much greater than for water. However, the total utility of lifegiving water is much higher than the total utility from diamonds no matter how beautiful they may be. 'The point is that, while margins are reliable means for maximizing totals, they are very treacherous means for evaluating totals, as any student who has pondered the diamonds-water paradox must realize.' Daly continues with the warning that,

> Any sort of economic numerology which… insists on glossing over this treachery deserves a thorough dunking in the satirical acid of Johnathan Swift's *A Modest Proposal*… in which, using exchange-value [price] calculations, Swift logically demonstrates the "economic desirability" of eating children.
>
> *quoted in Daly 1968a, p. 395*

Nearly half a century after he wrote these words, Daly found it necessary to dunk three prominent economists though without reference to Swift. His targets were Professors William Nordhaus, Wilfred Beckerman and Thomas Schelling.

Nordhaus and Schelling are both recipients of the Nobel Memorial Prize in Economics and Beckerman is a distinguished and influential British economist. They were commenting independently on the likely magnitude of climate change on the total value of an economy's output as measured by GDP or an equivalent statistic.[3] Using agriculture as an example they made identical calculations. They each started from the observation that the value of agriculture in the USA is about 3 per cent of national output.

Referring to climate change Nordhaus reasoned that, 'there is no way to get a very large effect on the USA economy', Beckerman said that 'even if net output of agriculture fell by 50 per cent by the end of the [21st] century this is only 1.5 per cent cut in GNP', and according to Schelling 'the cost of living would rise by 1 or 2 percent, and at a time when per capita income will likely have doubled' (as quoted in Daly 2014a, pp. 107–108).

It would be hard to find a better example of the diamond–water paradox. As Daly pointed out, the total value of food cannot be calculated by multiplying the total quantity of each food by its market price. Market prices reflect the value of the marginal unit which can be very low if supply is ample and available at low cost. But if food were to be in short supply because of climate change, market prices would increase dramatically as would the contribution of agriculture to GDP. 'True, agriculture accounts for only 3% of GDP but it is precisely the specific 3% on which the other 97 percent is based!' (ibid., p. 108) In a typically apposite analogy, Daly says, 'The foundation of a building may be only 3% of its height, but that does not mean that we can subtract the foundation if only we add 3% to the flagpole on top of the building' (ibid.).

Daly considers several possible reasons why these economists would make such a fundamental error. One is the assumption that all parts of national output are substitutes for each other so that more information services can make up for a lack of food. Another possible reason is that they overlooked the impact of a loss of agricultural output on all sectors that rely on it as an input: restaurants, hotels, hospitals etc. Yet another, not mentioned by Daly, is the assumption that any deficiency in the national production of food can be made up by imports from other countries who somehow remain unaffected by climate change. Daly concludes, 'In all three cases, the bad argument was part of a larger defense of economic growth' (ibid.).

This brings us back to Daly's perspective on economics as a life science. Had Nordhaus, Beckerman and Schelling understood economics as a life science it is doubtful that they would have concluded that because the market value of agriculture is such a small proportion of GDP, its loss would barely be noticed. Food is fundamental to all living species, including humans. It is essential, as Schroedinger explained, for combatting entropy. Throughout history, villages, towns and cities have depended on the surplus production of food on farms for their sustenance and survival. An ample and secure supply of food is the foundation of civilization. In tranquil times with good harvests, it is easy to overlook the fundamental importance and value of food in an economy. To suggest that the contribution of

agriculture to economic output is not much more than the statistical error in measuring GDP misses the point.

An economic plimsoll line

Many years before he published his ideas about scale, distribution and allocation in
1992, Daly had made an analogy between the Plimsoll line on ships and the scale
of an economy. The Plimsoll line, named after Samuel Plimsoll who proposed it in
the late nineteenth century (Plimsoll 1873), is a line drawn around the hull of a
ship. As long as the line is visible then the ship is not overloaded and in danger of
sinking in rough seas. Daly believed that something equivalent to a Plimsoll line was
needed for the economy, a simple, easy to understand, measurement that indicates
whether or not the scale of the economy is excessive given its ship-like dependence on nature. And just as moving the cargo around the deck of an overloaded ship
cannot solve the problem of overloading, so, Daly says, changing the allocation of
resources in an economy cannot overcome the problem of excessive scale. The scale
of the economy, like the cargo limit on a ship, should be determined first, and then
distribution and allocation can follow.

One difference between Plimsoll and Daly is that Plimsoll's proposal for a load
line was easy to understand, easy to implement and easy to enforce, providing
there are sufficient inspectors. In comparison, the scale of an economy is not
simple to define or explain, the information required to measure it is not readily
available, and imposing scale limits on an economy that can be easily enforced
is multi-faceted and complicated. These are not aspects of scale that Daly has
concerned himself with in great detail; but others who share his outlook on the
physical basis of economies, have made valuable contributions to its measurement. Two good examples are the material flow accounts developed by a group
of scholars based mainly in Austria at the Vienna School of Social Ecology, led
by Marina Fischer-Kowalski, Helmut Haberl and Fridolin Krausmann. Several of
the early publications of Marina Fischer-Kowalski, Helmut Haberl and others on
'social metabolism' refer to Daly's work. Estimates of global and national ecological
footprints were led by Mathis Wackernagel of the Global Footprint Network in
Oakland, USA and William Rees in Vancouver, Canada who were also inspired
by Daly.

Material flow accounting

The idea of analysing the material throughput of economies has quite a long history. In 1969 Robert Ayres and Alan Kneese modified the Walras-Cassel theoretical
model of a multi-sector economy (named after its originators Leon Walras and
Gustav Cassel), to be consistent with the principle of materials balance (Ayres and
Kneese 1969).[4] Materials balance is the application of the conservation of mass (the
first law of thermodynamics) to physical systems. Ayres and Kneese considered the
idea of 'consumption' of products misleading. Production, they said, is a process

of transformation of materials, some of which ends up in the products and some ends up as wastes, possibly after recovery and recycling. The products also end up as wastes though, depending on their type and use, they can remain in the economy for some time. This was the same idea that Daly wrote about when describing economics as a life science and described by him in an expanded input-output framework. 'There is also a balance equation of the life process in physical units, based on the law of conservation of matter–energy' (Daly 1968a, p. 395).[5]

Daly did not produce his own estimates of material and energy throughput but relied instead on the work of others such as Adriaanse et al. (1997) and Mathews (2012) at the World Resources Institute. Material flow accounting (MFA) has advanced considerably in recent years. The 'metabolism' of economic and social systems has become a focus of study and principles of material flow accounting have been established and codified (OECD 2008; Eurostat 2018;[6] UN 2020). Numerous empirical studies using MFA have been undertaken at global, national and sub-national levels. Krausmann et al. (2018) estimated global material flows and stocks for 1900 to 2015.

Figure 6.1 shows a material flow accounting framework in which stocks (rectangles) are distinguished from flows (arrows). The similarity between this figure and Daly's earlier conception (Figure 4.2) are easy to see and this is not a coincidence. Daly's work was very influential for the Viennese School of Social Ecology and their efforts to develop quantitative methods and indicators to measure the physical size of the economy.

Applying the framework shown in Figure 6.1, Krausmann et al. estimated in considerable detail the annual and cumulative flow of materials through the global economy from extraction to use and output of wastes and emissions from 1900 to

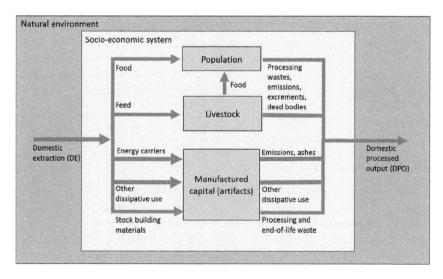

FIGURE 6.1 A material flow accounting framework

Source: Krausmann et al. (2018) p.132.

2015. They divided material flows into several use categories: food; feed; energy materials – fossil fuels, wood fuel and crops; other dissipative uses (e.g., seed, fertilizer minerals, salt), building materials; and several output categories: food/population, feed/livestock; energy wastes; building materials/manufactured capital; other dissipative uses; and processing and end-of-life waste. Figure 6.2 shows the estimated cumulative flow of materials through the global economy from 1900 to 2018 including the retention of materials in capital stocks (stock building). These capital stocks have been rising rapidly, doubling between 1980 and 2020, reaching a point where it is estimated that the 'anthropogenic mass' began to surpass all global living biomass having been only 3 per cent of it in 1900 (Elhacham et al. 2020).

Over years of dedicated research, Krausmann and his co-researchers have estimated the material throughput of economies as conceived by Georgescu-Roegen, Boulding, Daly and others, demonstrating the feasibility of measuring the scale of economies in purely physical terms. The progression from an empty to a full world, as Daly describes it, from 1900 to 2015 is indicated by the increase in the extraction of materials from 12 Gt/yr in 1900 to 89 Gt/yr in 2015. Outflows of wastes (58 Gt/yr in 2015) grew more slowly than inputs of virgin resources because of the accumulation of materials in stocks of infrastructure, buildings and equipment (Krausmann et al. 2018, p. 131).

BOX 6.1

His work definitely was influential for the Viennese school of social ecology and our efforts to develop quantitative methods and indicators to measure the physical size of the economy vis a vis the dominating monetary tools. Overall, Daly's thermodynamic perspective on the economy was an important background to the development of the whole concept of a social or industrial metabolism and the tools of material flow analysis.

Fridolin Krausmann

A powerful illustration of the transformation of the world from empty to full using a progression of 'Sankey' diagrams similar to Figure 6.2 but for annual data is shown in Figure 6.3.

The ecological footprint

A different but complementary approach to measuring the scale of economies is offered by the ecological footprint. It is a further development of Vitousek et al.'s estimate in the 1980s of the human appropriation of the products of photosynthesis which Daly referred to in several of his articles in the early 1990s as evidence for the growing economic subsystem in a finite global ecosystem (see Chapter 5).

Ecological footprint accounting can be thought of as the 'measurement of the *demand* on and *supply* of nature' (Global Footprint Network 2020). The ecological

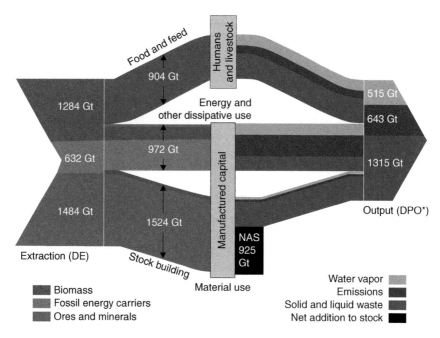

FIGURE 6.2 The cumulative flow of materials through the global economy, 1900 to 2018

Source: Krausmann et al. (2018) p. 137.

footprint of the global, national and regional economies provides a measure of the scale of these economies in terms of the demands they place on natural systems. The unit of measurement devised for this purpose is the standardized global hectare. A global hectare is a weighted, biologically productive hectare for comparing different types of landform according to their productivity.

A comparison of ecological footprints with estimates of biocapacity gives an estimate of the extent to which a region is full in terms of the demands it places on its own biocapacity. Some regions can draw on the biocapacity of other regions through trade, but this is not possible at the global level. The National Footprint Accounts are now published annually for over 200 countries. Daly often refers to the foundational work on the ecological footprint by William Rees and Mathis Wackernagel (Wackernagel and Rees 1996) to make the case that the human economy is exceeding the capacity of the planet to support it.

Figure 6.4 shows estimates of the global ecological footprint and global biocapacity for 1961 to 2017. Both increased but for different reasons. The ecological footprint increased because of the combination of an expanding human population and increasing levels of material consumption. Global biocapacity increased much more slowly, reflecting increases in productivity due to changes in production methods, especially in agriculture where output per acre grew with

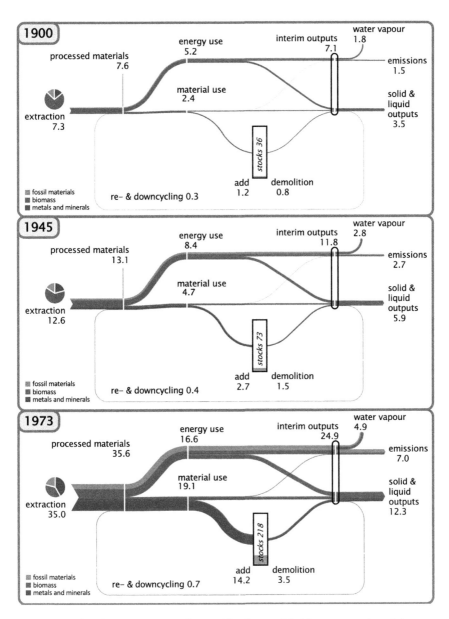

FIGURE 6.3 Global socioeconomic flows of fossil materials, biomass, metals and (non-metallic) minerals through the global economy between 1900 and 2015. All flows are shown true to scale in Gt/yr; stocks shown in the box in the centre as a grey bar are in Gt and scale differently than flows.

Source: Haas et al. (2020, p. 5).

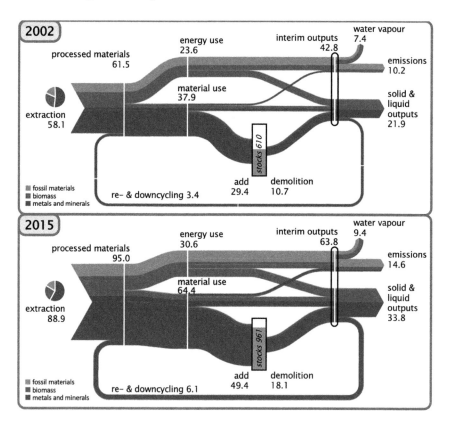

FIGURE 6.3 (Continued)

mechanization, and increasing use of energy, pesticides, fertilizers and genetically modified organisms. Figure 6.4 shows that in 1970, the global ecological footprint began to exceed global biocapacity, largely due to increasing emissions of carbon dioxide and the inability of natural systems to absorb them. This excess of human demands on what nature can provide comes at the cost of running down natural systems and cannot be sustained. The evidence that some planetary boundaries are being exceeded makes this increasingly clear (Steffen et al. 2015). As Daly insists, scale matters.

Is there an optimum size for an economy?

The fact that economies are growing while the biosphere is fixed or in decline owing to degradation caused by excessive extraction of materials and disposal of wastes, raises the question of whether there is an optimal size for the economy and if so, how can it be determined? We have seen that Daly distinguishes between growth as quantitative and development as qualitative (see Chapter 5) and that GDP is problematic as a measure of an economy's size because it conflates the two. Such is not

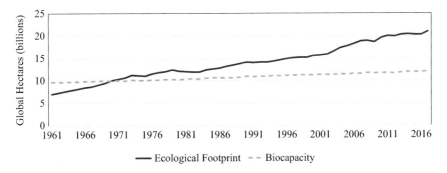

FIGURE 6.4 Global ecological footprint and biocapacity, 1961 to 2017

Source: York University Ecological Footprint Initiative & Global Footprint Network. 2020. National Footprint and Biocapacity Accounts, 2021 edition. Produced for the Footprint Data Foundation and distributed by Global Footprint Network. Available online at: https://data.footprintnetwork.org.

the perspective of most economists who are content to use GDP as a measure of an economy's size and as a normative indicator of a society's well-being. One example is *The Growth Report* written by 18 blue-ribbon contributors (including two winners of the Nobel Memorial Prize in Economics, Robert Solow and Michael Spence), from 16 countries which contains the following statement about GDP:

> Gross domestic product (GDP) is a familiar but remarkable statistic. It is an astonishing feat of statistical compression, reducing the restless endeavor and bewildering variety of a national economy into a single number, which can increase over time.…A growing GDP is evidence of a society getting its collective act together.
>
> *Commission on Growth and Development 2008, p. 17*

Daly's comment on this quote is instructive:

> Well, it may also be evidence of a society depleting its life-sustaining natural capital and counting it as current income, of asymmetric entries that count defensive expenditure on anti-bads (e.g., pollution cleanup) but fail to enter negatively the bads (pollution) that made the anti-bads necessary, or of shifting household production into the monetary economy because both spouses are now breadwinners (also add a further GDP increase from the extra salary). Also, since GDP counts gross rather than net investment, it increases with the depreciation and replacement of existing manmade capital. Correcting for these accounting anomalies can sometimes reduce countries in the 7 percent growth club to membership in the 0 percent growth club. The issue of distributive inequality also escapes the statistic.
>
> *Daly 2008a, p. 514*

These deficiencies in GDP are well-known to economists who understand how GDP is calculated, even though they may continue to interpret it as if they are of no real consequence. Growth in GDP remains the primary economic policy objective of most governments and economists, with growth qualified sometimes these days by adjectives such as inclusive, green, clean and sustainable. Daly sees it differently. To him, GDP is an easily misinterpreted statistic that provides false information about how well an economy is serving the interests of society in general. Growth in GDP, he argues, can become *uneconomic* in the sense that the benefits it brings, usually only to some, are exceeded by the costs it imposes disproportionately to others. Daly says that when economic growth, measured in the conventional way as an increase in GDP, becomes uneconomic, it has reached its optimal size. To illustrate this proposition, and in a largely unsuccessful attempt to persuade mainstream economists, Daly applies conventional economic thinking in an unconventional way.

Figure 6.5 shows how the marginal utility from production and consumption declines as an economy grows. Daly bases this decline on the simple idea that people as individuals and collectively generally satisfy their most important wants first and less important wants later. The marginal disutility of production and consumption is shown as increasing in Figure 6.5. This is due to shrinking ecosystem services as the growing economy expands into the non-growing biosphere. Daly uses his diagram to distinguish among three concepts of the limits to growth (Daly 2017).

The first is the 'futility limit,' which occurs 'when marginal utility of production falls to zero. Even with no cost of production there is a limit to how much we can consume and still enjoy it… in a given time period' (ibid., p. 94). Daly realizes that there are many people in the world living in considerable poverty and for whom the futility limit is far away. But the 'non-satiety postulate [of] neoclassical economics formally denies the concept of the futility limit [for people no matter how

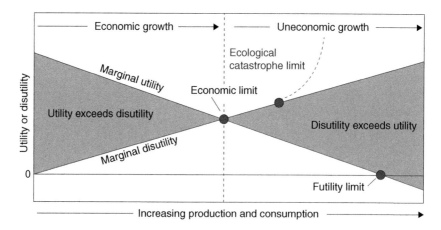

FIGURE 6.5 Three limits to growth

Source: Daly (2017) p. 93. Illustration by Jen Christiansen.

rich they are]… even though studies show that, beyond a certain threshold, self-evaluated happiness (total utility) ceases to increase with GDP' (ibid.).

Daly calls the second limit to growth the 'ecological catastrophe limit' where 'some human activity, or novel combination of activities, may induce a chain reaction, or tipping point, and collapse our ecological niche' (ibid.). This limit is represented by the sharply rising marginal cost curve, and as Daly notes, even that is optimistic: 'The marginal cost curve might in reality zig-zag up and down discontinuously… [because]… we may ignorantly sacrifice a vital ecosystem service ahead of a trivial one' (ibid.).

The third and most important limit is the 'economic limit', where marginal cost equals marginal benefit, maximizing net benefit from production and consumption. 'The good thing about the economic limit is that it would appear to be the first limit encountered. It certainly occurs before the futility limit, and perhaps as shown, before the catastrophe limit, although that is quite uncertain' (ibid.).

It is well known, especially among critics of GDP, that its originators never intended it as a measure of welfare. Simon Kuznets, who did much of the early work on GDP said to the US Congress in 1934 that, 'the welfare of a nation can scarcely be inferred from a measurement of national income' (national income is essentially the same as GDP). Thirty-eight years later, Robert Kennedy summarized his critique of GNP (GDP plus net income receipts from abroad) by saying, 'It measures neither our wit nor our courage; neither our wisdom nor our learning; neither our compassion nor our devotion to our country; it measures everything, in short, except that which makes life worthwhile' (Kennedy 1968).

The Index of Sustainable Economic Welfare (ISEW)

Not long after Kennedy made these remarks economists began to develop modified versions of GDP to make it align more closely with well-being. Daly joined with John Cobb and his son Clifford Cobb to develop the Index of Sustainable Economic Welfare (ISEW). They acknowledged the contributions of Zolotas (1981), though he did not consider sustainability, and Nordhaus and Tobin (1972), who did consider sustainability but without considering environmental issues. In 1989 Daly and John Cobb published the first version of the ISEW in their book *For the Common Good*, gratefully acknowledging Clifford W. Cobb's 'painstaking work and leadership' in preparing the Appendix on the ISEW. It was an implementation of the ideas sketched out by Daly in his contribution to a symposium organized by UNEP and the World Bank drawing on a series of workshops held over the previous six years (Daly 1989). The ISEW was revised for the second edition of *For the Common Good* in 1994 based on extensive reviews of their first attempt by several eminent economists.[7] In light of the reasons why GNP (and by extension GDP) is a poor indicator of economic welfare, Daly and Cobb set out to produce a statistic, comparable to GNP, that 'would give better guidance than the GNP to those interested in promoting economic welfare' (op cit., p. 443). Although Daly and Cobb have written independently and together on many aspects of welfare in the sense of well-being, they did not define 'economic welfare' for the purposes of their index. One definition was offered by the Cambridge economist Arthur Pigou in his classic *Economics of Welfare* (Pigou 1920) who said that economic welfare '... is that part of social welfare that can be brought directly or indirectly into relation with the measuring rod of money' (ibid., p. 11). The ISEW can be understood as an attempt to measure economic welfare defined in this way since all of the factors that enter into it are measured in terms of money.

BOX 6.2

I have always recommended Daly's work to my students and colleagues as I regard him as one of the clearest thinkers and writers in all of environmental science. I consider ISEW the most important alternative indicator to potentially replace the GDP.

Jeroen van den Bergh

Construction of the ISEW starts with annual expenditure on consumption, taken from the National Accounts.[8] An inequality weight is applied to consumption expenditure so that consumption expenditures by poorer people count more than by richer people. Estimated values for three variables related to economic welfare are added to this inequality-adjusted value of consumption:

- services of household labour;
- services of highways and streets;
- the portion of public health and education expenditures that contribute to welfare.

To account for the fact that consumer durables provide services over many years, expenditures on consumer durables are subtracted and an estimate of the annual value of their services is added.

The next step in the calculation of the ISEW is to subtract values for variables that detract from economic welfare. The Daly and Cobb list is extensive:

- defensive private expenditures on health and education;
- cost of commuting;
- cost of personal pollution control;
- cost of auto accidents;
- costs of water pollution;
- costs of air pollution;
- costs of noise pollution;
- loss of wetlands;
- loss of farmland;

- depletion of non-renewable resources;
- long-term environmental damage;
- cost of ozone depletion.

The final step in Daly and Cobb's calculation of the ISEW is to add the value of additional manufactured capital and the change in the net value of foreign investment. Summing all these values gives the ISEW, which divided by population gives the ISEW per capita, which, because it is measured in money, can be directly compared with GDP and GDP per capita, or as Daly and Cobb chose to do, with the almost equivalent GNP and GNP per capita. What they found when they made the comparison was discouraging. In the USA from 1951 to 1990 GNP per capita increased at an average annual rate of 1.89 per cent compared to 0.39 per cent for the ISEW per capita. As Figure 6.6. shows, when this is broken down by decade, both GNP per capita and the ISEW per capita declined in successive decades starting in 1960–70, turning negative in 1980–90 for the ISEW.

> Despite the year-to-year variations in the ISEW, it includes a long-term trend from the late 1970s to the present that is indeed bleak. Economic welfare has been deteriorating for a decade, largely as a result of growing income inequality, the exhaustion of resources, and unsustainable reliance on capital from overseas to pay for domestic consumption and investment… The most fundamental problem in terms of sustainable economic welfare is the decline in the quality of energy resources as measured by the ratio of energy output to energy input.
>
> *op cit., p. 507*

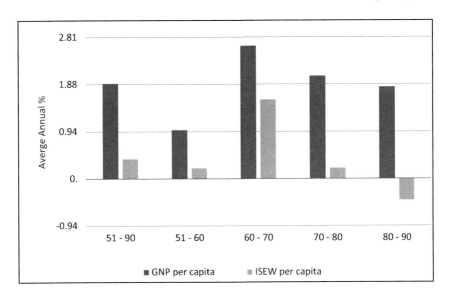

FIGURE 6.6 Annual per capita growth of per capita ISEW and GNP in the USA
Source: Based on data from Daly and Cobb (1974), Table A.13 p. 505.

Daly has used this negative growth in the ISEW per capita to support his contention that growth in the US economy has become uneconomic in the sense that it is reducing rather than increasing the economic welfare of Americans. If Daly is correct, then in terms of Figure 6.5 it means that the USA has passed the economic limit to growth or, as Daly might say, in the absence of a 'when to stop rule' the US economy has gone beyond its optimum size. 'Clearly the important question then becomes whether our nation is going to continue in its efforts to increase total output or whether we are going to redirect our focus towards the enhancement of sustainable economic welfare. Are the policies of our government going to be guided by GNP, or by some other measure of sustainable welfare?' (ibid.)

BOX 6.3

The ISEW created the foundation to nearly two decades of work on the GPI at the US state level, including an effort to pass GPI into law in Vermont. The scale/distribution/allocation framing has been highly influential in my teaching of ecological economics, as has his extensions of Georgescu-Roegen and the ends-means spectrum.

Jon Ericson

Daly and Cobb's questions remain pertinent even if their estimation of the ISEW could stand improvement, or if the ISEW itself were to be replaced with one or more metrics similar in scope showing a similar result but less reliant on monetary values. In any case, estimating the economic welfare of a nation over time is fraught with difficulties, which Daly and Cobb recognized and to which they drew attention. They pointed out that starting from personal consumption expenditures, as they did in building the ISEW, is problematic because it presumes that people are automatically better off if they consume more. Daly and Cobb questioned this assumption and noted research showing that after a certain level, a person's relative income or consumption matter far more to them than the absolute level. The income inequality weighting included in the ISEW goes some way to addressing this deficiency. Although the income inequality weighting reflects the law of diminishing marginal utility of income to the rich, there was no correction for diminishing marginal utility of income for an increase in total income to the entire population. If such a correction was made it would increase the divergence between ISEW and GDP.

Daly and Cobb remind us of a second problem with their estimate of the ISEW: variables, both positive and negative, that affect economic welfare were omitted because of a lack of data and the need for highly subjective judgment. Their examples of omitted variables include unreported income, changes in working conditions, expenditures on junk food, tobacco, pornography and 'innumerable other items that make questionable contributions to genuine economic welfare' (ibid., p. 460).

An additional limitation of their ISEW estimate is the 'heroic assumptions' that they had to make to include some of the variables such as the costs of long-term environmental damage stemming from human-caused climate change which was only beginning to be recognized as a major threat. Daly had been aware of the threat of climate change for a long time. In his paper on economics as a life science published in 1968, long before climate change was on the public agenda, he said that

> since the Industrial Revolution the tremendous consumption of carbon fuels has resulted in an increased concentration of carbon dioxide in the atmosphere. Since this gas increases the heat retention of the atmosphere, thus raising the average temperature, it may well be that the ultimate effect of the Industrial Revolution will be the melting of the polar ice cap and the inundation of large parts of the world.
>
> *Daly 1968a, p. 399*

Following the publication of the ISEW in 1989 and 1994, numerous studies have been undertaken for other countries comparing the ISEW or related metrics to GDP showing that the stagnation and decline in economic welfare reported by Daly and Cobb is not unique to the USA. A similar result has been found for many countries including Australia, Austria, Belgium, Germany, Italy, the Netherlands, New Zealand and Sweden (Pulselli et al. 2008; Kubiszewski et al. 2013). As of June 2020, Wikipedia listed 115 ISEW and GPI-type studies completed for countries and regions around the world.

Others have made suggestions for how to improve the ISEW (e.g., Dietz and Neumeyer 2006 who make several constructive criticisms of the ISEW). Clifford Cobb went on to develop the Genuine Progress Indicator based closely on the ISEW, and to found Redefining Progress, an NGO based in San Francisco. It was more than a coincidence that Mathis Wackernagel also spent some time in Redefining Progress working on the ecological footprint.

Interest in metrics other than GDP for assessing economies is spreading. Key international organizations including the United Nations, the World Bank, the OECD, Eurostat have undertaken work on alternatives to GDP, usually to complement it rather than supplant it, and there have been major studies for and by Commissions looking at better ways to measure progress than increases in GDP. The influence of Daly and Cobb's work on measuring the economy is clear, although it is not always given the recognition it deserves. One example is the Commission on the Measurement of Economic Performance and Social Progress established in response to President Sarkozy of France's dissatisfaction with the available statistical information about the economy and society. The Commission was led by Professors Stiglitz, Sen and Fitoussi (Stiglitz, Sen and Fitoussi 2009). Its task was to report on the measurement of economic performance and social progress. Daly was not invited to join the Commission, nor was he asked to make a submission to it. The first edition of *For the Common Good* is included in the references but not the second edition, and it is only mentioned once in the 300-page text as having 'further advanced' Nordhaus and Tobin's 1972 Measure of Economic Welfare (ibid., p. 239). The further advance was the inclusion of the environment, a serious omission one would think from a measure of economic welfare intended to take sustainability into account, but apparently not worth describing in the Commission's report.

A second example of how Daly's views on economic growth and how to measure it have been ignored by the mainstream is described by Daly himself in some detail (Daly 2014a, pp. 59–62). Earlier in this chapter we included a quote from Daly criticizing the interpretation of growth in GDP by the Commission on Growth and Development sponsored mainly by the World Bank. This quote comes from a review of the Commission's report by Daly which was studiously ignored (Daly 2008a). The main question about the report that Daly raised was 'Is growth still economic in the literal sense, or has it become uneconomic?', which as Daly says is surely not a trivial question. Daly expected a debate of his critical review, but no such debate transpired – even though sympathetic former colleagues of Daly's at the World Bank sent his review to authors of the Commission's report suggesting that a reply was in order. None came, even though the editor of the journal that published Daly's review offered for the second time to publish their reply.

Daly considered several possible reasons for the silence. One was that perhaps the report's authors were too busy responding to other reviews of the report, but there were few of these and most were descriptive, not critical, requiring no response. A second reason that Daly considered more likely was

> perhaps they made a political calculation of interest and advantage... A blue-ribbon panel of experts is presumed to be correct (especially if defending growth!), and a single critic is presumed to be wrong. Why risk upsetting that default proposition with a reply?
>
> *ibid., p. 60*

Clearly, Daly was disappointed, frustrated and angry at not evoking a response from those whose views he criticized. Never one to shy away from academic debate, and always ready to engage in a courteous manner typical of the southern gentleman that he is, Daly has been unable to induce most mainstream economists to address questions about economic growth that to ecological economists, heterodox economists more generally, other academics, and a very large number of activists and members of the public, seem worthy of debate. One unexpected voice of support for Daly's position came in 2008 when Solow, who had strenuously argued for substitution and technology as solutions to environmental pressures said,

> It is possible that the US and Europe will find that... either continued growth will be too destructive to the environment and they are too dependent on

scarce natural resources, or that they would rather use increasing productivity in the form of leisure....

Solow quoted in Stoll 2008, p. 92

Had Solow credited Daly for persuading him of this, we might have seen a greater interest among economists in the proposition that growth in these countries has become uneconomic, but that is yet to come.

Measuring distribution

In contrast to GDP, income distribution has an influence on the ISEW (and the GPI) through the inclusion of inequality weights applied to consumption expenditures. This is obviously relevant to economic welfare but is insufficient to fully capture the significance of inequality in the distribution of income and wealth, which is the second component in Daly's scale, distribution, allocation framework.

Compared with scale, where measurement especially in physical units is still comparatively new, income and wealth distribution has been the subject of empirical measurement and study for decades and even longer. Daly himself was influenced by Jan Pen's book on income distribution which included Pen's memorable parade of British income earners (Pen 1971). Pen imagined that every income earner's height was adjusted in proportion to their income. People with average income would have average height. Pen described a parade lasting one hour where people pass an observer in order from the shortest (lowest income) to the tallest (highest income). For most of the hour the people, having lower than the average income, are very short. It takes 45 minutes for someone of average height to pass by. Then people in the parade start to grow very tall. With six minutes to go people in the top 10 per cent of income earners appear; minutes later, 20-foot tall doctors, lawyers and senior civil servants pass, then corporate executives, bankers, stockbrokers quickly going from 50 to 100 to 500 feet. The final glimpse, lasting only seconds, is of a few entertainment stars, highly successful entrepreneurs, and members of the Royal family. The parade ends with John Paul Getty, the American oil magnate who also held British citizenship. He was so tall, no one knew how rich he was. His head was in the clouds.

In 2014 interest in income and wealth distribution received a considerable boost with the publication of Piketty's book *Capital in the Twenty-First Century* in 2014. The book contains extensive data on the distribution of income and wealth, which he and colleagues continue to update and make available on the World Inequality Database.

As an example of Piketty's work, Figure 6.7 from the World Inequality Database shows how the percentage of pre-tax national income in the USA accruing to the bottom 50 per cent and top 1 per cent changed between 1913 and 2019. Many commentators have noted that income distribution prevailing at the end of the second decade of the twenty-first century is comparable to what it was just before the Wall Street crash in 1929 that preceded the Great Depression of the 1930s.

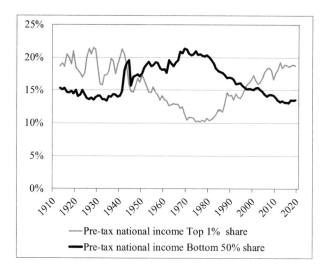

FIGURE 6.7 Income inequality in the USA from 1913 to 2018
Source: World Inequality Database.

Daly's point about distribution is that it is a proper subject for public policy and should not be left to the workings of the market to determine, a view that is widely, though not universally, shared. After World War II in the USA, as in many other countries, through taxes and transfers and other elements of the welfare state, the distribution of income became much less unequal, as is clear in Figure 6.7. In the 1970s income inequality began to increase and from 1980 did so sharply when neo-liberal economic policies were adopted in the Reagan-Thatcher years. This is strong evidence that policy does make a difference to income distribution, and to wealth distribution as well.

What sets Daly apart from others on distribution is his view that policy instruments designed to address unacceptable levels of inequality should be complemented by policy instruments to address excessive scale and inefficient allocation. A just distribution and efficient allocation, laudable as they are, are insufficient to deal with the question of scale, be that national or global. Furthermore, Daly points to the political tendency to deal with distribution indirectly by increasing scale. If everyone can become richer, so it is suggested, then we can avoid the conflict that is presumed to arise from redistribution. If scale is limited, we have to deal directly with distribution to ensure that it meets a reasonable standard of fairness with minimum and maximum income limits, a range of income differences sufficient for incentives but not sufficient for plutocracy (Daly 2017, p. 101). On occasion, Daly has suggested a factor of 4 to 1 for the highest to lowest income following Plato, a factor of 10 to 1 based on the range of incomes in government service, the military, universities and the Japanese economy, and mentioned a factor of 100 for those who favour inequality. These numbers are useful for starting a discussion of a maximum desirable degree of income inequality but for Daly it is the principle of a limit to income inequality rather than the exact range that has interested him the most.

When it comes to a just distribution of wealth, Daly draws on principles devised by John Locke and John Stuart Mill in the UK and Thomas Jefferson and the other Founding Fathers in the USA (Daly 1977a, pp. 53–56; Daly and Townsend 1993, pp. 332–335). The primary concern of all these men was with private property, and in their day, this was mainly land and farm implements (not forgetting the abhorrent institution of slavery). They believed strongly in the view that property rightly belongs to people who have applied their labour to the land making it and its product their own, but there were limits. Locke, for example, put the limit at whatever 'anyone can make use of to any advantage of life before it spoils, so much may he by his labor fix his property in. Whatever is beyond this is more than his share, and belongs to others' (quoted in ibid., p. 333). Daly comments, 'Clearly, Locke had in mind some maximum limit on property, even in the absence of general scarcity' (ibid.), something that today's most vociferous defenders of property rights have forgotten.

This view of private property has several implications. 'Property should be acquired through *personal effort*…, property implies *personal control* and individual responsibility…, property is relative to *human need*…' (McClaughry quoted in ibid., p. 333). It is obvious that what counts as property today is much broader than originally envisaged by the philosophers and activists of a bygone era and its distribution is wildly unequal. In the USA, for example, 'the top 1% wealth share rose from 22% in 1980 to 39% in 2014. Most of that increase in inequality was due to the rise of the top 0.1% wealth owners' (Alvaredo et al. 2020).

This degree of inequality in the distribution of wealth is not only unjust, it runs against the very purpose envisaged by those who built the case for private property on the avoidance of exploitation. If you owned your own land and tools, or just tools if a smithy or baker, then by virtue of your labour you owned what you produced and could consume or sell it. Exploitation under these conditions was not impossible but it was difficult. Daly reminds us that,

> according to Mill, private property is legitimated as a bastion against exploitation. But this is only true if everyone owns some minimum amount. Otherwise, private property, when some own a great deal of it and others have very little, becomes the *instrument* of exploitation rather than a guarantee against it.
>
> *Daly and Townsend 1993 p. 332*

In order to address the extreme disparities in the distribution of wealth, even more so than with income, its measurement is fundamental.

Conclusion

Daly's contribution to measuring the economy is first to argue that scale matters and that it is a separate and distinct matter for public policy. From there he has offered trenchant criticisms of GNP and GDP for measuring the size of economies. They fail to distinguish between quantity and quality, mix goods with bads, and omit many variables relevant to economic welfare such as the environment and income distribution. In close collaboration, John Cobb, his son Clifford Cobb, and Daly developed the Index for Sustainable Economic Welfare and estimated its value for the USA over many years. The estimates showed that in the USA the ISEW per capita plateaued in the 1980s and then declined, indicating that economic growth had become uneconomic.

Daly used his insights to criticize the report of the Commission on Growth and Development, but his criticisms were ignored, despite efforts by him and others to prompt a response. The Commission on the measurement of economic perform-ance and social progress was no better. This reluctance of mainstream economists to engage with Daly has been all too common throughout his career. We will encounter it again in later chapters. Fortunately, Daly's influence among researchers, students and activists has been considerable, and his ideas about the significance of the scale of economies have inspired others to develop new and different approaches to measurement, material flow accounts and the ecological footprint being two such examples.

Daly's interest in scale stems from his understanding of the economy as a sub-system of the biosphere. The one is growing the other is not. The limits to growth that follow from this reality raise questions of distributive justice that can no longer be ignored or shuffled aside on the assumption that everyone can gain from continued growth. Great strides have been taken in recent years in the measurement of the distribution of income and wealth, and there is renewed interest in policy measures such as a basic income and job guarantee. The emphasis that Daly has given to distribution as a pre-condition for allocative efficiency is very much in line with the increasing priority being given to distribution in this very unequal world.

Notes

1 Roland McKean was the first economist to use the term 'uneconomic growth' (McKean 1973, p. 207) in this sense but Daly, who was unaware of McKean's usage, introduced it independently and is credited with popularizing it and exploring its multiple dimensions and implications.

2 Summing final expenditures is one of three equivalent ways of measuring GDP. GDP is

> '[1] an aggregate measure of production equal to the sum of the gross values added of all resident institutional units engaged in production (plus any taxes, and minus any subsidies, on products not included in the value of their outputs); [2] the sum of the final uses of goods and services (all uses except intermediate consumption) measured in purchasers' prices, less the value of imports of goods and services, and [3] the sum of primary incomes distributed by resident producer units'
>
> (OECD 2001) i.e. value added.

3 Nordhaus referred to 'national output', Beckerman to 'GNP' which is GDP plus net receipts from abroad, and Schelling to 'national income' by which he probably meant GNP.

4 The Walras-Cassel model of a market economy is similar to the Arrow-Debreu model mentioned in Chapter 5.

5 A materials balance equation for production and consumption of a national economy was first included in environmentally-extended input-output analysis and used for empirical estimates by Victor (1972).

6 Eurostat has an online database of material flows in EU member countries.

7 As an example of exceptional courtesy and generosity to one's critics, John and Clifford Cobb sent the original version of the ISEW to a number of recognized experts asking for their criticisms, paying an honorarium to them, and publishing their collected criticisms in a volume, along with a new version of ISEW (the one in the second edition of *For the Common Good*), which would either take account of the criticisms and suggestions, or explain why not (Cobb and Cobb 1994). It included comments from Robert Eisner, E. J. Mishan, Jan Tinbergen and other national accounts experts, yet received little attention. Daly did not join in editing this book because at the time he was too busy arranging to leave the World Bank and move to the University of Maryland. He believed that Clifford and John at this stage did not need his help and also that his reputation as anti-growth might have dissuaded participation by some mainstream economists in what nevertheless was an objective technical review (Personal communication).

8 Full details of the calculation of the ISEW for the USA are provide in the Appendix to Daly and Cobb (1994).

7

WHAT'S WRONG WITH ECONOMIC GROWTH?

> At this point the growthmaniacs usually make a burnt offering to the god of technology: surely economic growth can continue indefinitely because technology will continue to 'grow exponentially' as it has in the past. This elaborately misses the point.
>
> *Daly*

Daly's disaffection with economic growth runs strong and deep. His critique of economic growth, both in theory in and practice, has been a common theme throughout his career as is apparent from the preceding chapters and will become even more so in the succeeding ones. It is hard to find anything that he has written in which he fails to criticize growth, often along with the individuals and institutions that promote it. In recent times his criticisms of growth have been reflected in the various adjectives used to describe different kinds of growth, but growth just the same – green growth, inclusive growth, sustainable growth, smart growth, broad-based growth, clean growth, shared growth, resilient growth and pro-poor growth, climate-friendly growth – presumably because growth in practice is none of these. Daly's alternative to economic growth is more radical. Since the 1970s he has made the case for a steady-state economy, one that does not grow in physical terms, which we will explore in the next chapter. As a prelude, it is useful to summarize Daly's critique of economic growth because it provides the rationale for his advocacy of a steady-state economy.

DOI: 10.4324/9781003094746-7

BOX 7.1

Daly's most important contribution has been his insistent critique of economic growth, starting at a time when this was unheard of among economists. His arguments have provided very important support for all the rest of us, when we engage in debates.

Inge Røpke

On three occasions Daly wrote his own summaries of what's wrong with economic growth: 'A Catechism of Growth Fallacies' (Daly 1977a), 'The Economic Growth Debate: What Some Economists Have Learned But Many Have Not' (Daly 1987) and 'A Further Critique of Growth Economics (Daly 2014a). All three accounts are comprehensive, but they differ in their orientation and emphasis. All are instructive. In the first one Daly describes fallacious aspects of different views about the feasibility and desirability of economic growth. In the second he frames his criticism of growth in terms of its biophysical and ethicosocial limits, and in the third he addresses confusions about economic growth. Naturally, there is some overlap among the three accounts and many of the specific points he raises come up in various parts of this book. Nonetheless, in order to fully appreciate his work on steady-state economics, it is essential to understand what has motivated him. First and foremost, it is his critique of economic growth.

Defining economic growth

The first thing that Daly finds wrong with economic growth is how it is defined. Differences in definition run through his critique of economics so we will address it first.

Daly is quite clear on how he defines economic growth:

> By 'growth' I mean quantitative increase in the scale of the physical dimensions of the economy; i.e., the rate of flow of matter and energy through the economy (from the environment as raw material and back to the environment as waste), and the stock of human bodies and artifacts. By 'development' I mean the qualitative improvement in the structure, design, and composition of physical stocks and flows, that result from greater knowledge, both of technique and of purpose. Simply put, growth is quantitative increase in physical dimensions; development is qualitative improvement in non-physical characteristics.
>
> *Daly 1987, p. 323*

Daly's definition of economic growth is at odds with how it is defined by most economists, governments and international institutions such as the World Bank and the OECD. They use increases in real GDP to measure economic growth, which is denominated in terms of money, not in physical quantities as Daly would have it. The many problems of using such a measure were discussed in the previous chapter and some will reappear in this catechism of growth fallacies. To avoid confusion due to differences in definition in the discussion that follows, we will make it clear if we are talking about increasing real GDP (GDP for short) or Daly's quantitative increase in physical scale when using the term economic growth, or just plain growth when the difference in definitions is of no consequence. While Daly considers the definitional difference between growth in GDP and growth in throughput to be very important for clear thinking about growth, he notes that in practice the two can be quite closely related: 'GDP growth correlates positively with throughput growth… while throughput per dollar of GDP has recently declined somewhat in some OECD countries, the absolute level of throughput continues to increase as GDP increases' (Daly 2007a, pp. 88, 90). An even stronger statement from Daly is, 'Real GDP… is the best index we have of total resource throughput' (Daly 2014a, p. 63).

A catechism of growth fallacies[1]

Daly always chooses his words carefully and in this instance the choice of the word catechism is significant. The Merriam Webster dictionary defines catechism as 'a summary of religious doctrine often in the form of questions and answers'. Daly's approach is to question the claims made about economic growth, noting the fallacies on which they are based. The choice of a religious term implies a commitment to these claims more akin to faith than to fact or rational argument.

Daly's catechism of growth fallacies can be conveniently divided between those that deal with the feasibility of growth and those that deal with its desirability.

The feasibility of economic growth

Consistent inconsistencies and avoiding the main issues (pp. 102–103)

Daly writes that, 'Growthmen are forever claiming that neither they nor any other economist worth his salt has ever confused GNP [or GDP] with welfare.' He gives an example of Nordhaus and Tobin (both recipients of the Nobel Memorial Prize in Economics) who say that 'Gross National Product is not a measure of economic welfare… maximization of GNP is not a proper objective of economic policy… Economists all know that…'. Then, in the same article they say, 'But for all its shortcomings, national output is about the only broadly based index of economic welfare that has been constructed.' As Daly says, 'Either GNP (or NNP[2]) *is* an index of welfare, or it is *not*'.

Despite protestations to the contrary, it is difficult to make sense of the widespread commitment to economic growth, as measured by increasing GDP and GNP, if economists and politicians do not think that it is a good thing. And by a good thing presumably they mean it is either a direct measure of economic welfare or that it is highly correlated with it. While there has been some movement away from this position by a few politicians (for example, the Wellbeing Economy Governments of Iceland, New Zealand, Scotland and Wales https://wellbeingeconomy.org) and by some economists since Daly wrote his catechism of growth fallacies in 1977, the commitment to increasing GDP remains a priority against which almost everything else relating to the economy is judged.

Admitting the thin edge of a big wedge (pp. 104–105)

Paul Erhlich's book *The Population Bomb* (1968) was one among several in the 1960s and 70s that expressed great concern about the rapidly expanding human population. It took tens of thousands of years for the global population to reach one billion by the beginning of the nineteenth century. Another billion people were added in the next 125 years, a third billion only 30 years later by 1960, a fourth billion in the next 14 years, the next 13 years for the next one billion, 12 years for the next billion by 1999 and again another by 2011, with the next billion forecast to be added by 2023 bringing the global population to 8 billion.[3]

Some critics say Ehrlich was wrong when he forecast imminent widespread famine in the 1970s and 80s but was he wrong to alert us to the difficulties of providing adequate food, clothing, housing, and all the amenities of a decent life for the rapidly expanding population? While there is plenty of room for debate about how many humans the planet can support in the short and long term, most people would agree with Nordhaus and Tobin who wrote in 1972 that 'We know that population growth cannot continue forever' (Nordhaus and Tobin 1972, p. 18).

Daly is happy to agree with Nordhaus and Tobin on population growth, but the matter does not end there. It is not just the growth in the human population that must come to an end, but the growth in the population of human artifacts as well, and for much the same reasons. It is the combination of more people with more artifacts that determines the material and energy requirements of the human economy. If growth in these requirements does not cease, they will inevitably run up against the biophysical limits of the planet, if they have not already done so.

Daly terms the population of humans 'endosomatic capital' (endosomatic meaning within the body of an organism). What differentiates humans from other species is the accumulation of 'exosomatic capital'. Daly explains:

> Cars and bicycles extend man's legs, buildings and clothes extend his skin, telephones extend his ears and voice, libraries and computers extend his brain and so on. Both endosomatic and exosomatic capital are necessary for the maintenance and enjoyment of life. Both are physical open systems that maintain themselves in a kind of steady state by continually importing low-entropy matter-energy from the environment and exporting high-entropy matter-energy back to the environment. In other words, both populations

require a physical throughput for short-run maintenance and long-run replacements of death by births. The two populations depend upon the environment in essentially the same way. The same biophysical constraints that limit the population of organisms apply with equal force to the population of extensions of organisms. If the first limitation is admitted, how can the second be denied?

This is a powerful argument which harks back to Daly's understanding of economics as a life science (see Chapter 4). Humans and the physical capital and consumer durables that they make are open systems dependent on inputs of low entropy matter-energy for their operation and maintenance. Their quantities are limited by the availability of low entropy matter-energy that they and their structures require. While it might be argued that the specific sources of low entropy differ for humans (food) and their artifacts (fuel and electricity) there are some obvious overlaps. Petroleum products, for example, are used in the production of human food without which many of the nearly 8 billion people alive today could not survive, and neither could most of the transportation equipment in the world function without diesel or gasoline to fuel it.

The parallel between the population of artifacts and humans is instructive but may not be as straightforward as Daly might have us believe. One difference is that humans are relatively homogeneous whereas our artifacts are extremely diverse. This complicates generalizations about their totals and their measurement. Another is that humans are both means, in the sense of endosomatic capital, and ends since they provide the purposes which the two types of capital are intended to serve. Still, his main point remains. Human artifacts cannot keep growing in aggregate in material and energy terms, any more than people can, without imposing increasing damage on the planet with adverse implications for all living species. And, as we saw in the previous chapter, from 1900 to 2015 the annual global extraction of materials increased over seven times between 1900 and 2015, and energy use grew even faster, so that now we are transgressing some planetary boundaries and are threatening to transgress others in the near future. Technological change, which has been rapid and widespread, does make a difference though, as we will see in the next fallacy – not always in the direction that its proponents expect.

Misplaced concreteness and technological salvation (pp. 105–107)

In Chapter 3 we saw Alfred North Whitehead's influence on Daly through his 'lurking inconsistency': the combination of science based on deterministic mechanisms and a belief in self-determining organisms. Another way in which Whitehead influenced Daly is through the Fallacy of Misplaced Concreteness, which Whitehead defined as the 'accidental error of mistaking the abstract for the concrete' (Whitehead 1925, p. 51). There are many instances where Daly uses this fallacy to criticize neoclassical economics. Nowhere is this more apparent than in *For the Common Good* (Daly and Cobb 1989, 1994). The five chapters of Part One of the book all have 'misplaced concreteness' in their titles: Chapter 1 is on the fallacy of misplaced concreteness in economics and other disciplines. This is followed by chapters on misplaced concreteness in the market, in measuring economic success, in *Homo economicus* (the abstract notion of humans much used in economics), and in land. Daly and Cobb give even more examples of the fallacy of misplaced concreteness when they discuss international trade, money, debt and wealth, and exchange versus use value.

According to Daly's teacher Georgescu-Roegen, 'It is beyond dispute… that the sin of standard economics is the fallacy of misplaced concreteness' (Georgescu-Roegen 1971, p. 320). The fallacy of misplaced concreteness awaits us whenever we mistake our abstractions for the reality we use them to represent. In his catechism of growth fallacies Daly uses the fallacy of misplaced concreteness in his critique of economic growth in relation to technology.

> Technology is the rock upon which the growthmen built their church… speaking of it as a thing that grows in quantity. From there, it is but a short step to ask whether this thing can grow exponentially, like many other things, and to consult the black art of econometrics and discover that it has!

Apparently, an exponential increase in technology (whatever that may mean) is just what is needed to negate the effects of exponential increases in resource depletion and pollution due to economic growth. Daly gives the following example of the view that technology does grow exponentially: 'technical progress shows no signs of slowing down. The best econometric estimates suggest that it is indeed growing exponentially' (Passell et al. 1972, quoted in Daly). To be fair to econometricians, Daly acknowledges that others are more cautious in their estimates of technology's contribution to production since it cannot be measured directly. It was Solow who propagated the idea that that whatever could not be statistically explained by labour and capital in a production function to account for economic growth could be attributed to technology. But as Daly has observed, many things that can account for economic growth (i.e. growth in GDP) are typically omitted from econometrically estimated production functions, materials and energy for example, so who is to say how much can be attributed to technology?[4]

Daly also questions the assumption, usually found in the pro-growth position, that 'technological change is exclusively part of the solution and not part of the problem'. Writing in 1977 Daly found this view 'ridiculous on the face of it', and cited Barry Commoner who wrote about the environmental hazards of technology in support (Commoner 1971). A more recent book with extensive documentation on the past and likely future failures of technology to 'save us or the environment' is *TechNo-Fix* by Huesemann and Huesemann (2011). It would appear that Daly's scepticism of what we can expect from technology stands on solid ground, and that we should pay more attention to the type of technology, its potential burden on the planet, and building an 'institutional sieve [that] will let pass the good kind of technology while blocking the bad kind'.

Two factor models with free resources and funds that all nearly perfect substitute for flows (pp. 107–108)

Well before his debate with Solow and Stiglitz in the 1990s, which need not be repeated here (see Chapter 4), Daly had been criticizing the standard two factor (capital and labour) representation of production. The essence of his critique, following Georgescu-Roegen, is the conception of labour and capital as funds that act upon material and energy flows to produce output. The funds do not end up in the output. Their job is to use energy to process the materials that do. From this perspective, capital and labour might be substitutes for each other, and materials and energy might be substitutes for each other but, as Daly and Georgescu-Roegen argue, capital and labour are complements, not substitutes, for materials and energy.

> Capital is a fund, material and energy resources are flows. The fund processes the flow and is the instrument transforming the flow from raw materials to commodities. The two are obviously complements in any given technology.

Daly is focusing on capital here because it is usually new technology embodied in new capital that is regarded, incorrectly, as a substitute for the flow of raw materials.
Daly continues:

> ... allowing for technological change does not alter the relationship. The usual reason for expanding (or redesigning) the capital fund is to process a larger, not a smaller, flow of resources, which we would expect if capital and resources were substitutes. New technology embodied in new capital may also permit processing different materials, but this is the substitution of one resource flow for another not the substitution of a capital fund for a resource flow.

Expanding or redesigning capital also has implications for labour, substituting for it in most cases.

When Daly wrote that 'allowing for technological change does not alter the relationship [between funds and flows]' he is correct in that a fund without a material flow cannot produce a physical output. He is also correct when a larger and/or more powerful technology increases the flow of materials. For example, a new oil refinery compared with an older one or a larger, faster piece of equipment in a car assembly plant, are both designed to increase throughput. With respect to resource extraction, new technologies are introduced deliberately to increase the rate of extraction – the flow – but as Daly points out, these technologies often reduce future flows. Commercial fishing and oil extraction are well-documented examples of where this has happened.

But these are not the only possibilities. Increased insulation is an addition to the capital stock of buildings specifically intended to reduce the flow of energy used for heating and cooling. In this case is it not reasonable to describe insulation as a substitution of capital for a flow of energy? And if a more efficient piece of equipment processes the material flow on which it operates such that the waste per unit of material flow is reduced, then can this not also reasonably be described as a substitution of capital for a material flow? Daly's objection seems to be that to do so obscures what is really happening: one piece of capital equipment is substituted for another, resulting in a reduction in the flow of material. The substitution is between capital. The result is a reduction in flow.

Whether or not the end result will be an absolute reduction in flow depends on what happens to the total output from the substituted equipment. If, by lowering production costs, the price of the product is reduced, sales might increase sufficiently so that even with the reduction in waste per unit of material flow, total material and energy flows increase. This is the familiar 'Jevons paradox' or rebound effect when improvements in efficiency lead to an increase in the use of energy and materials, not a reduction as might be supposed. Perhaps Daly's insistence on the more precise description of what happens when more efficient equipment replaces less efficient – a substitution of capitals with flow effect – reduces the likelihood that the Jevon's effect will be overlooked.

These examples suggest that Daly may have overstated the inability of capital to substitute for resources in the presence of technological change. He is not wrong in general that technological change has gone hand in hand with increases in material and energy flows in the past century or more. The data are very clear on this: global extraction of all four major categories: biomass, fossil fuels, industrial minerals and construction materials has increased many-fold during the most rapid period of technological innovation in history (Krausmann et al. 2018). That they did not increase as much as global GDP suggests that technology had a mitigating effect, but by no means a decisive one – certainly not sufficient to justify the exclusion of materials and energy from aggregate production functions of an economy.

Daly's view is that it is a serious misconception of production to neglect resource flows as justified by 'the tacit justification [that] reproducible capital is a near perfect substitute for land and other exhaustible resources' (Nordhaus and Tobin 1972, quoted in Daly). Not only does he consider it improper to compare factors with different dimensionality (funds versus flows), but reproducible capital is made from resources so that it cannot be a near perfect substitute for the very thing of which it is made.[5] As Daly says, underlying the neglect of material and energy resources in the standard aggregate production functions, is

> The assumption is that in the aggregate resources are infinite, that when one dries up there will always be another, and that technology will always find cheap ways to exploit the next resource… As long as economic growth models continue to assume away the absolute dimension of scarcity… current growth economics has decoupled itself from the world and has become irrelevant. Worse, it has become a blind guide.

But resources are such a small percentage of GNP (pp. 109–112)

Daly surmises that another justification for ignoring resources is the small value of GDP they represent. This is the argument made more recently and independently by Nordhaus, Beckerman and Schelling that even if agricultural output falls by 50 per cent as a result of climate change it would hardly be noticed since agriculture only accounts for about 3 per cent of GDP (see Chapter 6). In his catechism of growth fallacies Daly contests this view on the basis that resources are under-priced because 'labor and capital are two powerful social classes, while resource owners, for good reasons, are not.' This is different from, though not inconsistent with, his criticism of the Nordhaus, Beckerman and Schelling position on agriculture and climate change which rested on their neglect of the impact of declining food supply on food prices. Instead, Daly employs a class analysis, quoting Marx in support, to account for the underpricing of resources.

> Capital and labor are the two social classes that produce and divide up the firm's product. They are in basic conflict but must live together. They minimize conflict by growth and by throwing the growth-induced burden of diminishing returns on to resource productivity… Capital is the dynamic, controlling factor. It is not for nothing that our economic system is called 'capitalism' rather than 'resource-ism'… keeping the returns to capital high by keeping the accounting price of resources in vertically integrated 'internal operations' so low that even cheap and custom scrap is unattractive by comparison.

Present value and positive feedback (pp. 113–114)

Daly objects to the argument that the market automatically provides for conservation 'by offering high profits to farsighted speculators who buy up materials and resell them later at a higher price.' He bases his objection on two grounds. The first is that 'exponentially growing extraction leads to "unexpectedly" sudden exhaustion.' This proposition was popularized in *Limits to Growth* with the 'riddle' of a lily pond. If the lilies double in size every day and they would completely cover the pond in 30 days, when would it be half covered? The answer is on day 29. Since a constant rate of economic growth is exponential, Daly suggests the same logic applies to resource extraction and questions the farsightedness of even the most astute speculators: 'For linear trends, the past is a good guide to the future. For exponential growth, the past is a deceptive guide to the future.'

Daly's second objection to the effectiveness of the market for promoting conservation is more subtle. It concerns the present-value calculations of investors who must decide between leaving resources in the ground in the expectation of higher future prices and extraction today and investing the profits in something that offers a higher rate of return. He argues that high and increasing current growth rates, based on high and increasing current depletion rates, lead to high and increasing discount rates applied to future values. The last condition in turn leads to a low incentive to conserve, which feeds back to high current depletion and growth rates, high discount rates, and so forth. Present value calculations thus have an element of positive feedback that is destabilizing from the point of view of conservation.

This dynamic between the rate of economic growth, the rate of return to capital, the discount rate used in present value calculations to determine the profit-maximizing rate of resource depletion, and the excessive level of short-term extraction that results is another argument for why the market cannot be relied upon to regulate the scale of an economy in the face of absolute scarcity. It runs counter to the usual assumption of equilibrium that is relied on so heavily in neo-classical economics. Similar, self-generating instabilities are also characteristic of the financial markets, as we shall see in Chapter 10 on money and banking.

The fallacy of exponentially increasing natural resource productivity (pp. 115–118)

Any threat to endless economic growth posed by a diminishing resource base can be overcome if the productivity of the resource rises faster than the remaining stock declines, that is, providing the resource stock does not disappear altogether. Daly quotes Solow saying as much, 'there is really no reason why we should not think of the productivity of natural resources as increasing more or less exponentially over time' (Solow 1973, p. 45). If this were true, says Daly, it would mean that limits on throughput would be perfectly consistent with continual growth in GDP and 'therefore not such a radical proposal'. But Daly does not think it is true. His review of the evidence available in 1977 'supports no generalization about resource productivity at all' and certainly does not support Solow's view that, 'the world can, in effect, get along without natural resources' (Solow 1974, p. 11). More recent data on resource productivity supports Daly's scepticism. Although 'most countries have improved their material productivity [GDP/materials] over time', it has not been anything close to an exponential decline. 'Global material productivity has *declined* since about the year 2000 and the global economy now needs more materials per unit of GDP than it did at the turn of the century' (UNEP 2016, p. 26, emphasis added).

The ever expanding service sector and 'angelized GNP' (pp. 118–119)

Expenditures that make up GDP and GNP can be divided into expenditures on goods and expenditures on services. Since goods are by definition physical, they require resources and energy throughout the supply chain, with associated environmental impacts. After they have been used and reused, goods become wastes, some of which is recycled, but eventually end up degraded requiring disposal. Services are different. They are activities performed by some people for the benefit of others, hairdressers and mechanics for example, rather than the production of a physical good. Daly says that: 'Advocates of growth frequently appeal to the increasing importance of services, which, it is assumed, can continue to grow indefinitely, since such activities are presumably nonpolluting and non-depleting.'

Daly contests this view on two grounds. First, all services require some physical resources. Hairdressers use clippers, scissors, dryers, shampoos, conditioners, electricity etc., mechanics use many tools, replacement parts, oil etc. Some services (e.g., international air travel) are more resource intensive than some goods (e.g., locally grown food) and as Daly observes, people employed in the service sector spend their incomes on goods and services. No one can live on services alone.

The second flaw in the argument of eternal service-based economic growth is that the service sector has already grown significantly, in some countries now exceeding 80 per cent of GDP, and yet total materials used has continued to increase. This was already clear when Daly quoted a report of the US National Commission on Materials Policy written in 1973: 'In 1969 a dollar's worth of GNP was produced with one half of the materials used to produce a dollar's worth of 1900 GNP in constant dollars. Nevertheless, over the same period total materials consumption increased by 400%.'

Another way of looking at this is to recognize that all consumption is ultimately the consumption of services. 'Service (net psychic income) is the final benefit of economic activity' (Daly 1974, p. 36). Goods are wanted for the services they provide, whether this is food (for the service of nutrition), toothpaste (for the services of clean teeth and breath), cars and bicycles (for the service of mobility), luxury cars and bicycles (for services of status and mobility) and so on. The difference, therefore, between goods and services is more one of degree than of kind having to do with their respective material and energy requirements and the mode of service delivery. All commodities, whether classified statistically as goods or services, require throughput linking resource extraction to final disposal, governed by the laws of thermodynamics.

Misleading views on misallocation and growth (pp. 122–124)

Daly sees many problems with the view proffered by Beckerman (1974) and other economists that to prove the growth of GDP is excessive it is necessary to show that investment is excessive. As Daly explains, excessive investment in economic terms means that current consumption is being sacrificed for greater consumption in the future beyond what is optimal, with optimal being determined by equality between planned saving and investment. Daly calls this a *'behavioural equilibrium* without regard for any ecological limits that are necessary to preserve *biophysical equilibrium.*

Neoclassical economics through the sub-discipline of environmental economics teaches that ecological costs can be internalized by adjusting market prices. Daly describes this as 'an impossible task' if it is meant as a comprehensive strategy to deal with environmental degradation. 'Even granting the impossible task of internalization, all that means is that all *relative* scarcities are properly evaluated. Growth could continue and *absolute* scarcity could become ever greater.' The problem is made worse when it is recognized that

> the market is not able to allocate goods temporally over more than one generation. Indeed, when different generations (different people) are involved, the issue is one of distribution not allocation. Future people cannot bid in present markets... All interdependencies over time and space cannot be fit to the procrustean bed of an unrestricted price system.

What second law? (pp. 124–126)

We saw in Chapter 3 that from the perspective of Daly's Ends-Means spectrum, economics concentrates on the satisfaction of intermediate ends with intermediate means. Orthodox economists pay little attention to the Ultimate means – the fundamental matter-energy from which the intermediate means are derived, or to the Ultimate end – the purpose of one's life or life in general that the intermediate means are assumed to serve. Much of Daly's critique of mainstream economics stems from its unduly narrow focus resulting in the failure to appreciate the relevance of the second law of thermodynamics to economic growth.

As an example, Daly quotes Harvard economist, Richard Zeckhauser who, in 1973, in the midst of the oil embargo imposed by members of OPEC wrote: 'Recycling is not the solution for oil, because the alternate technology of nuclear power generation is cheaper' (Zeckhauser quoted in Daly). Daly comments,

> the clear meaning of the sentence is that recycling oil as an energy source is possible but just happens to be uneconomical, because nuclear energy is cheaper. The real reason that energy from oil is not recycled is of course the entropy law... This nonsensical statement is not just a minor slip-up we can correct and forget; it indicates a fundamental lack of appreciation of the physical facts of life.

Daly provides two even more egregious examples of the neglect by economists of the biophysical systems within which economic growth occurs and the physical laws by which they operate. A widely cited survey of the environment in economics from 1976 begins, 'Man has probably always worried about his environment because he was once totally dependent on it' (Fisher and Peterson 1976, quoted in Daly). It would not be difficult to find a similar sentiment today from many quarters. The implication of the statement, as Daly observes, is that we are no longer totally dependent on the environment, presumably because of technology. Such a view flies in the face of the fact that humans and our economy are open systems that obtain materials and energy from the environment, transform them, and return them, degraded, to the environment. 'Man and environment are so totally inter-dependent it is hard to say where one begins and the other ends. This total inter-dependence has not diminished and will not in the future, regardless of technology.'

The second example Daly chose to underline the fallacy that the second law can be safely ignored by economists, comes from the influential book *Scarcity and Growth* (Barnett and Morse 1963). 'Advances in fundamental science have made it possible to take advantage of the uniformity of energy/matter' (Barnett and Morse, quoted in Daly). Daly responds,

> It is, however, not the uniformity of matter-energy that makes for usefulness but precisely the opposite. It is non-uniformity, differences in concentration and temperature, that makes for usefulness. If all materials and energy were uniformly distributed in thermodynamic equilibrium, the resulting 'homo-geneous resource base' would be no resource at all.

So far, we have concentrated on growth fallacies outlined in Daly's catechism that are concerned primarily with the feasibility of economic growth. In various ways they come down to a lack of appreciation of the economy as a sub-system of the biosphere and its consequent dependency on material and energy throughput. These are key features of Daly's pre-analytic vision of economics. Now we turn our attention to fallacies in the catechism that are more about the desirability of eco-nomic growth than its feasibility.

BOX 7.2

Beyond Growth... was an exciting redemption of economics for me – I had previously written off economics after suffering through a year-long Economics 101 course... there was no relation of the economy to the environment or energy, no grounding of economics to contemporary real-world economic issues, a disregard of justice (writing off unemployment as 'natural'), and no indication that there are various brands of economics such that everyone should be able to find at least one that speaks to their values!

Eric Miller

The desirability of endless economic growth

Can't get enough of that wonderful stuff (pp. 99–100)

If something, anything, any stuff, confers benefits and costs, gains and losses, pros and cons, but people only attend to benefits, gains, and pros in making their assessment, then it follows that more is better, and even more is even better. This is the nub of Daly's critique of economic growth – especially when economic growth is defined as an increase in GDP which fails to distinguish systematically between benefits and costs.

> Growth in GNP [and GDP] should cease when decreasing marginal benefits become equal to increasing marginal costs ... but there is no statistical series that attempts to measure the cost of GNP [and GDP]. This is a growthmania, literally not counting the costs of growth. But the situation is even worse. We take the real costs of increasing GNP [and GDP] as measured by the defensive expenditures incurred to protect ourselves from the *unwanted* side effects of production and *add* these expenditures to GNP [and GDP] rather than subtract them. We count real costs as benefits. This is hypergrowthmania.

Following American economist Irving Fisher's definitions of capital and income (Fisher 1906), Daly suggests that the way to fix this problem is for statistical agencies to keep three separate accounts. These would be: a value index of material and energy throughput which is the cost of maintaining or adding to the existing capital stock; an account recording the value of services yielded over time by the total stock of human and physical capital; and an account of the capital stock itself.

The hair of the dog that bit you (pp. 100–101)

The hair of the dog that bit you, is a colloquial expression that refers to alcohol consumed with the aim of reducing the effects of a hangover. Just as more alcohol is a poor remedy for a hangover, more growth in GDP, says Daly, can be ineffective for dealing with the problems that growth creates. This is a standard argument: economic growth will solve the environmental problems it generates. As we saw in the previous chapter, growth in GDP is 'uneconomic' if the benefits that it brings are outweighed by the additional costs. Daly argues that the likelihood of this situation grows with increases in GDP since people satisfy their most pressing needs and wants first and less important ones later. Conversely, environmental costs rise with economic growth with the increasing quantities of materials and energy that growth entails. More growth, says Daly, is not likely to solve the problems that growth creates.

Crocodile tears from latter-day Marie Antoinettes (pp. 103–104)

'Let them eat cake,' said Marie Antoinette, the wife of King Louis XVI of France, when learning of peasants starving from a lack of bread during one of the famines that occurred during his reign. Except that she did not say it – to do so would have been out character, and had she said it she would not have been the first to do so, according to her biographer, Antonia Fraser. Still, it is a sentiment that resurfaces in different contexts to make a point. Daly uses it in his critique of 'orthodox growthmen who want to avoid the distribution issue'. For example, 'Growth is a substitute for the equality of income. So long as there is growth there is hope, and that makes large income differentials tolerable' (Wallich 1972, quoted in Daly). For Daly the addiction to growth is based on an addiction to large inequalities in income and wealth. 'What about the poor? Let them eat growth! Better yet, let them feed on the hope of eating growth in the future!'

We saw in the discussion of scale, distribution and allocation in Chapter 5 that Daly calls for separate and distinct policy measures for securing a just distribution. He finds the idea that growth is an acceptable substitute for justice ill-conceived and reprehensible especially when used by people of privilege to justify large inequalities in the distribution of income and wealth. Daly's position is that the solution to inequality in a rich country like the USA is redistribution not more growth.

Youth culture and frustrated pyramid climbers (pp. 114–115)

For an economist, Daly is unusually sensitive to demography. He observes that the average age in a growing population of given life expectancy is lower than in a stationary population. It follows that opportunities for advancement can be expected to arise sooner with population growth. Yet, 'a stationary population is a part of a steady-state economy.' Rather than see this as a justification for growth – economic and population – Daly points to the positive implications of a stationary population.

> More individuals will learn to seek personal fulfilment outside hierarchical organizations. Within such organizations, fewer people will be automatically promoted to their level of incompetence... Perhaps giant bureaucracies will even begin to dissolve and life will reorganize on a more human scale.

By saying 'will reorganize' rather than 'will be reorganized' Daly anticipated the work of system theorists and their use of 'emergent properties' to describe autonomous, unplanned systemic changes such as occurred as a result of the Covid-19 epidemic of 2020/21, in contrast to top-down decision making.

Pascal's Wager Revisited (p. 115)

The seventeenth-century French philosopher Blaise Pascal argued that a rational person should believe in God, because if God exists then the gains could be enormous – eternity in Heaven, and enormous losses – eternity in Hell, could be avoided. If God does not exist, then what losses there might be from a belief in God would be small (ignoring the trauma of living a lie and assuming God will not find out). Daly draws a parallel with Pascal's argument for a belief in God with 'the growthmania position [which] rests on the hypothesis that technological change can become entirely problem-solving and not at all problem creating and can continually perform successively more impressive encores as resources are depleted.'

As Daly says, just like Pascal's wager: 'We can err in two ways: we can accept the omnipotent technology hypothesis and then discover that it is false, or we can reject it and later discover that it is true.' Daly asks, 'which error do we most wish to avoid?' His answer to this Pascal-like question is that it is far better to act now as if technology will not solve depletion and other environmental problems by deliberately reducing our demands on the biosphere for materials and energy. If that turns out to be mistaken, and the techno-optimists were right, the losses would be manageable. 'If we later discover that the hypothesis [of omnipotent technology] is true we could always resume growth.' The alternative of acting on the assumption that technology will save us and pursue growth with full force only to discover later that we were wrong could well be catastrophic.

The more-is-better concept of efficiency (pp. 121–122)

Daly begins his critique of the more-is-better concept of efficiency with a quote from economist Arthur Okun, who stated that 'efficiency means getting the most out of a given input.' This, says Okun, implies that 'more is better, in so far as the "more" consists of items that people want to buy' (Okun 1975, quoted in Daly). Daly is quick to offer an equivalent definition of efficiency, 'getting the same output with less input', which implies that less is better when 'less consists of items that people would like to avoid buying if only they could.' In other words, when efficiency is defined as output/input it can be increased by increasing the numerator or reducing the denominator. What matters, says Daly, is what the denominator and numerator stand for.

Okun accepts the principle of consumer sovereignty: whatever people buy is good so the more the better. But Daly says that consumer 'choices do not reveal much about welfare unless we know the alternatives available. And economic growth often narrows the range of alternatives', especially if it comes with significant environmental and social costs. Maintaining output while reducing these costs, paid for or not, could be a better way to use an increase in efficiency to raise welfare. Doing so economy-wide is what the steady-state economy is all about. As we shall see in the next chapter, Daly has a lot more to say about efficiency in relation to steady-state economics than simply assuming that gains in efficiency translate into more-is-better.

Zero growth and the Great Depression (p. 126)

One of the most frequently repeated growth fallacies is the assumption that periodic economic recessions are a 'real life tryout of zero growth' (*Fortune*, 1976, p.116 quoted by Daly). '*Fortune* identifies SSE [a steady-state economy] with a failed growth economy.' Many more similar statements can be found expressing the same fallacious argument. As Daly says,

> no one denies that the failure of a growth economy to grow brings unemployment and suffering. It is precisely to avoid the suffering of a failed growth economy… that we advocate a SSE… Growthmania reigns supreme even when even the failures of a growth economy become arguments in its defense!

So ends Daly's catechism of growth fallacies written in the 1970s. Despite his best efforts, in the 1970s and 80s, Daly was unable to make much impression on the discipline of economics in which he was trained. Later that began to change and his continuing efforts to explain what is wrong with economic growth were part of the reason.

The economic growth debate: what some economists have learned but many have not[6]

Ten years after Daly's first summary of the case against economic growth, Daly offered a critique of economic growth in a paper originally written for a round-table meeting held in 1983 which asked the question, 'Limits to Growth: What Have We Learned? (Daly 1987). The other panellists were all distinguished economists from the USA: Robert Pindyck, Thomas Schelling, William Nordhaus and Allan Manne, who Daly tells us did, 'not necessarily share the views [he] presented,' which was probably a polite understatement. Being in a minority of one, Daly was on the defensive at the meeting, but he saw an opportunity for a breach in the positions of Schelling and Nordhaus which he tried to exploit. Schelling argued that the future is very uncertain, that much is unknown about climate change and its effects, good and bad, so let's wait. Nordhaus had used his DICE model to make estimates of the impact of climate change on growth over decades to an impressive level of precision. He used these estimates to assess how much it is worth spending now to mitigate climate change. Daly suggested that it would be interesting to hear them discuss their differences (meanwhile taking some of the pressure off him). He was disappointed that Schelling remained silent while Nordhaus told Daly that he was wrong about significant digits, entirely missing the point. The upshot was that even though Daly had made a considerable effort to couch his critique of growth in economic terms there was no real engagement, no points put forward to counter his, and no real attention paid to limits to growth. He came away angry and frustrated at the reluctance of these leading economists to take his arguments seriously and at his inability to persuade them to. This was a pattern that would mark much of his career.

What was it that the other roundtable economists found uninteresting, unconvincing or both? Daly began this roundtable paper with his now familiar definitions of growth (quantitative) and development (qualitative) that were quoted earlier in this chapter. The main substance of the paper concerns 'the two categories of limits to growth [biophysical and ethicosocial] and the nature of the welfare [well-being] losses that come about when each limit is stressed by growth.'

In the previous chapter we noted Pigou's definition of economic welfare as 'that part of social welfare that can be brought directly or indirectly into relation with the measuring-rod of money' (op cit., p. 11). Pigou's ultimate interest was in welfare as a whole and he recognized that increases in economic welfare could be at the expense of reductions in non-economic welfare, 'and if this happens, the practical usefulness of our conclusions is wholly destroyed' (ibid., p. 12). This observation by

Pigou is often overlooked. He went further: 'any rigid inference from effects on economic welfare to effects on total welfare is out of the question' (ibid., p. 20). To justify the study of the causes of economic welfare Pigou relied on a

> judgement of probability [that]… qualitative conclusions about the effect of an economic cause on economic welfare will hold good also of the effect on total welfare. This presumption is especially strong when experience suggests that the non-economic effects produced are likely to be small.

He went further to say that 'in all circumstances the burden of proof lies upon those who hold that the presumption should be overruled' (ibid.).

The essence of much of Daly's work is an acceptance of this challenge to provide proof. His argument that in rich countries, growth has become uneconomic rests squarely on the view that, at the margin, the costs of economic growth, exceed its benefits. He has made this argument in different ways and with varying emphasis on different considerations and supportive data. In his 1987 paper he makes this argument based on limits to growth.

> The nearer we are to limits the less can we assume that economic welfare and total welfare move in the same direction. Rather we must learn to define and explicitly account for the other sources of welfare that growth inhibits and erodes when it presses against the limits.
>
> *Daly, p. 324*

Biophysical limits to economic growth (pp. 324–327)

It is not uncommon for great teachers to have been students of great teachers themselves. Daly's great teacher was Georgescu-Roegen. Pigou's great teacher was Alfred Marshall also Professor of Economics at Cambridge University and author of the *Principles of Economics*, the leading textbook on economics at the turn of the last century in all its nine editions. Marshall showed a clear appreciation of the first law of thermodynamics and its relevance to economics when he wrote:

> Man cannot create material things… His efforts and sacrifices result in changing the form or arrangement of matter to adapt it better for the satisfaction of his wants… as his production of material products is really nothing more than a rearrangement of matter which gives it new utilities, so his consumption of them is nothing more than a disarrangement of matter which diminishes or destroys its utilities.
>
> *Marshall 1920, quoted in Daly*

These are the opening words of a chapter entitled 'Productive' in the first edition of Marshall's *Principles*; so prominently displayed that it is a wonder that they did not find their way into the body of neoclassical economic theory that Marshall did so much to construct.

Daly is generous when he says: 'Economists seem to have learned about the first law of thermodynamics (conservation of matter-energy) and the limits it imposes. Production functions are now sometimes required to respect a materials balance constraint.' He cites the important paper by Ayres and Kneese (1969) that added the materials balance principle to the Walras-Cassel general equilibrium model of an economy, but the extent to which economists in general have built this into their own work is questionable. In any event, as Daly points out, if economic growth was only constrained by the first law, the situation would be very different. 'We just keep rearranging and disarranging indestructible building blocks. Nothing is used up.'

But when second law considerations are introduced, the picture changes. Materials can be reused and recycled but not without some degradation at each stage so that new virgin supplies are required even in the absence of economic growth. Energy can be used more efficiently, but each time it is used to perform work its capacity to do further work declines until it is literally useless:

> Since the rearrangement of matter is the central physical fact about the economic process, we must ask what determines the capacity to rearrange matter? Is that capacity conserved, like matter-energy itself, or is it used up? Is all matter equally capable of being rearranged? The answers to these questions are provided by the second law... to apply conclusions derived from a model of the circular flow of money to issues dominated by the linear throughput of matter-energy is a classic case of the fallacy of misplaced concreteness.

Daly acknowledges that 'one may grant that biophysical limits are real but still doubt that they are near'. He cites population growth and refers to 'evidence that global per capita production of the basic renewable resource systems (forests, fisheries, croplands, and grasslands) have all peaked and begun to decline'. And before climate change was on the public agenda he observed that 'as fossil fuel subsidies [are]... withdrawn there will be an acceleration in the rate of productivity decline of renewable systems'.[7] This is likely to bring even more pressure on the habitat of wildlife and species loss, which is another aspect of biophysical limits that did not escape Daly's attention. 'The idea that biophysical limits to growth are near as well as real is not just the fabrication of "doomsayers."' Post-1987 evidence of biophysical limits to growth is even more compelling than when Daly gave this commentary on the state of nature, making his conclusion even stronger today (see Victor 2019, Chapters 5 and 6).

Ethicosocial limits to growth (pp. 327–336)

Writing in the early twentieth century Pigou may well have been correct that 'the non-economic effects produced are likely to be small', but by the end of the century times had changed. The world had become full. Even if growth is still biophysically possible, Daly contends that other factors may limit its desirability especially in rich countries. In his 1987 paper he named these ethicosocial limits to growth and describes four of them:

The desirability of growth financed by drawdown is limited by the cost imposed on future generations (pp. 328–30)

Daly believes that 'the basic needs of the present always should take precedence over the basic needs of the future… [but] the extravagant luxuries of the present [should not] take precedence over the basic needs of the future.' This is a moral judgement that he is quite prepared to defend and admonishes economists who 'avoid such judgments by appealing to the market where everyone's preferences count, weighted by their incomes [everyone that is who is alive today].' Daly follows Talbot Page who took a Rawlsian approach (Rawls 2005) to inter-generational distribution 'by imagining an intergenerational distribution that would be regarded as fair by a convention of representatives of each generation, who did not know in advance the place of their generation in the temporal sequence' (Daly).

Daly also mentions the argument that we encountered in Chapter 4, that 'provision for future people is partly a public good, and becomes more so the farther into the future one looks'. This follows from the fact that the number of shared descendants increases exponentially with each additional generation.

> Whatever responsibility we feel for [future generations] beyond, say, our grandchildren must be put into effect through collective measures rather than through individualistic market behavior. In sum, obligations to future generations provide a moral limit to the rate of drawdown, and indirectly to the rate of growth.

The desirability of growth financed by takeover [of habitat] is limited by the extinction or reduction in number of sentient sub-human species whose habitat disappears (pp. 330–331)

Daly recognizes that 'sub-human' species have intrinsic as well as instrumental value. He makes 'no apology for… the term "sub-human species" [and realizes that]… some ecological egalitarians object to the term.' Daly takes his lead from Bentham who said 'the question is not can they *reason*? Nor, can they *talk*? But, can they *suffer*?' He also draws on his religious convictions as a basis for the intrinsic value of other species interpreting the biblical statement that 'a man is worth many sparrows' to mean that a sparrow must be worth something (Matthew 10:31). What 'may be more subtle and inscrutable', says Daly, is 'the value God places on His creation and His purposes for it… than simply maximizing present value for the current generation of entrepreneurs.' Daly rejects the 'plainly ludicrous' idea that the limit on the destruction of habitat and demise of 'sub-human' species is already accomplished by the market: 'Preservation of sub-human species, like provision for the distant future, is a public good that must be served by collective action.' Determining the limit on the takeover of habitat and consequential limit on economic growth that might follow is 'a major philosophical task', not to be left to 'some clever econo-metrician [who] will not shrink from the task of imputing implicit shadow prices to sparrows'.

Limits from the self-cancelling effects of aggregate growth

The proposition that the desirability of aggregate growth diminishes with growth has been explored by many writers. Daly quotes John Stuart Mill, 'men do not desire to be rich, but to be richer than other men.' The significance of relative rather than absolute income being the main source of a person's well-being is central to what has come to be known as the Easterlin Paradox. In 1974 economist Richard Easterlin reported that 'in a given country at a given time one finds a positive cor-relation between income and happiness… But for different countries with very different income levels the differences in reported happiness are small.' Easterlin concluded that relative income rather than absolute income is what matters most to people's happiness, especially in rich countries. Since economic growth is growth in absolute income it contributes little or nothing to overall happiness, or welfare, well-being, satisfaction – all synonyms as used by Easterlin. This conclusion has been much debated with Easterlin answering his critics and essentially reaffirming his original position (Easterlin 2016).

Daly also drew on the work of economist Fred Hirsch who based the idea of self-cancelling effects of economic growth on the concept of 'positional goods' (Hirsch 1976). A positional good is valued according to the status it signifies. Since one person's higher status comes at the expense of others' lowered status, positional goods do not increase aggregate well-being. To the extent that economic growth consists of an increasing proportion of positional goods, it does not improve total welfare.

Another contributor to ethicosocial limits to growth that Daly cited is Stefan Linder (Linder 1970). Linder noted that consumption takes time. The more goods and services people try to consume as a result of economic growth the more stressful it becomes. Multi-tasking becomes the norm: 'As we become goods-rich we also become time-poor, and can afford fewer time-intensive activities such as personal care of the aged, the sick, and of children, as well as domestic service.' Daly concludes that: 'In sum, it would appear that aggregate growth just shifts the burden of scarcity on to time and relative position… these self-canceling effects [imply] that growth is less important for human welfare than we have heretofore thought.'

Depletion of moral capital as a limit to growth (pp. 333–336)

Daly's fourth ethicosocial limit to growth is the most ethical of all: 'The desirability of growth is limited by the corrosive effects on moral standards of the very attitudes that foster growth, such as glorification of self-interest and a scientistic-technocratic world view.' Daly credits Hirsch for pointing out Adam Smith's view that in addition to the invisible hand which guide self-interest so that the community's interests were also served, moral constraints derived from religion, custom and education also help ensure that the wider community benefits from trade. Ezra Mishan who was an expert in welfare economics, also recognized that economic growth resting on an individualistic ethos, undermines its social foundations. Mishan went further than Hirsch, arguing that the effect of a moral consensus is that much stronger if it rests on a belief in its divine origin. This comes very close to Daly's understanding of the Ultimate end in his Ends–Means Spectrum (see Chapter 3).

Daly has little sympathy for those like biologist E.O. Wilson who acknowledge the role played by a belief in 'objective transcendental value' but think that Darwin's theory of evolution precludes its existence. Regardless, theists and atheists both face dilemmas that remain unresolved. Theists have to contend with the dilemma presented by competing theistic teachings, each believed by their followers to be the word of God; and atheists have no unquestionable moral principle(s) upon which judgements can be based. But does this matter? All can agree that morality plays an important, even essential, role in sustaining society. In Daly's words,

> At a minimum the problem of sustainability requires maintaining intact the moral knowledge or ethical capital inherited from the past. In fact, sustainability really requires an increase in knowledge, both of technique and of purpose, to offset, in so far as possible, the inevitable degradation of our physical world.

The truth of this statement does not depend on the origin of the moral knowledge or ethical capital inherited from the past. Daly accepts that this is logically correct. Truth of a proposition does not depend on origins, but then, he wonders, what does its truth depend upon?

> The presumed moral knowledge could come from modern atheistic and materialistic philosophers. But do they base it on their own imagination or intuition, or on evolutionary genetic natural selection, or on unacknowledged, perhaps unconscious, vestiges of their culture inherited from a pervasive but fading theistic past? Which has more authority? Which is more receptive to the persuasive lure of purpose coming from the future? I am persuaded by the last, though I can't prove it any more than others can prove the other alternatives. I think there are no knockout arguments to such a deep perennial question. But there are to my mind important degrees of persuasiveness, even though good and reasonable people may differ. If I believed that creation was entirely the random result of multiplying infinitesimal probabilities by an infinite number of trials, or something close to that, I would find it hard to work for its preservation and care.

Daly, personal communication

This second attempt by Daly to summarize the case against economic growth was published in the year that the USA suffered its largest single-day percentage drop in the Dow Jones Index of share prices. The USA economy went into recession in 1990 so there was little appetite for criticisms of economic growth when growth was seen to be the solution, not the problem. But that was not the whole story. In 1989 the International Society for Ecological Economics was founded based heavily on Daly's work. The International Society, with 11 regional societies around the world, hosts a biennial conference and publishes the successful academic journal *Ecological Economics*. Two years later Daly was awarded the Grawemeyer Award for Ideas Improving World Order, the first of many that he received almost yearly in the 1990s and into the twenty-first century. His work was being recognized and his influence was spreading, though it was yet to breach the walls of orthodox economics.

Eleven confusions about growth[8]

Thirty-seven years after publishing his catechism of growth fallacies and 27 years after exploring the biophysical and ethicosocial limits to growth, Daly once again marshalled his arguments against economic growth. This time he set out to expose and clarify confusions about growth, covering much of the same ground as previously so the discussion here can be brief.

One can nearly always find something whose growth would be both desirable and possible (pp. 62–63)

To Daly this is a confusion between aggregate growth and reallocation. 'Aggregate growth refers to growth in everything… Reallocation, by contrast, means that some things go up while others go down, the freed up resources from the latter are transformed to the former… the fact that the reallocation remains possible and desirable does not mean that aggregate growth is possible and desirable.' In addition, Daly has made it clear elsewhere that there are no limits on anything non-material such as justice, peace, understanding, knowledge, wisdom and love. These can grow forever.

Since GDP is measured in value terms it is therefore not subject to physical limits (p. 63)

Daly says that this common confusion ignores the fact that GDP is a value measure of goods and services which have physical dimensions: 'Growth refers to real GDP, which eliminates price level changes. Real GDP is a value-based index of aggregate quantitative change in real throughput… The unit of measure of real GDP is not dollars, but rather 'dollar's worth'. A dollar's worth of gasoline is a physical quantity…'

A more subtle confusion results from looking at past totals rather than present margins (pp. 63–64)

Daly acknowledges the benefits of past growth, benefits that one can well argue outweigh the costs, at least until recently in historical terms. But along with wealth has come 'illth', a term Daly borrowed from John Ruskin, the leading art critic of the Victorian era. Daly gives as examples of illth the widely documented environmental degradation and burgeoning debt. Looking to the future, what matters is whether the benefits of further growth outweigh the costs. Daly's contention is no, they will not, especially in the USA.

Even if it is theoretically possible that someday the marginal cost of growth will become greater than the marginal benefit, there is no empirical evidence that this has happened yet (pp. 64–65)

Daly rejects this contention. He refers to the Index of Sustainable Economic Welfare (ISEW) and its successor the Genuine Progress Indicator (GPI): 'Both show that, for the United States and some of the wealthy countries, GDP and GPI were positively correlated up until around 1980, after which GPI leveled off and GDP continued to rise' (see Chapter 6).

The way we measure GDP automatically makes its growth a trustworthy guide to economic policy (pp. 65–66)

This idea is based on the assumption that willing participants to a transaction must both consider themselves to be better off or they would not trade. Daly claims that this assumption can lead some to draw the conclusion that an increase in the sum total of such transactions, GDP, automatically signals improvement. He gives several reasons why this conclusion is false. It overlooks all those affected, usually adversely, by transactions in which they are not included – environmental pollution being the most obvious example of these external costs. Then there is the failure to account for the consumption of resources as a cost to be deducted from the value of output. 'Cut down the entire forest this year and sell it, and the entire amount is treated as this year's income… Consuming capital means reduced production and consumption in the future' and this is not accounted for in GDP. A third problem is that GDP includes 'defensive expenditures' such as on pollution abatement and mental health treatment in response to deteriorating conditions. This is perverse says Daly, if rising GDP is taken to mean that circumstances are improving.

As natural resources become scarce we can substitute capital for resources and continue to grow (pp. 66–67)

This is the proposition that Daly debated with Solow and Stiglitz (see Chapter 4). It comes down to the lack on their part, and common to neoclassical economics more generally, of a 'realistic analytic description of production.' It fails to recognize '… that factors are of two qualitatively different kinds: resource flows that are physically transformed into flows of product and waste; and capital and labor funds, the agents or instruments of transformation that are not themselves physically embodied in the product.' Once this is realized then it becomes clear that 'the basic relation between resource flow on the one hand and capital (or labor) fund on the other is complementary [not substitution]'.

Knowledge is the ultimate resource and since knowledge growth is infinite it can fuel economic growth without limit (pp. 67–68)

Daly is 'eager for knowledge to substitute physical resources to the extent possible' and so he advocates several measures to make resources expensive (e.g. severance taxes) and knowledge cheap (e.g. patent reform). But knowledge has its limits. Daly says: 'If I am hungry, I want real food on the plate, not the knowledge of a thousand recipes on the Internet.' He is troubled that 'knowledge is naturally depleting while ignorance is renewable… mainly because ignorant babies continually replace learned elders.' Daly adds that '…vast amounts of recorded knowledge are destroyed… by decay, fires, floods, bombs, and bookworms,' and is concerned that digital storage is vulnerable too. Even then 'knowledge must exist in someone's mind… otherwise it is inert'. Daly does not say whether artificial intelligence and deep learning are bringing life to the inert, reminiscent, perhaps, of Dr. Frankenstein.

Knowledge does not always grow. 'Some old knowledge is disproved or canceled out by new knowledge, and some new knowledge is discovery of new biophysical or social limits to growth… New knowledge is not always a pleasant surprise for the growth economy … for example [new knowledge of] climate change from greenhouse gases … How can one appeal to new knowledge as the panacea when the content of new knowledge must of necessity be a surprise?'

Without growth we are condemned to unemployment (pp. 68–69)

To correct this confusion Daly reminds us that when the USA and many other countries, adopted full employment as a policy objective in the mid-1940s, 'economic growth was then seen as the means to attain the end of full employment'. Later this relationship was 'inverted': 'economic growth has become the end, and if the means to attain that end – automation, off-shoring, excessive immigration – result in unemployment, well, that is the price "we" have to pay for the supreme goal of growth.' It follows that 'if we really want full employment we must reverse this inversion of ends and means'. Daly suggests several policies to do just that: 'restricting automation, off-shoring, and easy immigration to periods of true domestic labor shortage… In addition, full employment can also be served by reducing the length of the working day, week, or year, in exchange for more leisure, rather than more GDP.'

Daly brings in a class analysis to explain the trends he adamantly opposes: 'corporations, hungry for cheaper labor, keep bleating about a labor shortage... What the corporations really want is a surplus of labor, and falling wages [so that] gains from productivity increase will go to profit not wages. Hence the elitist support for automation, off-shoring, and lax enforcement of democratically enacted immigration laws.' These particular comments, and others like it in other publications show that Daly is not trapped by a single, blinkered theoretical framework and that he is more than ready to draw on political economy – and ecology, physics, sociology, and theology – as and when he deems them relevant.

We live in a globalized economy and have no choice but to compete in the global growth race (p. 69)

Daly's response to this assertion follows from his previous one on growth and employment. Globalization, he says, is 'the engineered integration of many formerly relatively independent national economies into a single tightly bound global economy'. He sees it as a perversion of the objectives of the post-WWII Bretton Woods system 'aimed at facilitating international trade... [fostering] trade for mutual advantage among separate countries'. It was the World Trade Organization 'and the effective abandonment by the World Bank and International Monetary Fund of their Bretton Woods charter [that sought] free capital mobility and global integration [that] were not part of the deal.'

Daly has argued for some time that the theory of comparative advantage on which the mutual gains from trade is based assumes that capital is immobile between nations. When capital is internationally mobile specialization and trade become based on absolute advantage (i.e. where the rate of profit is greatest) and mutual gains from trade are no longer guaranteed.

Globalization is not inevitable: 'it was an actively pursued policy... to increase the power and growth of transnational corporations by moving them out from under the authority of nation states and into a nonexistent "global community".' Daly reminds us that globalization was and still is a choice. It is a choice that has begun to be questioned as a result of the undue dependence on imports of critical medical supplies made visible by the Covid-19 pandemic, and the vulnerability of nations to trade sanctions imposed for political purposes (see Chapter 13 for more on globalization and comparative advantage.)

Space, the high frontier, frees us from the finitude of the earth, and opens unlimited resources for growth (pp. 70–71)

Communications satellites oriented to Earth 'yes', space colonization 'no'. That, in short, is Daly's assessment of what outer space offers Earth's current and near future human dwellers. 'The numbers – astronomical distances and timescales – effectively rule out dreams of space colonization.' He asks rhetorically, 'if we are unable to limit population and production on earth, which is our natural and forgiving home...

then what makes us think we can live as aliens within the much tighter and unforgiving discipline of a space colony on a dead rock in a cold vacuum?'

Without economic growth all progress is at an end (p. 71)

'On the contrary', says Daly, 'without growth, now actually uneconomic growth if correctly measured, true progress finally will have a chance.' Daly reprises several of his key themes in responding to this final confusion about growth. Growth is quantitative, development is qualitative. 'Development without growth beyond the Earth's carrying capacity is true progress.' This means using 'technical improvement in resource efficiency' as an adaptive responsive to lower resource throughput enforced through limits. Development also means 'ethical improvement of our wants and priorities'.

Growth, whether measured in terms of GDP or matter-energy throughput has become uneconomic, 'but of course most economists do not admit that growth is, or even could be uneconomic. They seem determined to avoid discussion of arguments or evidence to the contrary.'

BOX 7.3

Daly was the first I learned from to oppose fundamentally economic growth, which impressed me as courageous. Today, I appreciate his pioneering role more than I did at that time in the mid-90s.

Marina Fischer-Kowalski

Conclusion

This chapter began with a catechism of growth fallacies, followed by a consideration of biophysical and ethicoscocial limits to growth, ending with 11 confusions about growth. Spanning three decades we see the considerable effort that Daly has made to wean a culture, a civilization even, away from the pursuit of economic growth as the long-term trajectory for humanity. He is pessimistic that this will be achieved without 'the will and inspiration to care for it', which he believes cannot come from a view that 'our lives are... a purposeless happenstance', which is his interpretation of neo-Darwinism (see Chapter 3).

Daly closes his 11 confusions about growth with reference to 'decision-making elites' such as 'the elite-owned media, the corporate-funded think tanks, the kept economists of high academia, and the World Bank... the Gold Sacks (sic) and Wall Street' [who] ... may already tacitly understand that growth has become uneconomic.' But why should they worry when they have figured out 'how to keep the dwindling extra benefits for themselves, while 'sharing' the exploding extra costs with the poor, the future, and other species?

Notes

1 Unless otherwise indicated all page references are to Daly 1977 which was based on an earlier version of the same arguments in Daly 1972.

2 NNP is GNP minus the depreciation of capital and GNP is GDP plus net receipts from abroad. In his earlier writings Daly used GNP which was more common then but more recently GDP has become more widely used. GNP is retained in the direct quotes and GDP is added in parentheses to show that the remark applies to it as well.

3 All population data from Worldometers: www.worldometers.info/world-population/#milestones, accessed 30 June 2020.

4 Ayres and Warr (2010) estimate at that exergy (the capacity of energy to do work) accounts for most, if not all, of the increase in GDP not attributable to capital and labour.

5 If a production function for capital with capital, labour and resources as inputs is substituted for capital in a production function with capital, labour and resources as inputs, capital disappears altogether. 'Any possibility for solving a problem of depleting resources through substitution with capital has evaporated' (Victor 1991, p. 198).

6 Unless otherwise indicated all page references are to Daly 1987.

7 Daly does not explain the connection between reduced fossil fuel subsidies and reduced productivity of renewable systems. Presumably, if subsidies to fossil fuels are withdrawn there will be a greater demand for renewable substitutes (biomass for fossil fuels for example), possibly exceeding sustainable yields.

8 Unless otherwise indicated all page references are to Daly 2014a.

8

STEADY-STATE ECONOMICS

Unless the underlying growth paradigm and its supporting values are altered, all the technical prowess and manipulative cleverness in the world will not solve our problems and, in fact, will make them worse.

Daly

Among Daly's many contributions to ecological economics none is likely to have a greater and more lasting significance than his analysis of and advocacy for a steady-state economy. As is typical of so much of his work Daly has built on the work of predecessors, most notably John Stuart Mill, Frederic Soddy and Nicholas Georgescu-Roegen. But he has also brought his own imagination and insights as well as his remarkable capacity for expressing complex ideas in simple terms. It is fair to say that Daly has virtually single-handedly ensured that the steady-state economy remains on the agenda informing discussions of the future of the economy, environment and society.

We begin this chapter by tracing the main strands of the history of the steady-state economy. This may seem like a digression because it describes the contributions of others, but it is essential for a full appreciation of the influences on Daly in this central aspect of his work. It is also necessary for assessing how much further Daly has taken steady-state economics, to the point where it now presents a serious and timely challenge to the growth orthodoxy. We will see that before Daly, most contributions to steady-state economics were descriptive rather than analytical, whereas Daly developed comprehensive principles of steady-state economics that can be compared and contrasted with economic principles found in other schools of thought.

DOI: 10.4324/9781003094746-8

A short history of the steady-state economy[1]

John Stuart Mill was not the first economist to write about the steady-state economy (he used the term stationary state), but he was among the first to contemplate it with pleasure rather than distaste as Adam Smith, Thomas Malthus and David Ricardo had done before. They established the foundations of classical economics which interpreted and promoted the new phenomenon of economic growth first experienced in Britain in the eighteenth century before spreading to other countries.

In his *Principles of Political Economy* (first published in 1848), Mill devoted an entire chapter to the stationary state, 'a stationary condition of capital and population' which he pointed out did not imply a 'stationary state of human improvement' (Mill 1848, p. 116). Daly quotes from this chapter in many of his writings (see, for example, Daly 1973a, Daly 1977a, 1991a, Daly 1996, Daly 2014a). According to Mill, 'in the richest and most prosperous countries' the arrival of the stationary state would soon follow 'if no further improvements were made in the productive arts, and if there were a suspension of the overflow of capital from those countries into the uncultivated or ill-cultivated regions of the earth' (ibid., p. 111).

Mill looked on the prospect of the stationary state as a positive rather than a negative development for several reasons that resonate today, and which have found their way into more current treatments such as Daly's. In his much quoted, eloquent language Mill writes:

> I am not charmed with the ideal of life held out by those who think that the normal state of human beings is that of struggling to get on; that the trampling, crushing, elbowing, and treading on each other's heels, which form the existing type of social life, are the most desirable lot of human kind, or anything but the disagreeable symptoms of one of the phases of industrial progress... the best state for human nature is that which, while no one is poor, no one desires to be richer, nor has any reason to fear being thrust back, by the efforts of others to push themselves forward.
>
> *ibid., pp. 113, 114*

Mill was careful to note that in the 'backward countries... increased production is still an important object' and argued that 'in those most advanced, what is economically needed is a better distribution, of which one indispensable means is stricter restraint on population' (ibid., p. 114). However, he gave few details of how such restraint is to be implemented.

Mill continued his case for the stationary state by stressing the disadvantages of there being too many people – even if they enjoyed a good material living standard:

> A population may be too crowded, though all be amply supplied with food and raiment [clothing]. It is not good for man to be kept perforce at all times in the presence of his species. A world from which solitude is extirpated, is a very poor ideal...
>
> *ibid., p. 115*

One can only wonder what Mill would say today if confronted with a world of some 8 billion people and still increasing, and with electronic communication and surveillance redefining, negating even, what it means to be alone.

Mill wrote about technology (which he called the 'industrial arts') and time spent working in relation to the stationary state. In doing so he anticipated later writers:

> there would be… as much room for improving the Art of Living, and much more likelihood of its being improved, when minds ceased to be engrossed by the art of getting on. Even the industrial arts might be as earnestly and as successfully cultivated, with this sole difference, that instead of serving no purpose but the increase of wealth, industrial improvements would produce their legitimate effect, that of abridging labour.
>
> *ibid., p. 116*

It would be a considerable stretch to say that Mill foresaw the environmental arguments for a steady-state economy that have become so central among more recent contributors. Yet we are reminded of such modern analytical tools as the ecological footprint (Wackernagel and Rees 1996) and HANPP, the human appropriation of the products of photosynthesis (Haberl, Erb and Krausmann 2007) when Mill wrote that there is not

> much satisfaction in contemplating the world with nothing left to the spontaneous activity of nature; with every rood of land brought into cultivation, which is capable of growing food for human beings; every flowery waste or natural pasture ploughed up, all quadrupeds or birds which are not domesticated for man's use exterminated as rivals for his food, every hedgerow or superfluous tree rooted out, and scarcely a place left where a wild shrub or flower could grow without being eradicated as a weed in the name of improved agriculture …
>
> *op cit., p. 116*

Mill concluded his remarkable chapter on the stationary state with a thought for the future, saying, 'I sincerely hope, for the sake of posterity, that they [the population] will be content to be stationary, long before necessity compels them to it' (ibid., p. 116).

Karl Marx is far better known for his analysis of capitalism and his prediction of its ultimate collapse than he is for what he had to say about steady-state economics. In the mid-nineteenth century, while mainstream economists were concerning themselves with the conditions for and implications of single and multi-market static equilibria, Marx devoted his attention to the dynamics of capitalism. He used the concept of 'reproduction', the process by which an economy, and more broadly, a society, recreates the conditions at the end of each period necessary for it to continue to the next. His analysis of capital accumulation and the declining rate of profit in a growing capitalist economy led him to conclude that eventually capitalism would fail to reproduce the conditions required for its ongoing existence.

As a prelude to this analysis, Marx analysed the requirements for 'simple reproduction', where workers receive a wage sufficient to reproduce themselves and the owners of capital replace worn out capital but do not expand it, spending all the surplus value generated in the economy on consumption. However, under capitalism Marx thought that simple reproduction would give way to expanded reproduction, i.e. growth, because of the capitalists' impulse to accumulate capital. Burkett argues that Marx was well aware of the 'natural conditions' required even for simple reproduction and he takes issue with those who claim that Marx was just as guilty of abstracting the economy from its dependence on the biosphere as mainstream economists (Burkett 2004).[2] Within the larger discussion of steady-state economics we learn from Marx that there is value in discerning which economic, social and environmental conditions must be reproduced, and which can be varied, without compromising the fundamental requirements of a steady-state economy. Such an economy was of no particular interest to Marx other than to explain why simple reproduction must give way to expanded reproduction. His main point in this respect was that a viable economic system should be capable of reproducing itself, but it must do so in a way that is consistent with reasonably stable social and environmental systems.

Daly also considered the institutions for a steady-state economy, devoting an entire chapter to the subject in *Steady-State Economics* (Daly 1977a, 1991a chapter 3), and though his views were not aligned with those of Marx, we do know that Daly read Marx, wrote about his views on population, poverty and development (see Chapter 12) and even taught Marxist economics: 'Marx for certain had many important and very useful insights' (Daly personal communication).

Like Marx, John Maynard Keynes contemplated the steady-state economy without naming it. Unlike Marx, Keynes thought that the steady-state was a very real possibility for those living in the second quarter of the twenty-first century, some 100 years after he wrote his essay: 'Economic Possibilities for our Grandchildren' (Keynes 1930). Considering economic growth in Britain since 1580, when Drake stole treasure from Spain, Keynes concluded that: 'assuming no important wars and no important increase in population, the *economic problem* may be solved, or be at least in sight of solution, within a hundred years' (ibid., pp. 365, 366 italics in the original). Keynes did not define what he meant by 'important' with respect to war and population, but World War II and more than a tripling of the world's population since 1930 likely qualify. Accordingly, we might extend his projection of when the economic problem could be solved somewhat further into the twenty-first century. But that is really not the point. Rather it is that Keynes anticipated the dramatic increases in economic output brought about by technological change and recognized that 'the economic problem is not – if we look into the future – *the permanent problem of the human race*' (ibid., p. 366, italics in the original).

Keynes' view of the steady-state economy was one of abundance and not in any respect a response to the need to bring economies into some sort of balance with the rest of nature – a theme that Mill had discussed nearly 100 years earlier and which is fundamental to Daly's position. But like Daly, Keynes was interested in the moral and social dimensions of economic life. His observations about the relationship between economic growth and morality and the opportunity for improvement that a steady-state economy would allow, though not guarantee, are not unlike those of Daly. Keynes wrote that 'all kinds of social customs and economic practices, affecting the distribution of wealth and of economic rewards and penalties, which we now maintain at all costs, however distasteful and unjust they may be in themselves, because they are tremendously useful in promoting the accumulation of capital, we shall be free, at last to discard.' In particular,

> The love of money as a possession – as distinguished from the love of money as a means to the enjoyments and realities of life – will be recognised for what it is, a somewhat disgusting morbidity, one of those semi-criminal, semi-pathological propensities which one hands over with a shudder to the specialists in mental disease.
>
> *ibid., p. 369*

The day has long passed since economics was called the 'dismal science', in part at least because Malthusian expectations that the human population would outrun food production have not materialized, at least to the extent envisaged by Malthus. These days it is fair to say that natural scientists are more readily persuaded than economists of the ultimate requirements for economic growth to come to an end because of resource and environmental constraints. This is especially true of those with a background in the life sciences where carrying capacity is a widely used concept, understood as a constraint on the growth of populations. Humans of course are a biological species, so the argument goes that we must also be subject to some sort of carrying capacity limit. Whether or not this applies to the growth of the economy as well as the human population is a complex question. As we have seen in the disagreements that Daly has had with conventional economists like Solow, Stiglitz and later with Julian Simon (Daly 1999b), the answer depends on: the definition of what is growing; possibilities and incentives for substitution among whatever may become scarce; and the role of technology in enhancing carrying capacity for humans.

One natural scientist who contributed to the discussion of the steady-state economy was geologist M. King Hubbert whom Daly cites in several of his publications (e.g. Daly and Cobb 1994, p. 408; Daly 2007, pp. 122, 123). Hubbert is best known for his work on peak oil and his prediction published in 1956 that oil production in the lower 48 states in the USA would peak in 1970 (Hubbert 1956). In 1974 Hubbert appeared before a Subcommittee on the Environment of the Committee on Interior and Insular Affairs in the U.S. House of Representatives. In his testimony he stated that 'a system is said to be in a steady-state when its various components either do not change with time, or else vary cyclically with the repetitive cycles not changing with time.' Hubbert contrasted the steady-state with the 'transient' state when 'various components are undergoing noncyclical changes in magnitude, either of increase or decrease'. He used these concepts to describe the historical transition of human societies from a steady-state to a transient state made possible by the utilization of fossil fuels (Hubbert 1974).

Taking the long view, from 5,000 years in the past to 5,000 years in the future, Hubbert argued that 80 per cent of all fossil fuels combined 'coal, oil, natural gas, tar sands, and oil shales' would be consumed within a span of about 300 years and that we were already well into this brief period: 'The epoch of the fossil fuel era can be but an ephemeral and transitory event – an event, nonetheless, that has exercised the most drastic influence so far experienced by the human species during its entire biological existence.'

Hubbert went on to argue that 'the exponential phase of the industrial growth which has dominated human activities during the last couple of centuries is drawing to a close ... [because] it is physically and biologically impossible for any material or energy component to follow the exponential growth phase... for more than a few doublings, and most of those possible doublings have occurred already.' Interestingly, in his testimony Hubbert admitted to having changed his mind about nuclear power based on fission as a substitute for fossil fuels since 'it represents the most hazardous industrial operation in terms of potential catastrophic effects that has ever been undertaken in human history'.

Hubbert concluded by saying that since 'our institutions, our legal system, our financial system, and our most cherished folkways and beliefs are all based upon the premise of continuing growth... it is inevitable that with the slowing down in the rates of physical growth cultural adjustments must be made.' However, he is not clear on whether he welcomed these adjustments, as Mill might have done, or whether he simply thought they were inevitable. Whatever the case may be, it is clear that his view on these matters coincided to a significant degree with those later expounded by Daly as later they were to discover. After attending a lecture given by Daly in Washington DC to an NGO in 1989, Hubbert introduced him- self – much to Daly's delight. They went for a drink at the Cosmos Club where Hubbert was a member. In their conversation Daly confirmed that fellow Texan Hubbert was the author of the textbook written for Technocracy, Inc., a movement after the Great Depression seeking to base economics on an energy theory of value and replacing economists with engineers. The movement, led by Howard Scott, a charismatic engineer, with scientific help from Hubbert, was popular for a while, given the eagerness for a better system induced by the Depression, but for good reasons did not last. Yet the well-written textbook was an early example relating thermodynamics to economics.

Another person, this time an economist, who had a direct influence on Daly is Kenneth Boulding. In 'Economics as a Life Science' Daly writes, 'among current theorists it would appear that only Kenneth Boulding... and Nicholas Georgescu-Roegen... reveal a disposition to take Marshall seriously [in preferring biological analogies to mechanical ones in economics]' (Daly 1968a, p. 393). Boulding made several contributions to our understanding of the dependence of economies on the biosphere in which they are embedded. His seminal essay 'On the Economics of the Coming Spaceship Earth' (Boulding 1966) is the most well-known, and deservedly so, since it provides a concise outline of what was later to become the framework of ecological economics. Boulding alluded to steady-state economics when he said that 'the closed earth of the future requires economic principles which are somewhat different from those of the open earth of the past' (op cit., p. 9). He explored some of these principles in his paper and developed his ideas further in a paper devoted specifically to a consideration of the 'stationary state' which he described as 'an integral part of the "economic imagination"' (Boulding 1973). Boulding stressed that 'the quality of the stationary state depends almost entirely on the nature of the dynamic functions relating the stocks to the flows...' and that '...all stocks, of course do not have to be stationary at the same time' (ibid., p. 92). He also distinguished among a number of 'quasi-stationary states in which some elements of the system are stationary while others are not'. Harking back to Mill, Boulding described one such state as having 'a stationary population and a stationary capital stock with... a change in the character of the capital stock' (ibid.), much like Daly.

Perhaps the most important point that Boulding made in his treatment of the stationary state is that 'no matter what element in the system is stationary... the critical question concerns the nature of the controlling mechanism which keeps it so' (ibid., p. 92). Such mechanisms may be draconian (e.g. forced population control); or more passive, even voluntary; or according to Boulding, they might engender 'mafia-type societies in which government is primarily an institution for redistributing income toward the powerful and away from the weak' (ibid., p. 95). This is a warning to be heeded as we move from discussing the rationale for a steady-state economy to its implementation: 'the problem of building political and constitutional defenses against exploitation may emerge as the major political problem of the stationary state' (ibid.). Anticipating Hubbert, Boulding concluded his comments on institutional considerations with a trenchant comment on existing institutions and their compatibility with the stationary state: 'precisely because existing institutions – political, economic, educational and religious – have exhibited survival value in a very rapidly progressing society, their survival value in a slow or stationary society is an open question' (ibid., p. 100).

One other contribution to steady-state economics that was published in Daly's formative years and which has had a lasting impact is *The Limits to Growth* (Meadows et al., 1972). This short book described a simulation model of 'the world system'. The scenarios that it generated include several in which the system collapses sometime in the twenty-first century. One such scenario, 'the "'standard" world model run, assumes no major changes in the physical, economic, or social relationships that have historically governed the development of the world system...[and] the behaviour mode of the system... is clearly that of overshoot and collapse' (ibid., p. 124). Other scenarios based on different assumptions showed how the system could be stabilized, at least over the duration of the model run (i.e. to 2100), approximating a steady-state.

The Limits to Growth has been subjected to an immense amount of criticism and is often dismissed out of hand today incorrectly as having been proven wrong (see Victor 2019, pp. 169–174 for more discussion). And yet when comparing what has actually happened in the world since *The Limits to Growth* was published with the scenarios described in the book, Graham Turner concludes with the sobering statement that 'the alignment of data trends with the model's dynamics indicates that the early stages of collapse could occur within a decade, or might even be underway' (Turner 2012, p. 124).

Daly takes leadership of the steady-state economics agenda

Daly began writing about the steady-state in the 1960s and has continued to the present day. In 1996 he wrote 'For over twenty-five years the concept of a steady-state economy has been at the center of my thinking and writing' (Daly 1996, p. 3). During an interview with Bernard Kunkel Daly related his interest in steady-state economics to his experience with polio as a child:

> I learned that somethings really are impossible. At the time, the popular idea was that if you had polio, you were supposed to get over it – if you just try harder, nothing's impossible. At a certain point I realized I was being fed a bunch of well-intentioned lies – somethings really are impossible – so I said to myself, the best adaptation when you come up against an impossibility is to recognize it and switch your energy to good things that are still possible … Now, you could make a big leap from that to my later economic theories: unlimited growth is impossible, so let's adapt to a steady state economy. That was never consciously on my mind, but looking back, if you were to put me on a psychiatrist couch, that might occur to the analyst.
>
> *Kunkel 2018, pp. 83–84*

Daly's first mention of a steady-state economy was at a talk he gave to the Yale chapter of Zero Population Growth where he presented the idea of throughput limits, entropy, limits to growth and so forth. John Harte, a Yale physicist was present, and he invited Daly to write a chapter for a book he was co-editing. Over lunch with Harte and his co-editor Daly agreed. He also wrote an edited version of the chapter for the *Yale Alumni Magazine* which was read by a lot of important people who graduated from Yale, one of whom happened to be Harrison Salisbury of the *New York Times*. Salisbury read Daly's article and called him, asking: 'Would you write an op-ed on the same thing?' Daly said, 'sure' and it was published under the title 'The Canary Has Fallen Silent'. It is the only piece that Daly has ever published in the *New York Times*, but it made an impact. One person who read it was Dennis Meadows, who at the time was working on *The Limits to Growth* (Meadows et al. 1972). Meadows' reaction was, 'Oh wow, this is very close to the kind of thing we're doing'. He wrote to Daly to suggest that, 'we should get together and talk about this'. They agreed to meet at the AAAS meetings in New York where Meadows asked Daly, 'What is your scenario for the…?' Daly said 'I don't have a scenario', but they talked about what one might look like. Daly followed up by sending Meadows some things he had written and in Chapter V in *The Limits to Growth*, entitled 'The State of Global Equilibrium', Daly is cited and quoted:

> For several reasons the important issue of the stationary state will be distribution, not production. The problem of relative shares can no longer be avoided by appeals to growth. The argument that everyone should be happy as long as his absolute share of wealth increases, regardless of his relative share, will no longer be available… The stationary state would make fewer demands on our environmental resources, but much greater demands on our moral resources.
> *Daly quoted in Meadows et al. 1972, p. 179*

In this quote Daly uses the term 'stationary state', which he took from the classical economists. He changed it to 'steady-state' in response to criticism that stationary sounded so static and people are not going to like it. Daly borrowed the term steady-state from science and demography, thinking it would resonate across different scientific fields without changing its meaning. What Daly overlooked was that in neoclassical economics, steady state is used to mean a state in which capital and labour are growing at the same rate, maintaining the same ratio, but there is positive economic growth. Daly's use of steady state to mean something else caused some confusion; in retrospect he thinks it might have been a mistake, and that he should have stayed with the stationary state or chosen something completely different like 'dynamic equilibrium' (Personal communication).

This sequence of fortuitous events shows how serendipity can play a role in a person's life. Had Daly not fallen under the influence of Georgescu-Roegen he might well have continued to concentrate on economic development in Latin America, unaware of biophysical limits to growth. Had he not spoken at Yale and had his ideas not attracted so much attention then perhaps he would have gone in a different direction. As it was, it did occur to him that if he was saying one thing and nearly everybody else was saying the opposite, maybe he was wrong, so he went over it again and again and tried to elicit criticism.

At first Daly thought that he was well within the fold of economics, but that he was saying something that was going to give economists something to work with which was important and for which he did not have all the answers. All he knew was that growth is not going to go on forever and that an economy predicated on this happening is going to fail in various ways. So, he thought, let's try to do something different, something that respects biophysical limits. Economics, he understood, is about limits. 'We all know scarcity is the basis of the science and so I'm just adding to the meaning of scarcity. We have scarcity in a broader sense. We're part of a larger system which is also containing it. And I thought that's not hard to understand. But that's not the way it happened.' (Personal communication)

Daly explains the resistance to his ideas as primarily stemming from the belief that growth is the solution to poverty and without growth the only solution is sharing which people are not prepared to do. Early on, Daly found that redistribution and population limitation are politically anathema and so is a steady-state economy. His view then, as now, is that there is a conflict between a political impossibility on the one hand, and a biophysical impossibility on the other hand; but it is the biophysical impossibility which is stronger and will prevail.

We can get a good indication of where things stood when Daly was beginning his inquiry into steady-state economics from the opening sentence in a paper entitled 'The Economics of the Steady-state': 'My title is somewhat pretentious since at present this "new economics" consists only of a definition of a steady-state economy, some arguments for its necessity and desirability, and some disciplined speculations on its appropriate institutions and the problem of transition' (Daly 1974, p. 15). In 1973 Daly published a collection of papers in a book entitled *Toward a Steady-State Economy* (Daly 1973a), which included contributions from Georgescu-Roegen on the entropy law, Ehrlich on population growth, Boulding on the spaceship economy and from several other authors whose names are still well-known today. The book also included a lengthy introduction by Daly describing 'the nature and necessity of the stationary state', followed later in the book by his paper on 'The Steady-State Economy: Toward a Political Economy of Biophysical Equilibrium and Moral Growth'.

Toward a Steady-State Economy went through two more editions (Daly 1980a and Daly and Townsend 1993). The 1973 edition was the precursor to *Steady-State Economics* (Daly 1977a), the first book devoted entirely to the steady-state economy and the second edition (Daly 1991a) still stands as the single most comprehensive treatment of the subject, made increasingly relevant with the passage of time.[3]

BOX 8.1

I had taken a few economics courses and was always bewildered how they failed to address the key issues I was wrestling with, such as environmental degradation, poverty, an economy that seemed to give little weight to well-being. Someone told me to read *Steady-State Economics* – it addressed all the issues that were missing from economics.

Tom Green

Daly's definition(s) of a steady-state economy

The subtitle of *Steady-State Economics* is 'the economics of biophysical equilibrium and moral growth', a subtitle that in eight words gives the essence of Daly's meaning of steady-state economics. He defines a steady-state economy (SSE) as:

> *an economy with constant stocks of people and artefacts, maintained at some desired, sufficient levels by low rates of maintenance "throughput" that is, by the lowest feasible flows of matter and energy from the first stage of production (depletion*

of low-entropy materials from the environment) to the last stage of consumption (pollution of the environment with high-entropy wastes and exotic materials). It should be continually remembered that the SSE is a *physical* concept. If something is non-physical, then perhaps it can grow forever.

op cit., p. 17. Italics in the original

More succinctly, Daly says, 'following Mill we might define a SSE as an economy with constant population and constant stock of capital, maintained by a low rate of throughput that is within the regenerative and assimilative capacities of the ecosystem' (Daly 2008b, p. 4). This definition focuses on keeping constant the stocks of people and artefacts with low rates of throughput that respect the limited capacities of the environment to generate resources and assimilate wastes. Counting people is easy; we do it on a regular basis through the census and so we know what is happening to the numbers of people, though the stock of people in terms of physical weight is complicated by changes in the prevalence of under nourishment and obesity. Counting artefacts is an altogether different matter. Statistical agencies do not keep systematic and complete inventories of artefacts and to the extent they do, they usually aggregate them in monetary units using market prices, a practice that is not without problems. Daly avoids the problem of aggregating capital in monetary terms through his insistence that SSE is a physical concept, but what does it mean to keep the stock of artefacts constant in physical terms? To simply add them up by weight or volume fails to allow for qualitative improvements in the stock and changes in its composition.

Another question that arises from Daly's definition of SSE based on maintaining a constant stock of capital is what is to be included as capital? In recent years it has become popular in some quarters to extend the concept of capital to virtually anything that endures over time: natural capital for nature, human capital for people, social capital for institutions, intellectual capital for ideas, in addition to the more common usage referring to financial capital, and manufactured capital. Daly refers quite often to natural capital in his writings and has defended it from criticism while recognizing some concerns, in particular the valuation of natural capital in terms of money (Daly 2020a). 'The danger comes in monetary evaluation [of natural capital] with its imposed assumption that different forms of capital, once valued in monetary terms, are substitutable in the same way that money is fungible' (Daly 2019b, p. 1). Notably, Daly's definition of a SSE based on minimizing throughput required to maintain a constant capital stock restricts the definition of capital, i.e. as human-made artefacts, to manufactured capital so, in this instance at least, problems associated with natural capital are avoided.

Partly as a way of avoiding the problems of measuring capital, Daly offers an alternative, more operational, definition of a steady-state economy that focuses on flows rather than stocks: 'We might define the SSE in terms of a constant flow of throughput at a sustainable (low) level, with population and capital stock free to adjust to whatever size can be maintained by the constant throughput that begins with depletion and ends with pollution' (ibid. 2008b, p. 3). Analytically Daly's two definitions of an SSE are equivalent: Definition 1 – minimizing throughput required to maintain a stock at a desired level; versus Definition 2 – maximizing a stock that can be maintained without exceeding a desired level of throughput. If the throughput arrived at based on the first definition is used in calculating a stock using the second definition, the stock will be the same regardless of which definition is used. This becomes slightly more complicated by the fact that Daly consistently mentions two stocks of interest: humans and artefacts, both requiring different throughputs for maintenance. Nonetheless, it is reasonable to describe his two definitions of an SSE as equivalent though they have different policy implications.

In terms of his second definition the measurement problem shifts from capital to throughput – the physical inflows and outflows to and from an economy. Unless we measure material throughput strictly in terms of weight and energy in some common unit, we will be unable to say whether material and energy throughputs are rising, falling or remaining constant. Such a high level of aggregation can be useful for assessing changes at the macro level, but risks overlooking the dramatically different resource scarcity implications and environmental impacts of quantitatively equal flows of materials and energy.

Daly does not express a preference for one or other of his two definitions of a SSE, other than that it may be easier to obtain more comprehensive information on the physical magnitude of throughput than on the size of the capital stock from which it is built. However, there is a case to be made that setting targets for throughput and allowing the stock(s) to change is preferable to setting targets for the stocks and minimizing throughput. An important aspect of Daly's rationale for an SSE is to bring the economy into some reasonable balance with the biosphere in which it is embedded. The main connection between the economy and the biosphere is throughput: the extraction of materials (including fossil fuels for energy) and the disposal of wastes. A larger capital stock requires a larger extraction of materials for its construction, and generally speaking a larger throughput to operate and maintain it. The environmental impact of the capital stock itself, though not trivial (e.g. loss of birds from collisions, loss of fertile land and habitat when built upon, disturbance to migration routes) is much less significant than the impacts from the extraction, use and disposal of material and energy, i.e. from throughput. It makes sense, therefore, to set policy targets for throughput as the best way to limit the environmental impacts of the economy and allow flexibility in the design and use of the capital stock, rather than to fix the capital stock and minimize the throughput required for construction, maintenance and operation, a point with which Daly agrees (personal communication).

One indication that an SSE defined in either of the ways suggested by Daly is as practical as the conventional reliance on GDP is that EuroStat, the statistical office of the European Union (EU), now reports the annual throughput of materials in total and by category and the accumulation of artefacts (but not humans) both measured in tonnes for the EU in total and for individual member countries. Figure 8.1 shows the flow of total materials through the EU economy in 2018. The sum of materials (as goods) imported and natural resources extracted within the EU equals

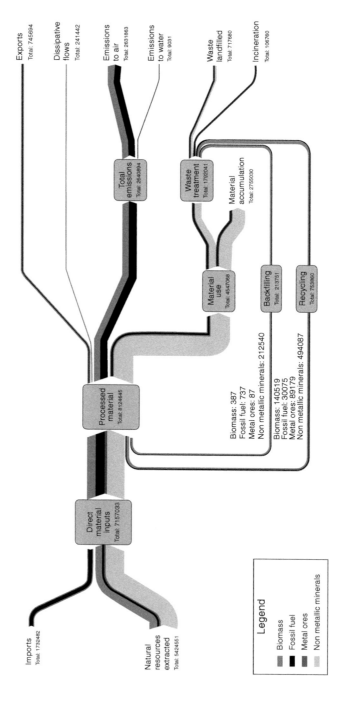

FIGURE 8.1 Flows of Material Resources (in Gt – billion tons) EU27, 2018

Source: https://ec.europa.eu/eurostat/web/products-eurostat-news/-/DDN-20200311-2.

the sum of exports, emissions to air, water and waste landfilled, dissipative flows, and material accumulation. Recycling and backfilling which reduce the requirement for virgin materials are shown feeding back into the supply of 'processed materials'. Another important feedback loop that is not shown in Figure 8.1 and which Daly has noted is the one between 'material accumulation', much of which is an expansion of the capital stock, and future maintenance requirements for 'direct material inputs'. Daly advocates limits on throughput in order prevent this feedback from disturbing the balance between the economy and the biosphere that the SSE is intended to achieve.

The emphasis that Daly places on the physical dimension of economies and the relevance of the first and second laws of thermodynamics to economics does not mean that he ignores the concepts that play such a central role in conventional economics. (We shall see in the next chapter that Daly has been criticized by some who think he makes too much use of conventional i.e. neoclassical economics.) One of these concepts is efficiency but in Daly's hands it takes on a rather different meaning.

BOX 8.2

The concept had enlightened me on two aspects. For one, a stark contrast to the fervent common belief that economic growth itself could cure all developing problems, Professor Daly was among the rarest economists who resolutely opposed the growth mania. For another, even among the objectors against blindly pursuing economic growth, few had actually taken a step forward from criticism to systematically constructing a solution framework, and that's why Professor Daly's Steady State Economy Theory stands out like a unicorn.

Ji Xi

Efficiency in the steady-state economy

The concept of efficiency figures prominently in neoclassical economics where it is a primary normative criterion for assessing alternative economic arrangements. More precisely, efficiency in welfare economics – the normative principles used in neoclassical economics – is 'Pareto efficiency'. Vilfredo Pareto (1884–1923) was an Italian engineer, economist, political scientist, sociologist and philosopher who gave his name to a particular meaning of efficiency. A Pareto efficient or Pareto optimal situation is one where it is impossible to make at least one person in an economy better off without making someone else worse off as judged by those directly concerned. A movement in the direction of Pareto efficiency that makes at least one person better off without anyone else being worse off is called a Pareto improvement. Pareto efficiency can be understood as a situation where all opportunities for mutually beneficial trades have been exhausted. The theoretical outcome of a competitive economy is that it is Pareto efficient. A well-known result in standard economic theory is that there is a different Pareto efficient outcome in a competitive economy for each distribution of income and wealth. Daly is not averse to using these insights into the allocative merits of a competitive economy (see Daly 1991a, pp. 82–86 and Chapter 8 in Daly and Farley 2011) but he is not content to leave it there. In a vain attempt to avoid value judgements or 'inter-personal comparisons of utility', neoclassical economists refrain from declaring one distribution better than another because different distributions of income and wealth involve gainers and losers. It follows that an imposed redistribution of income and wealth would not be a Pareto improvement since the gains of some come at the losses of others. As explained in Chapter 5, Daly takes a different view. It is precisely because there is a different Pareto efficient outcome from each distribution of income and wealth that Daly argues that a just distribution has to precede the determination of allocative efficiency. Contrary to those who think that a just distribution emerges from allocation determined in and by the market, Daly is firm in his view that a collective decision about distribution is a prerequisite for any market determined allocation to have normative significance and therefore proposes the establishment of minimum and maximum limits on incomes and wealth.

When it comes to a steady-state economy, Daly bases his analysis of efficiency on the relationship between inputs and outputs. A process that converts an input into an output is said to become more efficient if the same output can be produced with reduced input or if more output can be produced with the same input. Inputs and outputs can also be understood as 'the ratio of benefits and costs' (ibid., p. 78). Daly uses efficiency defined in this way to distinguish between quantitative growth and qualitative development by employing a simple identity (Daly 1974)[4]:

$$\underset{(1)}{\frac{service}{throughput}} \equiv \underset{(2)}{\frac{service}{stock}} \times \underset{(3)}{\frac{stock}{throughput}} \tag{8.1}$$

Daly holds that service is the final benefit of economic activity and 'throughput (an entropic physical flow) is the final cost… or rather the sacrificed ecosystem services provoked by the throughput' (Daly 1977a, 1991a, pp. 36, 37). Daly explains that 'the throughput flow does not yield services directly, it must first be accumulated and fashioned into a stock of useful artifacts (capital)' (ibid., p. 36). Here, it will be noted, that Daly is referring to material throughput since energy cannot be accumulated except in very small amounts. 'Stocks are in the center because they are the intermediate magnitude. It is stocks that directly yield services (ratio 2) and that require throughput for maintenance and replacement' (ibid., p. 37).

Each of the three ratios in identity (8.1) can be interpreted as a measure of efficiency. They are not independent of each other in that the value of ratio (1) is determined by the values of ratios (2) and (3). Daly uses the identity to amplify his definition of growth 'as an increase in total service resulting from an increase in stocks and throughputs, with the two efficiency ratios on the right-hand side of the identity held constant' (ibid., p. 77). In this definition of growth, service/stock and stock/throughput, two measures of efficiency, are held constant, which means that services increase with growth only by increasing throughput which automatically increases stocks. With development, there are three possibilities. One is to hold throughput constant. Stocks can still rise if ratio (3) increases. The second is to hold stocks constant in which case throughput will decline for ratio (3) to increase. The third is to hold throughput and stocks constant, then development occurs when ratio (2) increases.

This simple analytical framework brings clarity to Daly's distinction between growth and development. It also offers an alternative framework for analysing economies at all scales, from individual production units to entire macroeconomies since it can be applied at each of these levels. When thinking of a steady-state economy, Daly's main interest is at the macro level. He uses the following definition of efficiency to probe more deeply into the meaning of macro efficiency:

$$efficiency = \frac{benefit}{cost} = \frac{artifact\ service\ gained}{ecosystem\ services\ sacrificed} \tag{8.2}$$

Anticipating the objection that there are other costs besides ecosystem services sacrificed, Daly follows Irving Fisher and subtracts the disutility or 'disservices of labour from artifact services… [and treats]… the numerator as *net* artifact services gained' (ibid., p. 78). Daly's next step in the analysis of efficiency is to decompose 'artifact service gained/ecosystem services sacrificed' into four component ratios as in identity 8.3:

$$\frac{artifact\ sevices\ gained}{ecosystem\ services\ sacrificed} \equiv \frac{artifact\ sevices\ gained}{artifact\ stock} \times \frac{artifact\ stock}{thoughput} \times \frac{thoughput}{ecosystem\ stock\ sacrificed} \times \frac{ecosystem\ stock\ sacrificed}{ecosystem\ services\ sacrificed} \tag{8.3}$$

$$\qquad\qquad (1)\qquad\quad (2)\qquad\quad (3)\qquad\quad (4)$$

As Daly explains, each of the ratios (1) to (4) express a different dimension of efficiency (ibid., pp. 78–79):

(1) *Artefact service efficiency*: the efficiency of a given amount of stock in satisfying wants… depends on its *allocation* among different artefact embodiments and uses… and on the *distribution* of the stock among different people. [For example,

more housing services can be obtained from a housing stock comprised of many houses equitably shared than by a palace with one family, and the rest of the community living in shacks.]

(2)   *Artefact maintenance efficiency:* is essentially the turnover or renewal period of the artefact stock... Artefact maintenance efficiency is served by minimizing the throughput required to sustain a given stock. [This efficiency can be understood in terms of durability. A well-designed, well-constructed building can be maintained with less throughput than one than one that is in a constant state of repair.]

(3)   *Ecosystem maintenance efficiency:* reflects the degree to which the ecosystem can maintain a supply of throughput on a sustainable basis, that is, without a depletion of the natural stocks. It depends on the replaceability or renewability of the environmental sources and sinks. [For example, timber obtained from a sustainably managed forest is more efficient in this sense than timber obtained from clear cutting with no replanting.]

(4)   *Ecosystem service efficiency:* depends on allocation and distribution, as in ratio (1), but... on the allocation and distribution of *loss* rather than gain... These costs are allocated and distributed mainly through a web of ecological interdependence that lies outside the market. [When ecosystem stock has to be sacrificed to obtain throughput for the economy the associated loss of ecosystem services depends on what parts of an ecosystem are lost, whether it is concentrated in one location with little chance of recovery or spread around where it can be more easily replaced. For example, commercial fishing can destroy a fishery by depleting the stock but if the same fishing effort is spread among several fisheries, the impacts will be much less.]

Daly uses these four types of efficiency to explain the impact of technology in a steady-state economy, or indeed, in any economy. Technology increases efficiency by increasing the output (the numerator) from a given input (the denominator) or reducing the input for any given output, but there are limits. 'Ratio 1 is limited by diminishing marginal utility.' In the housing example, total utility is increased when the housing stock is more equitably shared, but for any given distribution, increases in the housing stock from the application of improved technology will add to well-being less and less with each increment. 'Ratios 2 and 3 are limited by the entropy law: nothing lasts forever.' With ratio 2, using the housing example, improved technology can allow a greater housing stock to be built and maintained from the same throughput. Technology can increase ratio 3 by improving timber harvesting methods and silviculture. 'Ratio 4 is limited by the law of increasing marginal costs, greatly complicated by the discontinuities arising from ecological thresholds and complex interdependencies' (ibid., p. 79). Technology, in the form of improved ecosystem management, can reduce the sacrifice of ecosystem services caused by a loss of ecosystem stock but a lack of knowledge about cause and effect and inherent uncertainty can make it difficult to identify the best opportunities.

The novelty of Daly's treatment of efficiency is his use of a commonsensical understanding of efficiency as a relationship between inputs and outputs linking the services obtained from artefacts to the related loss of services from the ecosystem. He proposes a steady-state economy as a way of finding the best balance long-term between the two, in sharp contrast to the growth paradigm which over values the expansion of artifact services and under values the loss of ecosystem services.[5]

Person in community

In the classification of academic disciplines economics is considered a social science. Social sciences can be thought of as the study of society and relations among people using scientific approaches and methods. Each social science focuses on a part of this broad terrain, with practitioners paying varying degrees of attention to ideas beyond their own discipline depending on their personal inclinations as well as the intellectual traditions with which they identify. In building his economics for a full world Daly drew heavily on physics and ecology from the natural sciences, philosophy and theology from the humanities; and various schools of thought in economics including elements of Hayek, Keynes and Marx, as well as classical economists Smith, Malthus, Ricardo and Mill, heterodox economists Georgescu-Roegen and Boulding, and mainstream neoclassical economists as well.

BOX 8.3

Daly and Cobb's more socially grounded construct of the 'economic being' seemed to me much more defensible than the standard neoclassical construct.

Rob Smith

In mainstream neoclassical economics the type of relations among people that are the primary object of study are those of buying and selling goods and services through markets. For much of mainstream theory these transactions are assumed to be operating among large numbers of anonymous individuals each seeking to maximize their own utility. The interactions may be face to face or they may be transactions involving institutions, particularly businesses established with the primary objective of maximizing profits. The aggregation of these individuals and institutions and their behaviour as individuals and in groups, make up a market economy, one that is capitalist when the land, labour and capital used in production are predominantly privately owned.

Of interest here is the particular characterization of the individuals participating in the economy on which so much of mainstream economics is based. 'The key assumptions in this case have to do with *Homo economicus*, that is, the understanding of the nature of the human being' (ibid., p. 5). '*Homo economicus*' or economic man, was introduced by critics of John Stuart Mill who used it in a pejorative sense, objecting to Mill's theoretical abstractions. 'Economic man also raised the indignation of Victorian moralists shocked at the postulation of such blatant selfishness' (Persky 1995, p. 222). Mill found it useful to abstract from many aspects of human behaviour in order to concentrate on the desire of people to accumulate wealth, with the added assumption that they are 'capable of judging the comparative efficacy of means for obtaining that end' (Mill, quoted in Persky, p. 223). A description of *Homo economicus* is fleshed out in Daly's textbook written with Josh Farley:

> *Homo economicus*... emerges from the discipline's foundations in utilitarianism and incorporates the following traits:
>
> 1. Insatiability... more is always better, and consumption is the major source of utility (i.e. well-being).
> 2. Perfect rationality. Individuals have stable, exogenously determined preferences... and make choices that best satisfy these preferences in the face of given constraints of time, income, and so on.
> 3. Perfect self-interest. Individuals (or at least families) do not care how their choices affect others and are not affected by the utility others experience. Social interactions matter only to the extent that they affect one's own consumption, leisure, and wealth.
>
> *Daly and Farley 2011, p. 233*

Daly and Farley comment, 'Though most economists recognize that the assumptions of *Homo economicus* are somewhat of a caricature of real human behavior, these assumptions nonetheless form one of the central pillars of conventional microeconomic theory' (ibid., p. 234). The extreme individualism that they describe became embodied in economics through the application of analytical methods borrowed from nineteenth-century physics that were based on interactions among atoms (Mirowski 1989). With *Homo economicus* as the conception of the human underpinning mainstream economics there is very little that is social in the fuller sense of the term. For example, there is no consideration of the social determinants of preferences such as the influence of the preferences of others, of education, of advertising (the primary purpose of which is to change peoples' preferences), and of culture in general. It is as if people are born as fully formed adults immune to the influence of others, with a lack of cultural, class and gender differentiation making them appear to be 'highly predictable across time and cultures' (ibid., p. 233) The problem with *Homo economicus*... is that she is an atomistic individual connected to other people and things only by external relations' (Daly 2014a, p. 159).

The unrealistic assumption of exogenously determined, fixed preferences, is fundamental to the principal normative criterion used in mainstream economics – the extent which people's preferences are met. The assumption of fixed preferences is built into the theory of welfare economics to assess improvements in economic efficiency. It is inherent in the practice of benefit–cost analysis used to assess proposals for projects and regulations. *Homo economicus* 'as a picture of human beings is profoundly erroneous. People are constituted by their relationships... The classical *Homo economicus* is a radical departure from social reality' (Daly and Cobb op cit., p. 161). If preferences are subject to change over time, the criterion of how well people are able to satisfy their preferences by making the economy more competitive or by building a new bridge becomes dysfunctional. This criterion can depend on whether preferences before or after they have changed over time are used to make the judgement. The same applies to any judgement based in mainstream economics about whether economic growth makes people better off when growth itself changes people's preferences.

Another limitation of *Homo economicus* is that an individual is regarded as a consumer, neglecting their role as a worker (or a citizen). Work is assumed to be a source of disutility for which the worker requires compensation. This is contrary to the obvious, that work, the kind of work, as well as working conditions physical and social, all contribute to a person's welfare. 'The goal of an economics for community is as much to provide meaningful and personally satisfying work as to provide adequate goods and services' (ibid., p. 165).

Despite their trenchant criticisms of *Homo economicus* and the economic theory and practices built upon it, Daly and Cobb say 'many principles of classical and neo-classical economics, with proper historical qualifications, will function in an economics based on the different model of *Homo economicus* as person-in-community'. In other words, rather than reject the concept of *Homo economicus* entirely, they retain an emphasis on the individual but within a broader and richer social context than simply the market. Daly and Cobb go on to explain that in this new model the well-being of a community, understood as the pattern of relationships is

> at least as important as the possession of commodities... This model of person-in-community calls for provision of goods and services to individuals, but also for an economic order that supports the pattern of personal relationships that make up the community.
>
> *ibid., pp. 164–165*

Rather than analysing economies exclusively as transactional, monetized interactions among self-regarding, rational, well-informed individuals, Daly and Cobb propose an alternative image of *Homo economicus* as 'person-in-community'. They explain that 'we are not only members of societies, but what more we are also depends on the character of these societies. The social character of human existence is primary' (ibid., p. 161). They continue:

> But what is equally important for the new model [of the human] – and absent in the traditional one – is the recognition that the well-being of a community as a whole is constitutive of each person's welfare. This is because each human being is constituted by relationships to others, and this pattern of relationships

> is at least as important as the possession of commodities… Hence this model of person-in-community calls not only for provision of goods and services to individuals, but also for an economic order that supports the pattern of personal relationships that make up the community.
>
> *ibid., pp. 164–165*

Communities are far more than an aggregation of individuals. They have

> boundaries that are both inclusive and exclusive. The relationships by which we are defined as persons in community are those with people and places we know, with whom we share some common history, language, and laws. They do not include all possible relations with all people all over the globe, except in a very abstract and tenuous way… The person-in-community understanding of who we are means that my welfare depends much more on the quality of all the relationships that define me than on my external relations to the commodities I buy or consume.
>
> *Daly 2014a, pp. 158–159*

Daly's view of the importance of community to people's sense of well-being has been corroborated by a considerable amount of social science research, much of which has informed the eight editions (as of 2020) of the World Happiness reports, the first one being published in 2012. The authors of these reports, all economists, state that, 'institutionally, building a government that is trustworthy and functions well, and culturally, building a sense of community and unity among the citizens are the most crucial steps towards a society where people are happy (Helliwell et al. 2020, p. 140). More specifically, 'On a cultural level, arguably the most important factor is to generate a sense of community, trust, and social cohesion among citizens. A divided society has a hard time providing the kind of public goods that would universally support each citizen's ability to live a happier life. In a divided society, people also tend to be less supportive of various welfare benefits because [they] worry they would benefit the "other" groups as well. When people care about each other and trust each other, this provides a much more stable base on which to build public support for various public goods and welfare benefit programs' (ibid.). Daly's sentiments exactly, expressed three decades earlier.

Principles of steady-state economics

By defining a steady-state economy in quantitative terms Daly has already gone well beyond the purely descriptive approaches of his predecessors. He has opened the way to empirical assessments of the extent which the economies of the world are moving towards or away from a steady state (O'Neill 2012). And although Daly is not credited sufficiently, his emphasis on the materiality of economies has been taken up by the recent interest in a 'circular economy' in which reductions in material throughput are achieved through increased sharing, reuse, product durability, recovery and recycling (Ellen McArthur Foundation 2013).

Defining a steady-state economy is an important first step but, as Daly appreciates, if steady-state economics is to challenge the mainstream growth paradigm, it must be fleshed out with a set of principles. Daly has written a great deal about these principles, but he has not assembled them in one place. In this section we do just that and in doing so provide an appreciation of the breadth and depth of Daly's contribution to steady-state economics. Some of the principles are drawn from previous chapters. Others are covered in later chapters including this one. What follows is a list of Daly's principles of steady-state economics with references to other parts of the book or to Daly's writings directly, where fuller discussions can be found.

Pre-analytic vision or paradigm

1. The economy is a sub-system of the biosphere and the constraints of that finite and complex biosphere must be included in economic reasoning (Chapter 4).
2. The economy and all economic activities are subject to the first and second laws of thermodynamics which define the biophysical roots of scarcity (Chapter 4).
3. The closer the economy approaches the scale of the whole Earth the more it will have to conform to the physical behaviour mode of the Earth. That behaviour mode is a steady-state – a system that permits qualitative *development* but not aggregate quantitative *growth* (Daly 2008b, p. 1).
4. *Person-in-community Homo economicus*, the self-interested, insatiable, atomistic, abstract human being on which neoclassical economic theory is based, is only connected to other people through external, transactional, relations. In steady-state economics the individual remains important but within the context on community where other types of social relations are recognized as important and 'whose very identity is constituted by internal relations to others in the community' (Daly 2014a, p. 159).

Analytic: Economic structure

5. The economy is comprised of three hierarchically ordered dimensions: scale, distribution and allocation (Chapter 5).
   a. There is a primary normative objective for each dimension:
      - optimum scale
      - just distribution
      - efficient allocation
   b. Target values for the three dimensions are to be determined in sequence: scale then distribution then allocation. The reverse sequence is invalid.

Analytic: Macroeconomics

6. The total scale of resource throughput should be limited to ensure that the scale of the economy (population times per capita resource use) is within the carrying capacity of the region avoiding capital consumption (Daly 1990).

7. Growth is quantitative, development is qualitative. In a steady-state economy there is development without growth (Chapter 8).

8. Optimal scale is reached when then the marginal cost of an increase in scale is judged to approximate without exceeding the marginal benefit. Growth beyond that is uneconomic (Chapter 6).

9. Just distribution, intra-generational and inter-generational should be judged according to Rawlsian principles based on the veil of ignorance (i.e. a collective decision-making process wherein the status of each person in the process is unknown prior to the establishment of the fundamental social rules and norms). Just distribution includes maximum and minimum incomes and maximum wealth (Chapter 7, Daly 1977a,1991a, pp. 80–82).

10. Institutional arrangements for efficient allocation depend on the characteristics of rival/nonrival/congestible and excludable/non excludable which can apply to any good or service. For example, bread is rival (if one person eats it there is less for someone else) and excludable (it can be owned which confers on the owner the right to exclude others from eating it). Congestible refers to a park or a road which are non-rival if uncrowded but become rival (congested) as more people use them. The light from streetlights is nonexcludable, and nonrival, as is clean air. Of the six possible combinations of these characteristics only the combination of rival/excludable meets the conditions for efficient allocation by competitive markets. Access to congestible services can be restricted using prices but only during periods of high use (Daly and Farley 2011, pp. 168–169). For all other combinations different institutional arrangements are required.

11. Renewable resources: harvest rates should equal regeneration rates (sustained yield) (Daly 1990).

12. Tinbergen's principle of at least as many instruments as there are policy objectives applies (Chapter 4).

13. Three separate accounts should be maintained: a value index of material and energy throughput which is the cost of maintaining or adding to the existing capital stock; an account recording the value of services yielded over time by the total stock of human and physical capital; and an account of the capital stock itself (Chapter 7).

14. Counting the consumption of natural capital as income should end (Daly 1996, p. 88).

15. Labour and income should be taxed less, and resource throughput more (Daly 1996, p. 88).

Analytic: Microeconomics

16. Production is undertaken by funds (e.g. capital and labour) that process flows of materials and energy (Chapter 4).
17. Funds can be complements or substitutes. Flows can be complements or substitutes. Funds and flows are predominantly complements (Chapter 4).
18. Efficient production requires maximizing the productivity of the limiting factor which has shifted from manufactured capital to natural capital (Daly 1996, pp. 78–80).
19. Manufactured capital should remain intact at the optimal level through investment (Daly 1990).
20. Investment should favour technologies that increase resource productivity (development) maximizing the amount of value extracted per unit of resource, rather than technologies for increasing the resource throughput itself (growth) (Daly 1990).
21. Investment in the exploitation of a non-renewable resource should be paired with a compensating investment in a renewable substitute (Daly 1990).
22. Waste emission rates should equal the natural assimilative capacities of the ecosystems into which the wastes are emitted (Daly 1990).

Money and banking

23. 100% reserve requirement for commercial banks. Creation of money is the sole prerogative of the Treasury and is aimed at price level stability, not at influencing the interest rate. Current investment should be paid for out of past savings, not out of money creation beyond the amount of new money that the public is voluntarily willing to hold as an asset (Chapter 12).

Internationalization not globalization

24. Internationalization is favoured over globalization. In internationalization the basic unit of community and policy remains the nation, even as relations among nations, and among individuals in different nations become increasingly necessary and important. In globalization many formerly national economies are integrated economically into one global economy, by free trade, free capital mobility, and though less significant, easy or uncontrolled immigration. Seek multilateral balance in trade flows, avoiding the accumulation of international debts (Chapter 13).

Population

25. Maximization of cumulative number of human lives ever to be lived at a level of per capita consumption sufficient for a good life. More people are better than fewer, but not if all are alive at the same time. Seek to balance births plus immigrants with deaths plus emigrants (Chapter 11).

Institutions

26. The guiding principle 'is to provide the necessary social control with a minimum sacrifice of personal freedom, to provide macrostability while allowing for microvariability, to combine the macrostatic with the microdynamic' (Daly 1977a,1991a, p. 51).
27. 'Maintain considerable slack between the actual environmental load and the maximum carrying capacity. The closer the actual approaches the maximum the less is the margin for error' (Daly 1977a,1991a, p. 51).
28. 'Making the transition to a steady-state economy... build the ability to tighten constraints gradually and to begin from existing initial conditions rather than unrealistically assuming a clean slate' (Daly 1977a, 1991a, p. 51).

Taken together these 28 'principles' are the core of Daly's steady-state economics. Strictly speaking there may be some redundancy (some principles derivable from others) but inclusiveness and clarity were chosen over strict logical parsimony in summarizing and classifying the principles.

BOX 8.4

It will be hard to overstate the potential influence of Herman Daly to move mankind to a new paradigm of global economic sustainability and avoid the collapse of civilization as we know it. Herman Daly's work and thoughts are built on a humane, just and equitable vision for humankind.

Robert Coish

Top 10 policies for a steady-state economy[6]

Thoughtful policies are based on principles and Daly's top 10 policies for moving to a steady-state economy are no exception. They are based on his principles for a steady-state economy.

1. Caps limit biophysical scale by quotas on depletion or pollution, whichever is more limiting. Auctioning the quotas captures scarcity rents for equitable redistribution.
2. Ecological tax reform – shift the tax base from value added (labour and capital) to 'that to which value is added', namely the entropic throughput of resources extracted from nature (depletion) and returned to nature (pollution).
3. Limit the range of inequality in income distribution with a minimum income and a maximum income.
4. Free up the length of the working day, week and year – allow greater options for part-time or personal work.

5. Re-regulate international commerce – move away from free trade, free capital mobility and globalization.
6. Downgrade the World Trade Organization, the World Bank, the International Monetary Fund.
7. Move away from fractional reserve banking towards a system of 100 per cent reserve requirements.
8. Stop treating the scarce as if it were free and the free as if it were scarce – Enclose the remaining open-access commons of rival natural capital (e.g., the atmosphere, the electromagnetic spectrum and public lands) in public trusts, and price them by cap-auction-trade systems, or by taxes. At the same time, free from private enclosure and prices the non- rival commonwealth of knowledge and information.
9. Stabilize population.
10. Reform national accounts – separate GDP into a cost and a benefits account.

Other alternatives to the growth paradigm

Sometime after Daly began writing about the steady-state economy other suggestions emerged for alternatives to the growth economy. They were motivated by many of the same concerns as Daly's, and contributors often cite his work, although they do not necessarily draw the same conclusions about what a different economy and a different economic paradigm might or should look like. To name some of the more prominent of these other alternatives to the dominant growth paradigm: there is sustainable development (World Commission on Environment and Development 1987), the circular economy (Ellen McArthur Foundation 2013), the well-being economy (wellbeingeconomy.org), degrowth (D'Alisa, Demaria and Kallis 2015), and the donut economy Raworth 2017), all of which draw on Daly's steady-state economics and will be mentioned again in later chapters.

Conclusion

It is more than 160 years since John Stuart Mill wrote favourably about the steady-state economy and over 40 years since Herman Daly literally wrote the book on the subject and began building a comprehensive set of definitions, concepts, principles, analyses and policy proposals based on a profound critique of the growth paradigm. In the meantime, economic growth has proceeded apace, and since the end of World War II it has been the over-arching economic policy objective of countries and governments around the world. Daly has been in the forefront of an expanding number of critics of economic growth and in proposing alternatives to the growth paradigm. His work has inspired many, including those unable to wean themselves entirely away from the pursuit of growth, but looking for greener or more inclusive growth. Looking back over his career of more than half a century and the mixed reaction his ideas about a steady-state economy has received, Daly can take some satisfaction that his influence through his work on steady-state economics has helped soften attitudes towards growth, while he remains fully aware that the change in paradigm for which he has fought long and hard, has only just begun. To that end he has provided a list of 'some issues that are not new, but which, in my opinion, constitute important unfinished business that I hope others will decide to at least further if not finish' (Daly 2019b, p. 1).

Notes

1 This brief history is a revised version of a section in Victor (2016).
2 See also Victor for a similar argument (1979) and the more thorough treatment in Foster (2000).
3 The second edition differs from the first in that a third section of further essays in the economics of sustainability has been added and the subtitle used in the first edition was dropped. The first two sections were unchanged from one edition to the next.
4 An identity is an equation which is always true, no matter what values are substituted.
5 'The growth paradigm… refers to the proposition that economic growth is *good*, *imperative*, essentially *limitless*, and the *principal remedy* for a litany of social problems. The growth paradigm appears ubiquitous, even natural, but it is uniquely modern' (Dale 2012).
6 These are summarized from Daly 2013b.

9

IN DEFENSE OF THE STEADY-STATE ECONOMY

> My problem with my fellow economists is not their frequent state of dis-
> agreement, but rather their near unanimous agreement in support of basic
> policies that are killing us. Instead of critical debates on vital issues, what
> resonates from academia is the unison snoring of supine economists in deep
> dogmatic slumber.
>
> *Daly*

The 1960s was a decade of change and transformation. The Cold War between the
USA and the Soviet Union was in full swing. The space race, which was very much a
part of the Cold War, was won by the Americans when Neil Armstrong became the
first person to walk on the moon. This accomplishment became a standard against
which the achievement of other societal goals came to be compared. 'If we can get to
the moon then surely we can…' or words to that effect. Technological optimism, the
idea that any problem can be overcome with the right technology, was already rampant,
especially in the USA. The successful moon-shot only added to this cultural meme.

Even before Armstrong made it to the moon and back, the images of the Earth
taken from space had begun to transform views of our place in the Universe. Ever
since Copernicus explained that the Earth travelled around the sun, we knew, at an
intellectual level, that we were not living at the centre of the solar system, let alone
the centre of the universe. But for the vast majority of people, the Earth remained
enormous, and our experience of it was flat rather than round. Only when we saw
our blue planet suspended in space, small and alone, did we begin to really appreciate emotionally as well cerebrally what it meant to live on spaceship Earth with
all our supplies on board but for the sunlight on which all life depends.

This transformation in worldview is still far from complete, partly because its
implications do not align well with the faith in technology engendered by the
technological successes that nourished it. The rapid expansion of the human

DOI: 10.4324/9781003094746-9

population brought to public attention by Paul Erhlich's *Population Bomb* (1968), combined with rising levels of consumption in many parts of the world, made some wonder whether and for how long this could continue. Daly, first as a student of Georgescu-Roegen, and then in his own right as a young professor at Louisiana State University was one of these. He was influenced more by the finitude of planet Earth and the constraining laws of thermodynamics than by technological wizardry, leading him to question the assumptions made about technology and humans in the neoclassical economics in which he was schooled. This led him to develop steady-state economics which is based essentially on an appreciation of the finitude of planet Earth and a more complete view of what it means to be human. In particular, it made Daly a critic of economic growth and a proponent of a steady-state economy.

BOX 9.1

I admire his courage to step outside of neoclassical economics and advance the steady-state economy argument when it was viewed as being extremely marginal. Thanks to Daly, people like me have had the work of an economist to draw upon to support our critique of capitalist-driven economic growth and its devastating consequences for ecosystems and communities... the logic is irrefutable and quickly grasped by students.

Laurie Adkin

Paradigm shift

For much of his career Daly, a warm, mild-mannered, strong-willed man willing to ask awkward questions, took the offensive. As we have seen, he wrote extensively on what is wrong with economic growth, debated its desirability and feasibility with all comers, and offered the steady-state economy as a preferable, even inevitable, alternative. When he read Thomas Kuhn's book *The Structure of Scientific Revolutions* (1962) Daly realized that what had begun for him as a simple query about economic growth had become a call for a paradigm shift in economics: a change in its fundamental presuppositions or, as Schumpeter called it, its 'pre-analytic vision' (Daly 1977a, 1991a, p. 14).

Paradigm change never comes easily, even in the natural sciences where theories that account for a wider range of observations eventually displace less successful ones. The emergence of a new paradigm inevitably represents a challenge to the existing, dominant one. In Kuhn's terminology, 'normal science' becomes threatened by 'revolutionary science' and is resisted. Paradigm change in the social sciences is attenuated by several differences between the social and natural sciences. One is that social systems, including economic and political systems, can change rapidly so that explanatory theories of a system's behaviour that once worked well can quickly become out of date because the system under study has changed. The economy studied by Adam Smith bears little resemblance to the economies of today, whereas Newton studied the same physical world as Einstein.

Controlled experiments are widely used in the natural sciences to establish causality. These are much more difficult to conduct in the social sciences. When Daly was a student, he was told that econometrics – the application of statistical methods to economic data to estimate relationships among variables – would help decide between different economic theories but it never lived up to its billing.

Another consideration that differentiates the social and natural sciences and which makes paradigm change more complicated in the social sciences is that people, who are the subject of study in the social sciences, are quasi-independent actors with minds of their own. Unlike inanimate atoms and molecules, people may change their behaviour in response to the theories devised to explain it. For example, predictions of stock market prices influence stock trades leading to different prices than those predicted. Does this mean the theory on which the predictions were based was wrong and should be changed? Weather forecasts affect behaviour too but the behaviour does not affect the accuracy of the forecast. It is much clearer that a series of bad weather forecasts indicates weaknesses in the theory and/or data on which they were based.

Notwithstanding these differences between the natural and social sciences, practitioners of both resist challenges to their prevailing paradigms. Often this is simply a matter of sensible precaution based on a lack of compelling evidence. At other times it can be due to perceived threats to individual scholars or groups of scholars with vested interests in the current paradigm. Daly learned of this resistance to paradigm change within economics when his ideas and influence began to spread. Soon after he began writing about the steady-state economy it became the subject of criticism, first from orthodox economists seeking to defend their views on economic growth (see Chapters 4 and 5), then from heterodox economists, including his mentor Georgescu-Roegen, and later from the left end of the ideological spectrum which includes advocates of degrowth and Marxists.

In this chapter and the next we will meet some of these critics, discovering what they had to say about Daly's steady-state economics, and his responses to them. As we shall see, both the criticisms and Daly's responses are of far more than passing interest. The financial crisis of 2007/2008, the Covid-19 pandemic, widening inequality, and the increasingly undeniable evidence of human caused climate change have led many to question the viability of the current economic system and the adequacy of mainstream economics to explain it. New ways of thinking about economics and the economy are being called for and many have been suggested. Some, such as doughnut economics, the wellbeing economy, postgrowth economics, degrowth, and regenerative economics are clearly inspired by Daly's proposals for a steady-state economy. If Daly's steady-state economy is fundamentally flawed, then all these attempts to find another way forward are also vulnerable. On the other hand, if the criticisms are invalid, or can be overcome by a different but related approach to Daly's, then so much the better for the future.

BOX 9.2

There is no better exemplar of how scientific discourse should be engaged than Herman Daly.

Robert Costanza

Daly's first defense of a steady-state economy[1]

Daly published a paper in 1972 entitled 'In Defense of a Steady-State Economy' (Daly 1972). It was written in the spirit of attack being the best form of defense. Sensing opposition to a steady-state economy, Daly launched into a critique of economic growth using many of the arguments that he later published as 'a catechism of growth fallacies' (Daly 1977a, 1991) (see Chapter 7). His first real defense of steady-state economics in which he responded to direct criticisms, appeared in 1980 under the title, 'Postscript: Some Common Misunderstandings and Further Issues Concerning A Steady-State Economy' (Daly 1980a). The issues Daly chose to write about then returned in different guises in the decades that followed.

Daly began his defense of a steady-state economy in 1980 with a section on problems of terminology. He explained that he 'adopted the term *steady-state* from the physical and biological sciences' to refer to 'an economy in which population and capital stock had ceased growing' (ibid., p. 358) and he emphasized the difference between quantitative growth and qualitative development. Then he turned his attention to repudiating the idea that to argue for a steady-state economy is a 'counsel of despair'. William Nordhaus, who made this charge, began working on environmental problems early in his career but from a decidedly neoclassical standpoint. As a member of the President's Council of Economic Affairs he acknowledged noted that 'the political and social impediments to metering or internalizing the undesirable consequences of our activities are becoming unmanageably high' (quoted by Daly ibid., p. 359). Nordhaus's view of those opting for a steady-state society as an alternative was 'a counsel of despair' (ibid.), an opinion that Daly vehemently disputed. He reached back to Mill's very positive description of a steady-state economy and all the reasons Mill gave for considering 'that a stationary condition of capital and population implies no stationary state of human improvement' (quoted by Daly, ibid.). 'The steady state', says Daly, 'is not an end in itself, but a means, a constraint imposed by the ends of justice, sustainability, and participation, as well as by the approximate steady state nature of the total ecosystem of which the economy is a part' (ibid., p. 360).

In his 1977 book, *Steady-State Economics*, Daly makes the fascinating observation that 'the most basic laws of science are statements of impossibility... [and that]... perhaps the success of science is due to its refusal to attempt the impossible' (op cit., p. 6). He mentions a couple of impossibility theorems in economics: Arrow's theorem of the impossibility of deriving social preferences from individual preferences through a voting system that satisfies a number of reasonable conditions, and the impossibility of having more than a single equilibrium price and quantity in a competitive market. In a speech on a panel at the Argonne National Laboratory on *The Limits to Growth* (Meadows et al. 1972), shortly after its publication, Daly mentioned the impossibility of maximizing more than one variable in a function – as in Bentham's 'the greatest good for the greatest number'. A Harvard economist on the panel said Daly was wrong and that in economics we frequently maximize more than one variable. During the coffee break a mathematician from Argonne remarked to Daly that 'someone should set that fellow from Harvard straight about maximizing more than one variable'. Daly agreed and suggested that he should do it since it would have more authority coming from a mathematician, and so he did which Daly found very satisfying to be right about a question of mathematics.

As part of his defence of a steady-state economy, Daly proposed another impossibility theorem: 'that a US-style high resource consumption standard for a world of 4 (sic) billion people is impossible... Even less is it possible to support an ever-growing standard of per-capita consumption for an ever-growing population' (op cit. p. 361). Daly's reason for this impossibility is that U.S. consumption levels are unattainable for all because

> the minerals in concentrated deposits in the earth's crust, and the capacity of ecosystems to absorb large quantities of exotic qualities of waste materials and heat set a limit on the number of person-years that can be lived in the "developed state", as that term is understood today in the United States.
>
> *Daly 1977a, p. 6*

Daly has little patience for those willing to rely on the demographic transition – the reduction in birth rates that can follow from a rise in incomes – to control population growth. He quotes the slogan often heard at the U.N. Population Conference in Bucharest in 1974 that 'development is the best contraceptive' (op cit., p. 363). Daly makes his point by extending the argument to literacy and public health programs. Why not rely analogously on a 'literacy transition' to solve illiteracy by automatic correlation with development asks Daly rhetorically. If those programmes make sense because they contribute to development then so does a programme of population control. Not to see this, he says, denies the impossibility of raising billions of people to rising US income levels by ignoring the trade-off between an 'increase in consumption per capita in exchange for a reduction in capitas' (ibid., p. 364). As we shall see in Chapter 12, Daly's position on population, and the related subject of immigration, has proved controversial even among his supporters.

One line of criticism levelled against a steady-state economy is often brought forward in times of recession – as the editors of *Fortune* did in the 1970s: 'The country has just gone through a real-life tryout of zero growth [the period 1973–75], which is remembered not as an episode of zero growth but as the worst recession since the 1930s' (quoted by Daly ibid., p. 365). Daly categorically rejects the charge that a recession or depression is equivalent to a steady-state economy. He considers it a serious mistake to confuse a 'failed growth economy' with a steady-state economy (SSE). 'The main reason for advocating a SSE is precisely to avoid the suffering of a failed growth economy' (ibid., p. 364). 'A growth economy and a SSE are as different as an airplane and a helicopter' (ibid., p. 365). An aircraft (like a growth economy) must move forward or crash but a helicopter (like an SSE) can remain stationary yet function well.

Daly rejects the argument that economic growth is essential for full employment. He reasons that limits to inequality – which is one feature of an SSE – would reduce aggregate savings and 'bolster aggregate demand and employment' (ibid., p. 366). Limiting matter-energy throughput 'would raise the price of energy and resources relative to the price of labor' (ibid., p. 366), encouraging a greater use of labour in production. A constant population and labour force, though taking time to achieve, would also help foster employment. In addition to these arguments made by Daly in 1980, we can add the possibility of shorter working hours to secure full employment in the absence of economic growth (Victor 2019, pp. 248–249, 327–330).

Much of the argument against Daly's steady-state economy relates to the functioning of the market economy. 'Only if we eliminate the market incentives for innovation and investment, or reduce the scope of market forces through further attenuation of private property in resources, must we face a real long-term resource crisis' (Economists G. Anders, W. Gramm and S. Maurice quoted in Daly op cit., p. 365). More recently, it was suggested that proponents of a steady-state economy are mistaken since people are 'problem-solvers, who can overcome resource constraints given the appropriate institutional setup: a system of secure property rights and the free formation of market prices' (Niemietz 2012). In other words, Daly says, synthesizing these critiques 'as long as we base our decisions on free market prices, resource constraints disappear as a long-run concern. We do not, in this view, need any ecologically or ethically determined limits on the total flow of resources' (ibid.).

Daly is quite well disposed to the market providing it is kept in its proper place within a larger scheme. 'Market prices are excellent means for efficiently allocating a given resource flow from nature among alternative uses in the service of a given population of already existing people with a given distribution of wealth and income' (ibid. 365). This is why he sees an important role for markets in a steady-state economy; where allocation, as in any economy, is a critically important function. But Daly does not think that markets and market prices are appropriate for other functions, in particular for deciding

> the rates of flow of matter-energy across the economy-ecosystem boundary or to decide the distribution of resources among different people (or among different generations, which, of course, are different people). The first must be an ecological decision, the second an ethical decision. These decisions of course will and should influence market prices, but the whole point is that

these ecological and ethical decisions are *price determining*, not *price determined*. Many economists simply fail to grasp this point.'

ibid., pp. 365–366

When Daly published, with Ken Townsend, the third edition of his collection of essays 13 years later he saw no reason to revise his defence of a steady-state economy. The criticisms remained much the same as did Daly's responses. Indeed, the criticisms have remained that way to the present day as far as mainstream economists are concerned. What they see as largely a problem of microeconomics best dealt with through a combination of adjustments to property rights and/or prices, Daly and like-minded ecological economists and natural scientists see as a system-wide macroeconomic problem of scale. This reflects a substantial difference in paradigms despite Daly's efforts to persuade other economists of the merits of a steady-state economy by applying the microeconomic concepts of marginal benefits and marginal costs to the size of an economy. In microeconomics the combination of diminishing marginal utility and increasing marginal cost is used to determine the optimal level of any activity, such as the level of output of a firm or the size of a dam. Daly thought the same logic should apply to the scale of the economy as a whole: 'the activity in question, growth, should be carried only to the point at which marginal costs equal marginal benefits' (Daly 1977a, 1991a, p. 28). If an economy grows beyond that level then Daly said the growth becomes 'uneconomic' (see Chapter 6). Later on Daly referred to the absence of a 'when to stop' rule in macroeconomics: 'The basic rule of microeconomics, that optimal scale is reached when marginal cost equals marginal benefit… has actually been called "the when to stop rule", — that is, when to stop growing. In macroeconomics, curiously, there is no "when to stop rule", nor any concept of the optimal scale of the macroeconomy. The default rule is "grow forever" (Daly and Farley 2011, p. 17).

Daly considered the absence of the idea of optimal scale in macroeconomics as a failure in logic when it is so central to microeconomics. Microeconomics studies components of the economic system that macroeconomics describes in aggregate. The logic of microeconomics is the logic of the part; macroeconomics is the logic of the whole. Since the macroeconomy is a part of the ecosphere it does not escape the microeconomic logic of the part. Its expansion into the finite ecosphere entails opportunity costs, just as when in microeconomics a firm or industry expands into the rest of the macroeconomy, raising the question of optimal scale. So Daly asks, at what point in the aggregation from micro to macro does the 'when to stop rule' of microeconomics no longer apply? Mainstream economics has no answer to this question.

Consequences

Daly's use of microeconomic thinking to make his macroeconomic case for a steady-state economy failed on two counts. It did not persuade orthodox economists and it became a point of criticism from heterodox economists that Daly relies too much on neoclassical thinking (see Chapter 10). It also failed to impress members of the Department of Economics at Louisiana State University (LSU) where Daly was working and making a name for himself as the champion of the steady-state economy. He had joined LSU in 1965 as an assistant professor in economics and progressed rapidly through the ranks, becoming associate professor with tenure in 1970 and a full professor in 1973. His promotions were based on an impressive publication and citation record, his popularity as a teacher and as a well-liked and affable contributor to departmental administration. Daly got along well with most of his colleagues at a personal level and with the university as a whole with friends in several departments.

The University recognized Daly's work and flourishing reputation by awarding him the Distinguished Research Master Award and later an Alumni professorship, which did not make the economics department happy. Over the years, both Daly and the department had changed, Daly in one direction and the department in the opposite direction, becoming more and more committed to neoclassical economics and less and less tolerant of dissenters. Daly increasingly became *persona non grata* in the department, which affected him and his family. His wife Marcia was at a social occasion when she was approached by one of the wives of a faculty member who said, 'Gee, it must be difficult for you to live with Herman with all his crazy ideas'. Shocked, Marcia replied, 'What do you mean?' 'Well, his book, what he says in it.' 'Have you read his book?' asked Marcia. 'Well. no'. 'Then maybe you should', said Marcia and turned away.

BOX 9.3

His contribution has been massive – even though he has been ostracized by his own academic community.

Mathis Wackernagel

It wasn't just at social gatherings that Marcia felt the displeasure of others. When Herman was travelling, she received intimidating phone calls from people who would not identify themselves, though Daly had his suspicions. This kind of social tension was increasingly frequent, as was Daly's isolation in the department which made it virtually impossible for him to supervise graduate students. It became increasingly obvious to him that any students he supervised would have a very hard time completing their programme to the point where Daly was obliged to resign from chairing the PhD supervisory committees of some very bright students just so the department would pass them. As the situation deteriorated, Daly became frustrated and angry. He had already decided to take early retirement from LSU when in 1988 an opportunity to join the World Bank as a Senior Economist in the Environment Department presented itself, so he resigned his position as a tenured professor at LSU and left.

BOX 9.4

His persistence in the face of entrenched orthodoxy is inspiring and courageous – I know how hard the system bites back when it is challenged and
how vicious people can be to those they disagree with.

Katherine Trebeck

It was sometime later that Daly came to recognize that he had been naïve. He had
thought that his critique of economic growth and his proposals for a steady-state
economy were ideas grounded in classical economics that could be discussed and
debated much like any other ideas that interested economists. He had not realized
that by challenging the paradigm of standard economics, especially its growth
emphasis, he was threatening the identity and self-esteem of economists who had
devoted their lives to it, so they took it personally. Unfortunately, what happened
at LSU is a microcosm of a larger story which continues to play out today. There
are too few opportunities for students to study ecological economics and where
economists critical of economic growth can ply their trade.

BOX 9.5

I wish to grow up to be an honest and committed intellectual of rigour like
him. I want to thank him for picking a battle with economists that must not
have been easy, and he has kept his ground.

Giorgos Kallis

Georgescu-Roegen, Daly and the steady-state economy

You might have thought that someone who argued in print that economics should
be a life science and that the laws of thermodynamics were highly relevant to economics would be assumed to know that nothing lasts forever, not an individual
animal or ecosystem, not an individual company or economy, not a planet, not a
universe. You might also have thought that of all the people who would make this
assumption, it would be the person from whom these lessons were well and truly
learned. And yet, when it came to the steady-state economy favoured by Daly over
the pursuit of uneconomic growth, his teacher and mentor Georgescu-Roegen
did not see it that way. The exchange of views between them might seem of little
relevance today were it not for the attention it has received from some proponents
of degrowth who see sharp differences between Daly and Georgescu-Roegen (e.g.
Bonauiti 2011), favouring the latter 'the posthumous patron saint of degrowth'
(Kunkel p. 25) and under-appreciating Daly's contributions. A careful examination

of their disagreement also helps clarify aspects of Daly's position on a steady-state economy. In the following chapter we will consider critiques of a steady-state economy from several heterodox perspectives and Daly's responses to them.

In 1972 Georgescu-Roegen gave a lecture at Yale University which, a year later, he wrote as a working document for the Commission on Natural Resources and the Committee on Mineral Resources of which Daly was also a member. Later, in 1975, Georgescu-Roegen published a third revised version under the title 'Energy and Economic Myths' (Georgescu-Roegen 1975).[2] One of the myths, or 'topical mirage' as Georgescu-Roegen called it, concerns the longevity of the steady-state economy: 'The crucial error consists in not seeing that, not only growth, but also a zero-growth state, nay, even a declining state which does not converge toward annihilation, cannot exist forever in a finite environment' (ibid., p. 367). Georgescu-Roegen goes on to say, that 'contrary to what some advocates of the stationary state claim, the state does not occupy a privileged position vis-à-vis physical laws' and cites Daly's lecture on The Stationary-State Economy in 1971 at the University of Alabama: 'The proponents of salvation through the stationary state must admit that such a state can have only a finite duration' (ibid.). Georgescu-Roegen objected to Daly's contention that 'the stationary-state economy is, therefore, a necessity' (ibid.) and he considered Daly's definition of a steady-state economy based on 'Mills vision' deficient because 'as Daly explicitly admits… [it] offers no basis for determining even in principle the optimum levels of population and capital' (ibid., p. 368). Georgescu-Roegen restated similar objections to the steady-state economy in other papers written shortly after his 1975 paper (e.g. Georgescu-Roegen 1977a). He acknowledged: 'Daly's insistence on the distinction between 'stationary' and 'static'… 'the former allowing for change as in Daly's distinction between growth and development, and describes it as 'the pivot of the rationalization of ecological salvation through the steady state'. Georgescu-Roegen continues,

> The stationary state as conceived by the Classical economists, especially by Mill, [and by extension to Daly] is so elastic that it may be adjusted with almost no conspicuous ado to almost any necessity of an argument… The duty of academia is to help attenuate… mankind's struggle with the environment and with himself… and not to delude others with ideas beyond the power of human science.
>
> *ibid., 1977a, p. 267*

These are withering criticisms coming from as celebrated an economist and former teacher as Georgescu-Roegen. But they were ill-founded. Daly made it abundantly clear that a steady-state economy is not eternal when he wrote: 'The SSE cannot last forever because of the entropy law.' In just nine words Daly shows that there is nothing that separates him and Georgescu-Roegen on this question. Then he adds, 'In the very strict and inclusive sense, a steady-state economy is impossible… [it] cannot last forever, but neither can a growing economy, nor declining economy' (Daly 1980a, p. 369). Daly makes an analogy between an economy and a candle and says 'after lighting and before going out, we can describe the greater part of the flame's life as a steady state, without implying that it will last forever' (ibid.). If we wish

> we can turn our resource candle into a Roman candle and burn it rapidly, and extravagantly, or we can seek to maintain a steady flame for a long time, or we can put out the flame before the candle has burned down. The steady-state view advocates the middle choice.
>
> *ibid.*

These unambiguous statements should have put an end to any claim that there was a fundamental difference between Daly's and Georgescu-Roegen's view on the longevity of a steady-state economy. In fact, if Georgescu-Roegen had read more closely Daly's chapter 'The Steady-State Economy: Toward a Political Economy of Biophysical Equilibrium and Moral Growth' (Daly 1973a) or in its original form as the lecture Daly gave at the University of Alabama in 1971 which Georgescu-Roegen cites in his 1975 paper, the dispute might have been avoided altogether. In his 1973 chapter Daly asks, 'What is the optimal time horizon or accounting period over which population and wealth are required to be constant?' (ibid., p. 155). Simply by asking the question we can see that Daly is not assuming that population and wealth remain constant forever in a steady-state economy. Then Daly says, 'Once we have fixed an accounting period, one may then ask how many accounting periods the total system should last' (ibid.), again implying that its lifetime is finite. His answer to the last question is that

> the current capacity of the ecosystem depends not only on the size of the stocks [to be maintained] and the rate of maintenance throughput [to be minimized], but also on the length of time over which the stocks are to be carried… [and] … the given endowment of non-renewable resources.
>
> *ibid., pp. 155–156*

If further confirmation of Daly's acceptance of the finite nature of a steady-state economy is required it is provided in a letter by Daly published in 1977 where he states, 'that a true steady-state economy in the strict thermodynamic sense is impossible.' Daly adds, 'I do not believe that the goal of a steady state economy offers ecological salvation – only that it offers a better target and a better paradigm for altering our policies than the alternatives' (Daly 1977b, p. 770). He goes on to defend his use of the concept of a steady state against Georgescu's criticisms by pointing out that the concept of a steady state population has been long used by demographers without implying that the human species is immortal. And Daly is clear on how far thermodynamics and economics can take us in determining what this steady-state population should be and what its standard of living should be: 'these are ethical questions, not to be derived from positive analysis, either economic or thermodynamic' (ibid.).

It is striking that Georgescu-Roegen opened his reply to Daly's letter by saying, 'My admiration for Herman Daly's cogent arguments and his enthusiasm for the steady state is second to none' (Georgescu-Roegen 1977b, p. 771). He continues not by disputing anything said by Daly, but takes issue with 'Mill's idea that it suffices to have a stationary society for men to cease being wolves to one another' (ibid.). Georgescu-Roegen then makes a point about demography and the importance of asking, 'How long can such a [stationary] population survive?' Again, this does not set him apart from anything Daly had written. Finally, Georgescu-Roegen argues against technological optimism which, Micawber-like, assumes that something will turn up to counteract the declining quality of natural resources that threatens the longevity of a steady-state economy. This too, is quite consistent with Daly's views and comes back full circle to his agreement with Georgescu-Roegen that 'a true steady-state economy in the strict thermodynamic sense is impossible' (op cit., p. 770).

In the years when the back and forth between Georgescu-Roegen and Daly about a steady-state economy was at its peak, *Scarcity and Growth Reconsidered* (Smith 1979) was published. It provides further evidence, should any be needed, that Georgescu-Roegen's differences with Daly on the implications of the entropy law for a steady-state economy were small. *Scarcity and Growth Reconsidered* was based on papers presented at a conference on 18 October 1976 organized by Resources for the Future (RFF) to reconsider the classic study *Scarcity and Growth* (Barnett and Morse 1963). RFF is a nonprofit organization 'for research and education in the development, conservation, and use of natural resources and the improvement of the quality of the environment' (ibid.). Most of the research supported by RFF in environmental and resource economics is an application of neoclassical economics but this conference drew from a wider field and included contributions by Daly. At the conference Barnett defended the original study emphatically, but Chandler Morse, who was present but not a presenter, backed away from its more extreme neoclassical position. Morse's former student from Cornell, Talbott Page, who then worked at RFF and was at the conference, was sympathetic to Daly and Georgescu-Roegen's position. Daly and Page became friends and kept in touch for years. Also present was Bruce Hannon, energy theorist and a kindred spirit to Daly who was disappointed not to get support from Alan Kneese, another attendee and a leading environmental economist in the 1960s and 70s.

Scarcity and Growth Reconsidered included comments by Georgescu-Roegen on Daly's paper and also on one by Stiglitz, whose 1974 paper we discussed in Chapter 4. Georgescu-Roegen's comments make it crystal clear that if he had differences with Daly they pale in comparison with his sharp critique of Stiglitz who gave a neoclassical analysis of the economics of natural resources. Georgescu-Roegen wrote:

> Given my own stand on the crucial role played by natural resources in the economic process, it may be superfluous to say with which of the two papers

I am in substantial agreement. Yet I deem it necessary to state from the outset that I am entirely out of sympathy with the manner in which J. E. Stiglitz dealt with his topic.'

Georgescu-Roegen in Smith (ed.) 1979, p. 95

After taking Stiglitz to task for 'a line of multifarious but ineffective fires in defense of a position to which many standard economists still cling with the tenacity of original sin' (ibid., p. 95), Georgescu-Rogen turns his attention to Daly. His first observation is that: 'This paper represents an improvement on Daly's earlier pleas' (ibid., p. 102), creating an expectation that Daly had recanted his assumption [which he never made] that a steady-state economy promises eternal life, or admitting some other error, or offering a new argument in favour of a steady-state economy. Instead, what we find is Georgescu-Roegen's laudatory description of Daly's Ends-Means Spectrum that Daly first published in 1973, which was not new, just previously overlooked by Georgescu-Roegen. Still, his positive assessment of the Ends-Means Spectrum shows the closeness of their views on the correlation of the hierarchy of ends and means with academic disciplines. However, Georgescu-Roegen parts company with Daly on three reasons that he claims Daly offers 'for his blueprint [for a steady-state economy], and it is with respect to these reasons that I entertain some still unshakable doubts [about a steady-state economy]' (ibid., p. 103).

The first of these reasons is the assumption he attributes to Daly that 'stationariness alone suffices to clear away all substantial conflicts between individuals or, especially, social classes. Quasi-stationary societies of the past proved that they were as vulnerable from this standpoint as the growing ones' (ibid.). Daly does not make this assumption in the paper on which this comment is being made and Georgescu-Roegen does not cite any other publication by Daly where he is supposed to have said this. What Daly does say in the paper is that 'Since aggregate growth can no longer be appealed to as the "solution" to poverty, we must face the distribution issue directly by setting up a distributist institution which would limit the range of inequality to some justifiable and functional degree' (Daly 1979, p. 85). Daly makes no assumption that an equitable distribution of income and wealth, which he regards as an essential aspect of a steady-state economy, will be any easier to achieve than the objectives for scale and allocation.

Georgescu-Roegen's second 'unshakeable doubt' about a steady-state economy is that 'there is nothing in Daly's setup to help us determine even in broad strokes the proper standard of living – "the good life" – and, worse still, the proper size of the population' (op cit., p. 103). Again, this is an overstatement or at least an inconsistency. Georgescu-Roegen complemented Daly for including his Ends-Means Spectrum in the paper yet he gives him no credit for locating the intermediate ends such as 'health, education, comfort etc.' served by intermediate means (stocks of artifacts, labour power) in relation to ethics, religion and the Ultimate end, which taken together give an idea of Daly's conception of the good life.

As for the proper size of the population, we can refer to Daly's statement of 'the old [question] of what is the optimum population? So far no one has given a definite answer, and certainly I cannot' (Daly 1973a, p. 154). Then anticipating Georgescu-Roegen's later criticism he says,

> it is sometimes argued that it is vain to advocate a stationary population unless one can specify the optimum level at which the population should become stationary. But I think that puts it backwards. Rather it is vain to speak of an optimum population unless you are first prepared to accept a stationary population – unless you are able and willing to stay at the optimum once you find it. Otherwise knowing the optimum merely enables us to wave goodbye as we pass through it.
>
> *ibid.*

Daly goes further than this in his 1976 paper when he calls for a population 'sufficient for a good life and sustainable in the long term… This means that birth rates are equal to death rates at low levels so that life expectancy is high' (op cit. p. 79).

The third point 'in great need of clarification is [Daly's] basic concept of a capital stock that remains constant in amount but may nevertheless undergo qualitative changes' (Georgescu-Roegen op cit., p. 103). This is the most valid of Georgescu's misgivings about the steady-state economy and we discussed it at some length in Chapter 8. There we drew attention to Daly's alternative definition of a steady-state economy based on constant throughput with stocks permitted to vary which avoids the problem of a constant capital stock. To be fair to Georgescu-Roegen, Daly did not include this alternative definition of a steady-state economy in his 1976 paper and he had not published it anywhere else at that point. We can only wonder whether, if Georgescu-Roegen had known of this alternative definition of a steady-state economy, he would have retained his unshakeable doubt. Perhaps most significant of all is that the reasons Georgescu-Roegen gave for his 'unshakable doubts' about a steady-state economy do not include any assumption by Daly that a steady-state economy is somehow immune to the entropy law and can endure forever, which was the original basis of their dispute.

This dispute, if it ever really existed, should have been laid to rest when Daly wrote: 'in the very long run of course nothing can remain constant, so our concept of an SSE must be a medium run concept in which stocks are constant over decades or generations, not millennia or eons' (Daly 1976, p. 80). A year after he had written his critique of Daly's steady-state economy, Georgescu-Roegen gave Daly a copy of *Evolution, Welfare and Time in Economics. Essays in Honour of Nicolas Georgescu-Rogen* (Tang, Westfield, Worley eds. 1976). It contained the handwritten inscription: 'To Herman Daly, my only follower, whom I recognize with pride, with my warmest thoughts.' It was signed Nicolas Georgescu-Roegen, October 22, 1976. Anyone who doubts the closeness of their positions would do well to think again.

Sustainable development and the steady-state economy

The only reasonable interpretation of Georgescu-Roegen's inscription is that, despite his disagreement with Daly about the steady-state economy, well-founded or not, he considered Daly to be a follower – his only follower, no less – of whom he was proud. Nonetheless, towards the end of his life Georgescu-Roegen again criticized Daly's steady-state economy but this time the context was the then newly popular concept of sustainable development.

The UN World Commission on Environment and Development established in 1983 published its report in 1987 which came to be known as the Brundtland Report after the Commission's chair. The term sustainable development pre-dated the Commission, but it was the Brundtland Report that was responsible for putting it into widespread use. Although the Commission defined sustainable development in several ways, only one definition emphasizing intergenerational equity became widely known: 'development which meets the needs of and aspirations for the present generation without compromising the ability of future generations to meet their own needs' (World Commission on Environment and Development 1987, p. 43).

In the absence of a single, precise definition, it became possible for people and organizations to define sustainable development in ways that best suited their interests. Twenty years after the Commission's report was published, Jim MacNeill, Secretary-General of the Commission and chief architect and lead author of the Commission's report, commented:

> I remain stunned at what some governments in their legislation and some industries in their policies claim to be "sustainable development." Only in a Humpty Dumpty world of Orwellian doublespeak could the concept be read in the way some would suggest.
>
> *MacNeill 2006, p. 167*

However, none of the Commission's definitions of sustainable development are a repudiation of growth. 'We see… the possibility for a new era of economic growth, one that must be based on policies that sustain and expand the environmental resource base' (World Commission on Environment and Development 1987, p. 1).

In Chapter 2 we gave an account of Georgescu-Roegen's misguided accusation that Daly was an architect of sustainable development, that it would be bad for Romania and that Daly should be fired from the World Bank. In fact, Daly's contribution to discussions and debate about sustainable development were quite different from what Georgescu-Roegen believed. Robert Goodland, who was Daly's mentor and boss at the World Bank, realized that a great deal of attention would be given to the Brundtland report at the 1992 United Nations Conference on Environment and Development to be held in Brazil. Goodland led a group consisting of Daly,

Salah El Serafy also at the World Bank and Bernard von Droste at UNESCO in an attempt to influence discussions of sustainable development at the conference. They co-edited a collection of papers including two by winners of the Nobel Memorial Prize in Economic Sciences: Trygve Haavelmo and Jan Tinbergen, and a Preface by two Ministers of the Environment, the Honourable Emil Salim, Minister of State for Environment and Population for Indonesia and Jose Lutzenberger, Secretary of State for Environment in Brazil (Goodland, Daly, El Serafy and von Droste 1991). The authors read and discussed each other's chapters and reached a consensus that the contributions were 'not only compatible with each other, but also mutually reinforcing' (ibid., p. 9). They agreed that:

> it is no longer tenable to make economic growth, as conventionally perceived and measured, the unquestioned objective of economic development policy… The planet will transit to Sustainability: the choice is between society planning for an orderly transition, or letting physical limits and environmental. damage dictate the timing and course of the transition… It is neither ethical nor helpful to the environment to expect poor countries to cut or arrest their development… therefore the rich countries… must take the lead in this respect. Poverty reduction will require considerable growth, as well as development, in developing countries. But ecological constraints are real and more growth for the poor must be balanced by negative throughput growth for the rich.'
>
> *ibid, pp. 10–14*

Daly's influence is clear throughout the book. Six out of the eight papers cite him directly and his name occurs 40 times in a book of only 100 pages. In his own paper, 'From empty-world to full-world economics: Recognizing an historical turning point in economic development' he emphasized the complementarity (not substitutability) between human-made and natural capital, and that natural capital has become the limiting factor in economic development so it should be favoured for investment. Natural capital liquidation should no longer be permitted to count as income (Daly 1991a).

Adjustments to GDP to avoid the mistake of counting resource depletion as income had been the focus of research by El Sarafy for some time. He was the author of a simple formula for dividing the income from the sale of non-renewable resources into true income that could be maintained indefinitely and capital depletion that could not, based on the ratio of reserves to current extraction and the rate of interest (El Serafy 1989). Daly often referred to El Serafy's formula in his own writings. El Serafy described his proposal in his contribution to the 1991 book as would be expected, but he also wrote about the steady state ideas borrowed from Daly, no doubt with Daly's blessing.

> If we are serious about saving our planet, we must seek a steady state for the economies of the rich, while the poor grow and develop so that poverty is

eradicated and income disparity, which is the source of so much environmental damage, reduced.

ibid., p. 66

Continuing with this Daly-inspired theme, El Serafy said, 'Clearly something drastic has to take place in social and industrial organization and in the modalities of international relations if a steady state of economic activity, involving constant level of throughput, is to prevail in the developed countries' (ibid., p. 67).

It is very clear from El Serafy's statements endorsed by Daly that they differentiated between the implications of sustainable development for rich and poor countries. Hence, it is difficult to understand what motivated Georgescu's harsh criticism of Daly in relation to sustainable development expressed in some of his personal letters though not, apparently, in print. In a letter to economist Kozo Mayumi, Georgescu-Roegen wrote that the steady-state 'for Bangladesh would mean a lasting condemnation to its present misery' (Letter from Georgescu-Roegen July 18, 1992 in Bonauti 2011 p. 232). He says much the same thing in a letter to James Berry (Letter from Georgescu-Roegen 20 May 1991 in Bonauti 2011, p. 232). In several of the letters selected by Bonauiti, Georgescu-Roegen describes sustainable development as 'snake oil… more beguiling than the steady state', a clear reference to Daly according to Bonauiti (ibid., p. 42)

In so far as Georgescu-Roegen's concern about sustainable development was related to his view that a steady-state economy would leave so many in endless poverty, we have seen that this could not be further from Daly's intention or of the other contributors to the Goodland et al. 1991 book. Perhaps Georgescu-Roegen had not seen the book at the time he wrote his letters, but he should have known that Daly had been differentiating between rich and poor countries in his work on a steady-state economy as far back as 1973. In the Introduction to *Toward a Steady-State Economy*, Daly wrote:

> Any discussion of the relative merits of the steady, stationary, or no-growth economy, and its opposite, the economy in which wealth and population are growing, must recognize some important quantitative and qualitative differences between rich and poor countries, and between rich and poor classes within countries.
>
> *Daly 1973a, p. 10*

Daly describes the differences and concludes that,

> the upshot of these differences is that for the poor, growth in GNP [GDP] is still a good thing, but for the rich it is probably a bad thing… In what follows we shall be concerned exclusively with a rich, affluent-effluent economy such as that of the United States.
>
> *ibid., pp. 10–12*

Unlike many others, the Brundtland Commission included, Daly was not content to leave sustainable development inadequately defined as a way of achieving a somewhat spurious consensus. He sought more clarity which he thought could be achieved by distinguishing between quantitative growth and qualitative development. He had hoped that 'the glaring contradiction [in the Commission's report] of a world economy growing by a factor of 5 or 10 and at the same time respecting ecological limits... would be resolved in future discussion' (Daly 1990, p. 1). Daly credited the Commission's chair, Gro Harlem Brundtland for providing 'a political opening for the proper concept of sustainable development to evolve', but he was disappointed that subsequent remarks by her veered towards the self-contradictory notion of 'sustainable growth' urging 'economic growth by a factor of 5 or 10 as a necessary part of sustainable development' (ibid. pp. 1–2). In contrast to Brundtland who was 'encumbered by the political necessity of holding together contradictory factions', Daly saw himself free to 'take up the challenge of giving the basic idea of sustainable development a logically consistent and operational content' (ibid.). In a brief paper that has since been cited over two thousand times, influencing numerous individuals and organizations, Daly proposed a number of 'operational principles' of sustainable development (Daly 1990). In Daly's words (ibid., pp. 3–5):

For the management of renewable resources... regenerative and assimilative capacities should be treated as natural capital... and maintained... at the optimal level:

1. Harvest rates should equal regeneration rates (sustainable yield);
2. Waste emission rates should equal the natural assimilative capacities of the ecosystems into which the wastes are emitted;
3. Maintain [manmade and natural] capital intact... at the optimal level... [Manmade and natural capital]... are basically complementary and only very marginally substitutable;
4. The quasi-sustainable use of non-renewable requires that any investment in the exploitation of a non-renewable resource must be paired with a compensating investment in a renewable substitute [by dividing] the net receipts from the non-renewable into an income component that can be consumed currently each year, and a capital component that must be invested in the renewable substitute;
5. Regarding technology the rule of sustainable development would be to emphasize technologies that increase resource productivity (development), the amount of value extracted per unit of resource, rather than technologies for increasing the resource throughput itself (growth);
6. From a macroeconomic perspective the scale of the economy (population times per capita resource use) must be within the carrying capacity of the region in the sense that the human scale can be maintained without resorting to capital consumption... this implies limits on and a trade-off between population size and per capita resource use in the region;
7. Poor countries which cannot afford any reduction in per capita resource use will perforce have to concentrate their efforts on population control. Countries

that have higher rates of per capita resource usage frequently have low rates of demographic growth and consequently must aim their efforts more at consumption control than population control, although the latter cannot be neglected in any country.

Daly concludes his proposals for operational principles of sustainable development by saying: 'Fighting poverty will be much more difficult without growth. Development can help, but serious poverty reduction will require population control and redistribution aimed at limiting wealth inequality' (ibid., p. 5).

Although Daly does not say so in the paper, his principles for sustainable development are derived from his larger set of principles of steady-state economics (see Chapter 8). Perhaps, as an employee of the Word Bank when he wrote the paper, this is a connection that he decided not to make explicit being less unencumbered by political considerations than he may have supposed.

BOX 9.6

This is why Herman Daly's work matters so much: despite the flaws of the particular economic paradigm in which we are currently embedded, he refuses to give up on economics, on an economic way of thinking about society. He applies a ruthless rigor to it, which is often sadly missing in his more orthodox counterparts, and in doing so, he points us towards solutions… When it comes to a comparison between Daly's steady-state economics and conventional economics, there is no contest.

Tim Jackson

Conclusion

Daly did not set out to challenge the paradigm of economics, but he was led to do so by his observation that the world had changed from empty to full and his belief in the necessity of a steady-state economy. His work on a steady-state economy soon attracted criticism from mainstream economists which helped Daly sharpen his ideas and also from his mentor Georgescu-Roegen. In the 1980s and 90s interest in sustainable development provided Daly with an opportunity to promote a steady-state economy under a different guise. Daly's widely cited principles of sustainable development helped clarify conditions for truly sustainable development though they were incomplete in a number of respects: for example, they did not include anything about limiting land use changes to protect habitat, and they were silent on the metrics to be used for implementing the principles and monitoring outcomes. But from the title of the paper where Daly first expressed his principles, 'Toward Some Operational Principles of Sustainable Development' (Daly 1990) it is obvious that he knew he was leaving plenty for others to do. As we have already seen, for example with respect to metrics, much has since been done. (See Chapter 6)

A further aspect of Daly's work on a steady-state economy that has attracted criticism especially from heterodox economists relates to a lack of attention on his part to the political economy of a steady state economy. This criticism and Daly's response to it is the focus of the next chapter.

Notes

1 Unless otherwise indicated all page references are to Daly 1980.
2 Since this is the most widely read of Georgescu's three versions and presumably the one most informed by his knowledge of Daly's position on a steady-state economy, it is the one that the comments in this chapter are based on.

10

HETERODOX CRITIQUES OF THE STEADY-STATE ECONOMY

> Steady-staters are used to being attacked by right-wing neoliberals. Attacks from left-wing neo-Marxists are new and require a reply. To put the matter simply, Marxists hate capitalism, and they mistakenly assume that steady-state economics is inherently capitalist.
>
> *Daly*

Daly's proposal for a steady-state economy has not been well-received by mainstream economists still committed to the growth paradigm. Heterodox economists, even some who share Daly's environmental perspective, have been critical too though for different reasons. These critics include proponents of degrowth, Marxists, and various individuals who object to specific aspects of Daly's prescription of a steady-state economy though are in general sympathy with him on the need for radical change. We consider this 'friendly fire' in this chapter and Daly's responses to it.

Degrowth and the steady-state economy

Giorgos Kallis, one of the most prolific writers on degrowth, opens a section on Defining Degrowth with the statement that, 'sustainable degrowth can be defined from an ecological-economic perspective as a socially sustainable and equitable reduction (and eventual stabilization) of society's throughput' (Kallis 2011, p. 874). He then cites Daly to explain the meaning of throughput and Georgescu-Roegen for the relevance of the laws of physics and eventual decline. Kallis, sounding very much like Daly, says, 'throughput reduction is incompatible with further economic growth and will entail in all likelihood economic (GDP) degrowth' (ibid.) Kallis agrees with Daly that a decline in GDP is not the objective but is a likely outcome of degrowth.

DOI: 10.4324/9781003094746-10

What then is the difference between Daly's steady-state economy and degrowth? Kallis gives three ways in which sustainable degrowth goes further than the '"old wine" for ecological economists familiar with steady-state economics' (ibid., p. 875). First is the 'concern whether the descent to a steady-state can be achieved primarily through economic reforms.' Second, 'degrowth opens up the discussion of selective downscaling of manmade capital', and third:

> selective degrowth opens up a political debate about which extraction-pro-duction-consumption activities need to degrow and which ones need to grow. This choice cannot be left to market forces alone… it is not only an ethical consideration, but also a politically pragmatic one, as popular support is required.…
>
> *ibid.*

The one element that stands out in this amplification of the meaning of degrowth that distinguishes it from Daly's steady-state economy is the political dimension of change that is fundamental to degrowth. Unlike Daly, whose main focus is on reform through changes in policy to achieve a steady-state economy, advocates of degrowth stress political aspects relating to power dynamics coupled with changes in culture and ideology. This is brought out clearly by Serge Latouche, one of the founders and intellectual leaders of the modern degrowth movement who says that 'degrowth is not a concept… it is a political slogan with theoretical implications' (Latouche 2010, p. 519). One might say, by way of comparison, that Daly's steady-state economy is not a slogan, but an economic concept with practical implications.

It does not follow from this difference in emphasis that degrowth and a steady-state economy are in conflict. On the contrary, it is easier to see them as natural partners complementing each other to the benefit of both. This is the position taken by Giorgos Kallis, Christian Kerschner and Joan Martinez-Alier in their discussion of the economics of degrowth. They say, 'nobody in the DG [degrowth] literature is preaching degrowth forever. As Kerschner (2010) has shown, the debate between DG and proponents of a SSE (going back to Georgescu-Roegen's excessive strictures against Herman Daly) is false: degrowth is the path of transition to a lower steady state' (Kallis, Kerschner, Martinez-Alier 2012, p. 173). Daly sees the relationship between degrowth and a steady-state economy in much the same way.

> There is really no conflict between the steady state economy and degrowth since no one advocates negative growth as a permanent process; and no one advocates trying to maintain a steady state at the unsustainable present scale of population and consumption. But many people do advocate continuing positive growth beyond the present excessive scale, and they are the ones in control, and who need to be confronted by a united opposition!
>
> *Daly 2014a, pp. 234–235*

It is with respect to the opposition of those in control that the difference between Daly and proponents of degrowth is most pronounced. Daly's approach is to develop and promote ideas for policies and institutions that if adopted would move an economy towards a steady-state and maintain it there in what he came to call a state of 'dynamic equilibrium' (ibid., p. 234). This is not enough for proponents of degrowth. Kallis for example, writes:

> Policies like these [i.e. the kind of policies proposed by Daly] that threaten to 'harm' the economy, are less and less likely to be implemented within existing market economies, whose basic institutions (financial, property, political, and redistributive) depend on and mandate continuous economic growth. An intertwined cultural and political change is needed that will embrace degrowth as a positive social development and reform those institutions that make growth an imperative. Sustainable degrowth is therefore not just a structuring concept; it is a radical political project that offers a new story and a rallying slogan for a social coalition built around the aspiration to construct as a society that lives better with less.
>
> *Kallis 2011, p. 873*[1]

Seen this way, degrowth can be regarded as an extension of Daly's steady-state economy rather than a critique in the usual sense of the term. Both seek economic and societal transformation. Daly and those like him see it being led by changes in policy and institutions. Others give priority to societal change through political and social activism and insist on an appreciation of the dynamics of societal change and a deliberate intervention in the political process. Without it, they say, policy alone will be insufficient.

Bonauiti thinks that the differences between degrowth and Daly's steady-state economics goes deeper than this, asserting, unconvincingly, that 'the different positions held by Daly on the one side, and by G-R [Georgescu-Roegen] and Latouche on the other are rooted in divergent pre-analytical views' (Bonauti op cit., p. 43). We have already seen that the differences between Daly and Georgescu-Roegen on the steady-state economy come down to a difference in time horizon, with neither thinking that a steady-state economy is ever lasting and both sharing the fundamental pre-analytical view of the economy embedded in the biosphere and subject to the laws of thermodynamics.

Yet even if Bonauiti is correct, the specific proposals in the degrowth literature and Daly's proposals for a steady-state economy are remarkably similar. Kallis compared policies proposed by himself, Daly, Latouche and the Spanish member of the Green Party in the European Parliament. He concluded that, 'there is broad agreement about the public policies that could have an effect in a degrowth direction' (Kallis 2018, Table 5.1). This conclusion was reinforced by Parrique (2019) who, in an 860-page treatise on the political economy of degrowth, undertook a more comprehensive comparison of degrowth policy proposals but excluded lists such as Daly's that do not include 'degrowth' in their name even though they are 'more or less in line with the idea of degrowth'. As one example, Parrique mentions Daly's 10 policies for a steady-state economy (Daly 2013b).[2]

On the question of the feasibility of fundamental societal change coming from changes in policy rather than being a necessary pre-requisite, Parrique's view is that 'a small reform is better than none, and one should think in terms of *Trojan horse policies* and *stepping-stone policies*, and not as an all-or-nothing policy agenda' (op cit., p. 511). No doubt, Daly would agree.

BOX 10.1

I was and I am enthusiastic about his capacity to explain the embeddedness of the economy in the social metabolism. He was able to explain Georgescu-Roegen in easier terms. His book of 1973, Steady State Economics, is the foundation for the Degrowth debate today, not always acknowledged.

Joan Martinez-Alier

Marxist critiques of the steady-state economy

It used to be that courses on the history of economic thought, economic history and comparative economic systems were included as standard fare in graduate programmes in economics. Now these subjects are considered specializations, if they are taught at all. They have been replaced by courses where the emphasis is on mathematics and statistics. As a result of his more traditional training in economics Daly is quite well versed in Marx though, as is he the first to admit, less well-trained in mathematics than he would have wished. He even a taught a course on Marxist economics at Louisiana State University for several years in the early 1970s. Daly draws on Marx's analysis of capitalism quite frequently in his own work though not always in a totally positive way. For example, he quotes a famous passage from Marx describing in surprisingly modern terms how production under capitalism exploits the soil as well as the worker:

> Capitalist production... disturbs the circulation of matter between man and the soil, in other words, prevents the return to the soil of its elements consumed by man in the form of food and clothing; it therefore violates conditions necessary to the lasting fertility of the soil... Moreover, all progress in capitalistic agriculture is a progress in the art, not only of robbing the laborer, but of robbing the soil; all progress in increasing the fertility of the soil for a given time is a progress toward ruining the lasting sources of that fertility. The more a country starts its development on the foundation of modern industry, like the United States, for example, the more rapid is the process of destruction! Capitalist production, therefore, develops technology,

and the combining together of various processes into a social whole, only by sapping the original sources of all wealth – soil and laborer.

Marx, Capital, Volume I, quoted in Daly 2014a, p. 31

Daly does not entirely agree with Marx's characterization of capitalist agriculture. 'Marx sees capitalists exploiting the soil as well as the laborer. Our analysis sees capital and labor maintaining an uneasy alliance by shifting the exploitation to the soil and other natural resources' (ibid.). However, Daly goes on to say in terms that Marx would have appreciated that if resource prices were to be raised as a matter of public policy effected through some new institution 'the labor-capital conflict would again become severe; hence the radical implications of the ecological crisis and hence the need for some distributist institution' (ibid.).

In his extensive published work Daly's references to Marx can be found alongside references to many other profound thinkers such as Aristotle, Adam Smith, John Ruskin, Wendell Berry, J.S. Mill, Charles Darwin, Karl Popper, Erwin Schrödinger to name a few. One of the strengths of his work, particularly on the steady-state economy, is the breadth of intellectual foundations on which it is built. This does not mean that he is perfectly aligned with any of them. Daly credits Marx's analysis of class struggle and inequality and agrees that 'exchange between the powerful and the powerless is often only nominally voluntary and can easily be shown to be a mask for exploitation, especially in the labor market' (Daly 1977a, p. 54). But Daly rejects out of hand Marx's economic determinism which he thought 'has now collapsed both intellectually and politically', and the implication of material determinism that 'economic growth is crucial in order to provide the overwhelming material abundance that is the objective condition for the emergence of the new socialist man' (Daly 1991a, p. 196). Such views leave Daly open to criticism from those more favourably disposed to Marx's analysis of capitalism. Some of this criticism can be described as friendly and some decidedly unfriendly and accusative.

An example of a friendly Marxist critic of Daly's steady-state economics is John Bellamy Foster, an American professor of sociology and editor of the socialist *Monthly Review*. Foster has written extensively and with great insight on the political economy of capitalism and the environment from a Marxist perspective. In his writings Foster often quotes and cites Daly on the steady-state economy and they both draw on some of the same earlier writers in their critiques of capitalism. One example is the classical economist James Maitland Lauderdale writing in 1819,

who made a critical distinction between "public wealth" and "private riches". Lauderdale called attention to the paradox that private riches could expand while public wealth declined simply because formally abundant objects with great use value but no exchange value became scarce, and therefore acquired exchange value and were henceforth counted as riches.

Daly and Cobb 1989 1994, p. 147

As the world becomes full, public wealth becomes scarce, 'turning formerly free goods into scarce resources' (ibid.), generating positive exchange value (prices) and increasing GDP. Bellamy Foster also wrote about the 'Lauderdale Paradox': the 'inverse correlation between public wealth and private riches such that an increase in the latter often served to diminish the former' (Foster and Clark 2009, p. 1).

Where Bellamy Foster parts company with Daly is on whether socialism offers a better alternative to a reformed capitalism, such as Daly proposes:

> First and foremost, among those who seek a Lesser Transition [a reformed capitalism] I would include Herman Daly, whose work has enormously impressed me over the years and from whom I have learned a great deal and have enormous admiration. Daly insists, in a powerful critique of business as usual, on the need for a steady state economy, which means an economy with no net capital formation. But he believes this can be done within a capitalist free-market institutional context.

Foster sees Daly's position as 'utopian reformism' and attributes Daly's unwillingness to embrace socialism, even in its ecosocialism form, as the way forward because of its abject failure in the Soviet Union. Daly's views, Foster says, are 'stuck in the old Cold War divide' (Foster 2018).

These remarks of Foster's were made in response to an article by French Brazilian Marxist sociologist and philosopher Michael Löwy on ecosocialism (Löwy 2018) to which Daly also responded. Foster is correct that Daly does draw lessons from the experiences of twentieth-century socialism but which he says Löwy asks us to ignore and

> focus on Marxist *theory* instead. This reminds me of the growth economists who ask us to focus on the elegance of neoclassical optimization theory and downplay the massive external costs and monopoly concentration of wealth and power of actual corporate capitalism.
>
> *Daly 2018*

Foster accepts Daly's case for a steady-state economy but insists that 'We can't afford a Lesser Transition that begins and ends with the quantitative notion of "no growth," as if this in itself is enough, and that does not address substantive equality, while pretending to address ecological sustainability – as if the two were not insep-arable. The goal has to be sustainable human development, which must necessarily make room for the poorest countries to develop' (op cit.). Put this way it is hard to see a difference between his views and those of Daly, who also argues for 'sub-stantive equality' and the need for a 'distributist institution', and a shift of the tax burden from labour to resources. But there is a difference between them and it is what Foster and other Marxists refer to as 'the logic of capital' – the imperative of capital accumulation considered fundamental to capitalism and the concentration of ownership and power in a few hands based on the exploitation of labour and nature that it necessitates. Foster calls for a 'movement toward socialism... to go beyond the pursuit of profit and capital accumulation and the reliance on commodity markets... to have any hope of coming out of the tunnel' (ibid.).

Daly, however, remains unpersuaded by such pleas because

> Marxists in general, take a very dim view of markets and a very rosy view of central planning... [they are reluctant to discuss] limiting population growth... a topic which Marxists, as well as the capitalist cheap-labour lobby, go far out of their way to avoid.
>
> *Daly, op.cit.*

Daly is sceptical of the 'democratic ecological planning' with all its attendant voting that Löwy counts on so heavily: 'We have all citizens spending absurd amounts of their time "democratically" voting, mostly about things they don't understand, while those with the most information about actual use-values... are "disenfranchised" by the absence of markets' (ibid.). Daly also questions the viability of 'overwhelming material abundance and the new socialist man... along with the abolition of scarcity. But with little discussion of growth or its costs' (ibid.).

The line of criticism from a Marxist perspective of Daly's steady-state economics based on the incompatibility of a steady-state economy and capitalism has been most vigorously argued by Richard Smith. He criticizes 'steady-state capitalism' as impossible and a distraction for 'a broad conversation about what the lineaments of a post-capitalist ecological economy could look like' (Smith 2010, p. 30). Daly, in his brief response to Smith, points out that he has never used the term 'steady-state capitalism'. Furthermore, in light of the cold war contest between capitalists and socialists each claiming that their system would grow faster, Daly's view of a steady-state economy 'is something different from both capitalism and socialism' (Daly 2010, p. 103).

It is different from capitalism because a condition for a steady-state economy is a limit on its material and energetic size (either constant stocks of capital and people or constant throughput flow at levels compatible with the limits of the biosphere). No such limits exist in capitalist economies or in the theory of capitalism propounded by neoclassical economists. They would have to be imposed as constraints on the economy, whether capitalist or socialist or something else. Smith may well be right that capitalists (including landowners) would oppose such limits as they did in the past in defense of slavery, child labour, and opposition to minimum wage, progressive taxation, a social safety net, socialized medicine, and a wide range of environmental legislation, all of which came to pass in capitalist countries. Importantly, Daly does not propose that the scale of economies be left to the market, or the distribution of income and wealth, and he proposes new institutions with these responsibilities, subject to democratic control. With these institutions and limits on scale and inequality in place, only then does Daly say that the market has a useful role to play in allocating available resources to alternative uses guided by the purchasing decisions of buyers who are in sole possession of information about what matters to them backed up by their ability to pay from their equitable share of income and wealth.

Smith ought to have understood all this, simply because Daly has repeated this message time and time again as Smith is aware. 'Daly's critique of the neoclassical defence of growth is probably the most devastating critique to come from within the profession' (op cit., p. 33). But it was easier for Smith to launch into a frontal assault on capitalism rather than address Daly's carefully circumscribed role of markets in a steady-state economy. Daly said in response to Smith that 'it was a change to be attacked by a socialist' but found that Smith does not 'say anything specific or helpful about moving to a steady state economy, whether a capitalist or socialist, or neither' (op cit., p. 103).

In his critique of Daly's steady state economy Smith makes uncalled for *ad hominem* attacks on Daly: 'even shareholders who are environmentally-minded professors investing via the TIAA-CREFF accounts, are seeking to maximize returns on investment' (op cit. p. 33) and 'under capitalism workers don't have job security like tenured professors' (ibid., p. 35). Again, Smith should have known better. In 1991 when Daly published the second edition of *Steady-State Economics*, the book which Smith cites frequently in Smith 2010, Daly had resigned his university professorship and was an employee of the World Bank (Smith 2010, p. xiv).

Yet, despite their differences, Smith's agenda for redesigning the economy as an 'eco-socialist economic democracy' has much in common with Daly's proposal for a steady-state economy. For example, Smith says,

> We're going to have [1] to find ways to put the brakes on out-of-control growth…, [2] … to physically ration the use and consumption of all sorts of specific resources like coal, oil, gas, lumber, fish, oil, water, minerals, toxic materials,… [3]… to sharply increase investments in things society *does* need like renewable energy, organic farming, public transit, public water systems, public health, quality schools… and many other currently under funded social and environmental needs.
>
> *op cit., p. 38*

The list continues but the point is made. When it comes down to what needs to be done, Smith has borrowed a lot from Daly. If there is a substantive difference between them, it is that Smith wants to replace capitalism first and then take the necessary actions to reduce humanity's burden on the environment. Daly, on the other hand, thinks capitalism can be changed from within through institutional re-design to limit scale and bring a just distribution of income to promote well-being without economic growth.

Elke Pirgmaier is another Marxist economist who finds fault with Daly's incorporation of the market in his proposals for a steady-state economy. She accuses Daly of attempting 'to squeeze neoclassical economics into a biophysical and ethical corset' (Pirgmaier 2017, p. 52). In a typically respectful email to Pirgmaier Daly writes,

Your 'corset analogy', – the basis on which you dismiss the importance of the macro constraints on scale and distribution in order to focus your criticisms on allocation – misses the mark. A corset changes appearance, not reality. Limiting scale by cap-auction-trade or throughput taxes is more like a diet than a corset – it actually reduces weight. Also, maximum and minimum limits on income changes real distribution of diet-limited throughput. So yes, a corset squeezes, but that is all that it does, and by focusing only on squeezing I believe your analogy misleads.

Daly 2016

In response to Pirgmaier's concern that Daly relies too much on neoclassical economics Daly says:

Regarding reliance on markets for allocation (once determination of scale and distribution have been taken from the market), it is important to remember that even neoclassical price theory recognizes the evils of monopoly, and from the beginning restricts market allocation to rival and excludable goods. Rival goods that are non-excludable give rise to the 'tragedy of open access commons'; and excludable goods that are non-rival (e.g. intellectual property) give rise to the 'the theft of the commons'. These concepts are neoclassical orphans that ecological economics has adopted in connection with markets. And if markets are not acceptable how do you solve the allocation problem? Central planning? You mention 'social provisioning' without explanation.

ibid.

Pirgmaier's criticism of Daly's inclusion of markets for allocation purposes notwithstanding, Pirgmaier accepts 'a certain role of markets as coordinating production, distribution and exchange decisions ' which if anything is broader than the role envisaged by Daly (ibid., p. 54).

Daly did not publish a response to Pirgmaier but one was provided by his co-author economist Josh Farley and environmental scientist Haydn Washington. They gave a detailed analysis of Pirgmaier's paper, with which Daly agreed, and concluded that her 'critique of SSE [the steady-state economy] is misinformed and muddled. She misrepresents both the SSE and its relationship to NCE [neoclassical economics]' (Farley and Washington 2018, p. 442).[3]

Other criticisms of Daly's steady-state economy

In addition to criticism from mainstream economists, proponents of degrowth and Marxists, Daly's steady-state economy has drawn responses from commentators who, broadly speaking, accept his critique of economic growth and the need to reduce the burden that humans place on the environment but find fault with Daly's steady-state solution. We will briefly mention a few.

Considering the concept of degrowth ambiguous and rather confusing, economist Jeroen van den Bergh argues for 'a-growth', which means being indifferent about growth (van den Bergh 2011). By economic growth van den Bergh means growth in real GDP or GDP per capita, accepting that both are imperfect indicators of social welfare. He bases his argument on the belief that it will be easier to get support for a progressive policy package by avoiding an explicit degrowth strategy. Economist Kate Raworth takes much the same position in her *Doughnut Economics* (2017), in which she explores the question of reconciling social goals inspired by the UN's Sustainable Development Goals with the planetary boundaries identified by Rockstrom et al. (2009).

Van den Bergh and Raworth's attempt to side-step the debate on environment versus growth is well-intentioned but problematic. It is predicated on growth meaning growth in GDP rather than Daly's definition of growth meaning growth in physical stocks or throughput or both. Since they agree with Daly that growth defined quantitatively should end, at least in the rich countries, and that physical stocks and throughput should decline until sustainable levels are reached, their agnosticism on growth in GDP admits the possibility that GDP (and GDP per capita) can continue to rise over the long term while physical stocks and/ or throughput decline. This is the position of mainstream economics and some advocates of green growth but most likely not a position with which they are in sympathy.[4]

Economist Clive Spash claims that 'Daly has never addressed the contradiction of maintaining the institutions of a capital-accumulating social and economic structure while deconstructing the central supports of economic growth, namely energy and material throughput and resource exploitation' (Spash 2015, p. 378). This is very similar to criticisms put forward by degrowthers and Marxists and to which Daly has responded though not, it seems, to their satisfaction. One difference between Spash and these others is that Spash identifies as an ecological economist and has written several papers on epistemological and philosophical aspects of ecological economics. In one of them he distinguishes among three camps in ecological economics: new environmental pragmatists, new resource economists and social ecological economists. He classifies Daly as a new resource economist which, in Spash's terms, means his approach is multidisciplinary rather than inter or transdisciplinary, a description with which Daly would agree. Spash acknowledges in a more recent paper that Daly rejected the 'growth paradigm' in 1972 giving him credit for introducing the term (Daly 1972), commenting that 'by the 1980's, amongst economists, Herman Daly was holding the anti-growth fort almost single-handed until the rise of modern ecological economics' (Spash 2020, p. 2). But Spash, remains troubled by Daly's inclusion of the market in his proposals for a steady-state economy (Spash 2013, p. 356, and 2020 p. 8) even though Daly would restrict its role to the allocation of rival, excludable goods and services, and only when it is constrained by publicly set limits on scale and inequality and competition among sellers prevails.

Economist John Gowdy, a long-time friend and colleague of Daly's and also a student of Georgescu-Roegen, shares Spash's concerns. In contrast to Georgescu-Roegen, 'there is no notion of how inequality, growth, accumulation, and the dynamism of a capitalist economy are intertwined' (Gowdy 2016, p. 81). Gowdy refers to a paper by Philip Mirowski, historian and philosopher of economic thought, to distinguish between Daly's 'mechanical' analysis and Georgescu-Roegen's 'complexity/evolutionary' one. This distinction maybe valid but it is not one that Mirowski makes (Mirowski 1988).

In their textbook on ecological economics Daly and Farley devote several pages at the start of the book to coevolutionary economics which they define as: 'The study of the mutual adaptations of the economy and environment. Economic activity induces changes in the environment, and changes in the environment in turn induce further changes in the economy in a continuing process of evolution' (Daly and Farley 2011, p. 462). They refer to Karl Polanyi's *The Great Transformation*, in which he explained that the economy is a component of human culture, 'and like our culture, it is in a constant state to evolution' (ibid.) They continue with a concise description of the coevolution of human economies from hunter-gatherer to industrial and give economist Richard Norgaard credit for 'many of the basic ideas' that they draw upon and endorse.

From Daly's perspective steady-state and coevolutionary economics are complementary since it is the coupling of human-natural systems that underlies the justification for a steady-state economy.

> Evolution and coevolution are the way the natural world works, and the economy, as a subsystem, must coevolve as the world evolves. But our old coevolutionary strategy of growth into an empty world has become maladaptive in the full world that continues to evolve, but not to grow. A steady-state economy looks like a more promising coevolutionary strategy.
>
> *Daly 2015b*

Norgaard sees it somewhat differently.

> Our disagreements were always subtle, but like his professor Nicholas Georgescu-Roegen, I was frustrated with [Daly's] appeal to the second law to justify a steady state and at other times he seemed to infer that scientists could define the limits within which the economy must stay. His view reinforced the idea that there is a nature out there apart from people, and most natural scientists were comfortable with this while my coevolutionary framing very much accepts that people have been interacting with and transforming nature for a long time. Steady-state is a better guiding principle than eternal growth but thirty-five years of debate over what constitutes sustainable development and finally a broader interest in coupled human-natural systems thinking has taken us into deeper more thoughtful and scientifically valid thinking.'
>
> *Personal communication*

Economist Inge Røpke a makes a strong case for a transformation of how introductory courses on economics should be fundamentally restructured 'in order to provide the foundation for a sustainability transition' (Røpke 2020). This is not a critique of Daly's steady-state economics so much as it is a critique of how economics should be taught. Røpke is not against neoclassical economics to the same extent as some others but she thinks that neoclassical approaches can be presented without endorsing them, something she implies Daly and Farley do in their textbook (Daly and Farley 2004). She says it was understandable that they did so 'because the first steps to get ecological economic ideas into the mainstream curriculum went through the teaching of mainstream environmental economics' (ibid). Now Røpke says, the situation has changed and 'we should try to avoid endorsing the basic thought patterns of neoclassical economics, even though it may be difficult for economics teachers, even heterodox economists, to give up teaching something that they struggled to learn as students' (ibid.). If Daly's experience is anything to go by this will be no easy task but one which he would no doubt support.

Conclusion

If it is true that it is better to be criticized than ignored, then Daly's steady-state economy has been quite a success. Three-quarters of the respondents to the online survey on Daly and his influence identified the steady-state economy and thermodynamics and economics, as the two aspects of Daly's work that have had the greatest impact. Apart from mainstream economists who see no reason that economic growth defined as increasing real GDP cannot and should not continue indefinitely, all of the critics of Daly's steady-state economy agree with his position that the economy is a subsystem of the biosphere which has grown too large. They also agree that very significant changes are needed to the economy in order to rectify this situation, taking into account wide disparities in incomes, wealth and culpability, among nations. Their differences with Daly lie in one or more of the following, real or imagined: Daly's lack of appreciation of the economy as a complex system; his failure to take into account the growth imperatives of capitalism; too much reliance placed on neoclassical economics as an explanation and source of solutions; and too little recognition of ideological, cultural and political factors. There are also those who take issue with Daly's proposals for limiting population growth, constraining lending by private banks, and pushing back against globalization and 'free' trade. These issues, which are as relevant today as when Daly first wrote about them many years ago, are covered in the following chapters.

There is another difference too between Daly and his critics. Some of the critics of Daly's steady-state economy see their objections as undermining the whole steady-state enterprise, without necessarily having anything, let alone anything better, to put in its place. Daly sees it differently. To him weaknesses in his analysis of problems and proposed solutions are opportunities for improvement and collaboration. Even to his harshest critics, taking Pergmaier as an example, he writes, 'In spite of our differences I found your article stimulating, and hope that you will continue to develop the better alternative for which [your] article would be the prelude' (Daly 2016).

The 'wild facts' that Daly refers to so often in his writings showing the extent to which the world has become full, demand a response other than growth and more growth. In the steady-state economy, Daly has provided one that some enlightened critics and contributors have chosen to build on while others look elsewhere. They may find something that is more complete, more fleshed out, than Daly's steady-state economy, but it is most unlikely that they will find something viable that is not consistent with the pre-analytic vision from which it springs.

Notes

1 One area where Kallis clearly disagrees with Daly is on immigration (see Chapter 12).
2 Given the emphasis in the degrowth literature on the importance of political context, it is surprising that Parrique found very little difference among the degrowth agendas proposed for France, Spain, Belgium, Germany and Finland.
3 Another, even less well-argued critique of Daly's steady-state economy from a Marxist standpoint by Troy Vettese (2020) met with a sharp response from Daly who at the same time acknowledged that he 'benefitted from reading, regardless of differences, some other Marxists and eco-socialists', naming John Bellamy Foster and Paul Burkett as examples (Daly 2020b).
4 Latouche also wrote about agrowth but gave it a different meaning: 'Rigorously, it would be best to speak about "agrowth", as one speaks about atheism [rather than agnosticism]. It actually means quite precisely, the abandonment of a religion: the religion of the economy, growth, progress and development' (Latouche 2010, p. 59).

11

POPULATION, MIGRATION AND IMMIGRATION

More people (and other creatures) are better than fewer – but only if they are not all alive at the same time!

Daly

Daly's definition of a steady-state economy based on maintaining constant capital stocks with minimal throughput includes the human population as well as the stock of produced artifacts: the built infrastructure, capital equipment and consumer durables. He explains that this definition of a steady-state economy is 'an extension of the demographer's model of a stationary population to include non-living populations of artifacts, with production rates equal to depreciation rates, as well as birth rates equal to death rates' (Daly 2015, p. 106).

The only requirement for a stationary population, assuming no net migration, is that the birth and death rates are equal. They can be equally high or equally low – the population will be stable in both cases – but there is a big difference in what it means for the individual members of the population. Average lifetimes are shorter when birth and death rates are high, longer when they are low and the difference in average lifetimes can be considerable. Since people have a natural preference for longer rather than shorter lifetimes, Daly favours low birth and death rates. He says the same should be true for artifacts. Their useful lives can be extended, requiring less frequent replacement and lower throughput, if they are made more durable, easier to repair, and less vulnerable to the whims of fashion; but this can be anathema in an economic system that seeks a high rate of growth in GDP. It would and should be different, Daly says, in a steady-state economy where the minimization of throughput to sustain humans and their artifacts at levels sufficient for a good life is the goal.

None of this is especially controversial. It all follows logically from the first of Daly's two definitions of a steady-state economy in which stocks are held constant

DOI: 10.4324/9781003094746-11

and throughput minimized (see Chapter 8). Daly does not discuss population in the context of his second definition of a steady-state economy where throughput is held constant based on resource and environmental considerations, and stocks are allowed to vary. Under these conditions the size of the human population in a steady-state economy so defined could fluctuate over time, though without an upward or downward trend. Again, this is hardly controversial. It only becomes so when policies designed to influence the size of the human population are broached; in particular, policies relating to contraception, abortion and immigration. As is characteristic of Daly, he does not shy away from discussing such policies and making his own position clear. We shall see in this chapter, his views have not always been received favourably, even by some who share Daly's views on the economy and the environment. However, before we come to that we will examine Daly's early work on population, pre-dating even his first publications on the steady-state economy, but which doubtless influenced his later thinking.

The population question in North Eastern Brazil

Daly's interest in population began early in his career. In 1967–68 he was a Ford Foundation visiting professor of economics at the University of Ceará in North Eastern Brazil. While there he wrote a highly original paper on the population question in North Eastern Brazil (Daly 1969). The paper was written in Portuguese, a language in which he had become fluent, and presented at a conference at the Fundação Getúlio then published (Daly 1968b). Daly produced an English version when he spent a year at Yale as a research associate after returning from Brazil. The paper was published in 1970 in an academic journal (Daly 1970a) and was also reprinted in Spanish for publication in Mexico, so its reach was considerable.

The paper's appearance in English as a discussion paper for the Economic Growth Centre at Yale University is a reminder that Daly started out as a growth economist. He believed then, as many economists still do today, that economic growth is desirable in itself and if not that, then essential for the achievement of other objectives. For this reason, arguments to the contrary are resisted, even in the richest countries, and in the face of mounting evidence that endless economic growth is infeasible.

When Daly wrote about population in North Eastern Brazil in 1969 the global population was 3.6 billion, an increase of 56 per cent since Daly's birth in 1938. Furthermore, the annual rate of increase had been increasing for 20 years, giving rise to considerable concern about whether food supplies could keep up with continuing population growth (Ehrlich 1968). In North Eastern Brazil, which covers 18 per cent of Brazil and where Daly was living in the late 1960s, the population was 27 million, growing at around 3.1 per cent annually, and per annual capita income was $150 (equivalent to about $1,000 in 2020 dollars). Daly realized that unless population growth in the region slowed, there would be little prospect of rising living standards for millions of very poor people, so he set about analysing the demographic situation in North Eastern Brazil in detail.

In order to understand the relationship between population growth and income growth, Daly used the

> simplest analytical framework for relating output growth to population growth... the Harrod-Domar relationship upon which the body of contemporary growth theory is based... the formula simply says that the per capita income growth rate equals (approximately) the total income growth rate... minus the population growth rate.
>
> *op cit., p. 540*[1]

Daly used this relationship to distinguish three ways to increase per capita income growth rate: by increasing the rate of savings and investment, increasing the productivity of capital, and reducing the rate of population growth. Drawing on the best data available at the time, Daly concluded that

> the Northeast is already operating at its capacity to absorb capital and technology (resulting in a high rate of growth) but is a long way from its capacity to reduce population growth. Furthermore, *the high population growth is the main social limit holding down the capacity to absorb capital and technology, and the savings potential.*
>
> *ibid., p. 542, emphasis in original*

This led Daly to propose a 'third development epoch of "population policy"' (ibid., p. 537) for Northeastern Brazil, to complement, not displace, the first two which were a major emphasis on fighting the periodic droughts followed in the 1950s by the prevailing 'capital-technology-planning' approach.

Daly summarized his reasons for a population policy for Northeastern Brazil, emphasizing fertility control in 10 main points:

1. Without it modal income is unlikely to increase (likely to decline), which means that the benefits of growth will not extend to the masses, who will form an ever larger percentage of the population, and who already compose about 90 per cent of the population.
2. It will greatly reduce the burden of educating the masses, a precondition for modern technology as well as democracy. Indeed, without population control the percentage of illiterates will probably continue to increase.
3. It will increase the savings potential by lowering the dependency ratio.
4. It will facilitate the introduction of high-productivity methods in agriculture which are not feasible with surplus farm population and low wages; the same is true for industrial automation. Alternatively, it will mitigate the problems of technological unemployment when automated methods become more economical than even subsistence labor.
5. It is necessary to keep the swollen cities from bursting.

6. The lower class wants birth control, although not articulately, and the upper class already practices it.
7. A dollar spent on birth control is, at the current margin, vastly more productive in raising per capita income than a dollar spent on conventional development projects.
8. Ultimately birth and death rates will be brought inexorably into equality. Without some form of birth limitation, the death rate will eventually rise to equal the birth rate at around forty-eight per 1,000, implying a life expectancy of only twenty-one years.
9. The region is unable to support its present population, as evidenced by high out-migration, illiteracy, disease, and infant mortality rates.
10. A high birth rate gives an impulse to structural inflation, which reacts unfavorably on k [the marginal capital-output ratio] and s [the annual savings rate as a percentage of annual output] (ibid., pp. 550–551).

Most economists would have been satisfied with these conclusions and the theoretical and historical analysis on which they were based, as a contribution to the burgeoning literature on economic development in poor regions, but not Daly. He followed the historical and analytical part of his paper with a discussion of the 'ideology of the population question', making comments that still resonate today. This willingness on Daly's part to explore the broader, social and cultural implications of economic analysis in this paper and in much of his later work, sets Daly apart from most of his contemporaries. Having arrived at what he considered a compelling economic case for population policy Daly wondered why it had not been adopted and why it was still strongly resisted. He knew that part of the answer was 'simply lack of information and lack of understanding of the economic arguments, plus natural inertia', but he thought this was 'far from being the whole story. To understand the rest, we must turn to the realm of ideology' (ibid., p. 554).

After analysing statements on population from a variety of Brazilian sources Daly realized that the case for population control that he found so compelling was based on a goal of maximizing per capita income. For others the goal was maximizing national power and prestige which is better served by total, not per capita, income. A larger population reduces per capita income unless total income increases even more. This is not a consideration when only total income matters. As long as more people means more total income, population growth is considered good.

To understand the ideological underpinnings of the different perspectives on population Daly found it useful to contrast the views of 'laissez faire libertarians' with those of Marx and 'Marxians'. The libertarians favour leaving 'as much as possible of economic life to automatic regulation [but] have usually made an exception for population…' (ibid., p. 559). Marx and the Marxians 'have generally considered socialist reform and population control to be incompatible alternatives for lifting the proletariat' (ibid.). Daly referred to 'Marx's contempt for Malthus' and later to the 'Left's denial of [the complementarity of population control and social reform for lifting the masses as] the clearest possible indication that their immediate aim is the building up of pressures for revolution… regardless of the human cost' (ibid., p. 568). Marx rejected anything Malthusian. Daly found it curious that Marxists want to centrally plan and control production but are quite laissez faire on reproduction (as long as it is enough to maintain the reserve army of unemployed). Later in this chapter we will look at Daly's highly innovative attempt in a different paper to reconcile Marx's views on population with those of Malthus more generally, but the seeds for these ideas were sown in his work on population in North Eastern Brazil.

Another of Daly's insights into the population issue stimulated by his work in Brazil was the distinction he drew between Marx's theory of exploitation and what Daly called 'Roman exploitation'. The Marxist theory of the exploitation of the working class by property owners is well-known. It is based on the ownership of capital (i.e., the means of production) by a few, obliging the 'dispossessed industrial worker – members of the proletariat' to sell their labour for wages. Wages are kept low by the existence of 'the army of the unemployed' who are willing to work for subsistence level payments. Daly compared Marx's idea of the industrial proletariat with the original Roman meaning of proletariat – 'producers of offspring for the republic'. He though the Roman proletariat was 'far more accurate in characterizing the lowest class of Northeastern society' (ibid., p. 561) since they were agricultural not industrial workers.

Observing that the upper class in Brazil practised birth control and had smaller families than the poor, Daly wrote

> the ownership by the oligarchy of both the means of production and the means of controlling reproduction leads to Marxian and Roman exploitation simultaneously, with the result that the proletarian population, like animal populations, is limited by its aggregate wealth, while the upper class population has the distinctively human characteristic of being limited by its standard of living.
>
> *ibid., pp. 561–562*

Daly's view was that 'the idea of Roman exploitation [which accounted for the denial of contraception to the poor] should be stressed as a cause of backwardness and social injustice along with Marxian exploitation' (ibid., p. 568).

To complete his discussion of ideology and population control in Northeastern Brazil Daly referred to the deep division that existed, and still does, in the Catholic Church with regard to birth control. Daly was more optimistic than others that the Church would change its position despite the then recent encyclical of Pope Paul VI *Humane Vitae* in 1968 that considered deliberate contraception as 'intrinsically wrong'. In all likelihood, the Pope felt it necessary to speak out on contraception at that time because of the arrival of the birth control pill that first came available in the USA in 1960 and spread far and wide very quickly, making contraception simpler than ever. Striking a tone reflective of the Cold War that dominated the late 1960s and early 1970s, Daly said:

> Catholics as well as the far left and the oligarchy should be vigorously challenged on the issue [of population control], since radical economic-demographic policy is the only viable individualistic alternative to collectivist revolution in the third world and the Communists know this very well.
>
> *ibid., p. 568*

It is not possible to ascertain whether Daly's paper had any real influence on population policy in Brazil in general or in North Eastern Brazil in particular. However, in 1974 in a statement at the World Population Conference in Bucharest the Brazilian government led many to believe that the government's attitude to population growth had changed. The statement recognized that all couples had the right to have the number of children they wanted, and the government's responsibility was to ensure that the poor also had this right. This statement was followed in 1978 by President Geisel's expression of concern over the consequences of continued population growth and later by President Figueiredo's recognition that progress in family planning was a prerequisite for the continued social and economic development of the country (Sanders 1984). Access to family planning services by the Brazilian government fell short in the years that followed these presidential pronouncements but their close resemblance to the views expressed by Daly a few years before suggests his voice, and those of others with similar views, was at last being heard.

Fifteen years after the publication of his 1970 paper Daly revisited the situation in North Eastern Brazil, this time armed with much better data on fertility and mortality by income group than was available for his initial study. Comparing 1977 to 1970 he found

> a clear narrowing of extreme class differences in fertility, and a consequent weakening of the effect on distribution of income per head compared to 1970. However, lower class total fertility was still over twice that of the upper class, so the effect remains highly significant.
>
> *Daly 1985, p. 333*

Yet, even in 1985 when Brazil was returning to democracy after 21 years of military rule, there was still a 'certain "disinclination" to study fertility in its socio-economic context.' Daly explained this 'taboo' in terms of 'an injustice too blatant to defend openly, but too important to the interests of the status quo to challenge openly'. He said that this reminded him of his childhood in segregated Southern United States where 'it was obvious to an unexceptional child that, by and large, the blacks did the dirty work and were poor, while the whites did the more interesting work and we're better paid. And segregation laws were intended to keep it that way' (ibid., p. 334). Some lessons you never forget.

The basic conclusions of Daly's second look at population in North Eastern Brazil were that:

> 1) North-east Brazil had probably the highest class difference in fertility of any society…, 2) Between 1970 and 1977 fertility fell significantly in all classes… [though] class differences… remained very high by international standards, [and] 3) a Marxian-Malthusian definition of social class, in terms of control versus non-control of both production and reproduction, fits the North-east and offers a possibility for integrating the valid insights of both traditions.
>
> *ibid., p. 338*

Reflecting years later Daly said that reproductive rights in North Eastern Brazil have much improved since the late 1960s, while ironically in recent years in the United States they have regressed. Even Planned Parenthood in the US is under attack. Neither Daly's 1968 article nor the 1985 sequel received many citations, but they were noticed favourably by demographer Kingsley Davis at Stanford and a few scholars in Southern Brazil. They also led to Daly's friendship with Jose Lutzenberger (see Chapter 2).

Marx, Malthus and population

The Marxian-Malthusian definition of social class that Daly mentioned in his 1985 paper had its origins in his 1970 paper just discussed on population in Northeastern Brazil. It was presented with more theoretical rigour a year later (Daly 1971a). Daly's intent was to show that 'the two historically dominant theories of poverty, the Marxian [exploitation of wage workers by profit seeking, owners of capital] and the Malthusian [population growing faster than the food supply], are not inconsistent, but complementary' (ibid., p. 25).

Daly is not and has never been a Marxist. For one thing, Daly considers Marx's labour theory of value as a far inferior explanation of relative prices to neoclassical price theory based on the interaction of producers (supply) and consumers (demand) in markets. Also, at a more philosophical level he could not accept Marx's dialectical materialism and historical determinism. But there are other issues on which he thought Marx had much to offer. This includes Marx's analysis of class conflict based on the exploitation of workers by the owners of capital.

> The word "exploitation" has been virtually banished from polite economics… The class of capitalists exploits the class of labourers by appropriating the entire surplus (national product excess over 'subsistence'). The national product is produced by labour. Capital and land enhance the productivity of labour, regardless of who owns them.
>
> *ibid., p. 28*

Daly drew a parallel between Marx's explanation of the exploitation of labour through wage labour by pointing out that workers also produce a surplus of new workers for the 'reserve army of the unemployed' (Marx's term). To put it in Marxist terms, just as labour power in the factory can produce more than its subsistence, so in the home it can reproduce more than its replacement. Daly notes that this Malthusian dimension of exploitation was certainly not emphasized by Marx, who hated Malthus, even though in this respect Malthus' theory supported Marx's argument by providing an extra source for the reserve army of the unemployed that is necessary to keep wages at subsistence. But for Marx this would have been an unacceptable concession to Malthus, so he based the reserve army only on technological and cyclical unemployment.

Daly began the combination of Malthusian and Marxist principles from the observation that average of GNP per head is 'meaningless' because it hides wide disparities among people and social classes. His remedy was to create a four-fold categorization, based on differences in population and income. Following Marx, Daly distinguished between two kinds of income: income to workers largely as wages (Yw) and income to capitalists largely as returns to property (Yp). Then he distinguished between two segments in the population: those that control their reproduction (Pc) and those who do not (Pn). These distinctions give rise to four possible types of social class and associated incomes per capita (income Y divided by population P):

1. Yp/Pc – the combination of property ownership and controlled fertility is characteristic of the *upper class*...
2. Yp/Pn – the combination of property ownership with uncontrolled fertility is characteristic of the *middle class*... a group that may be considered 'Catholic capitalists'...
3. Yw/Pc – the combination of labour income with controlled fertility is also a *middle class* characteristic ...this group could be called *neo-Malthusian labourers*...
4. Yw/Pn – the combination of labour income and uncontrolled fertility is characteristic of the lower class... the *classical Malthusian proletariat*

ibid., pp. 32–34.

Having created these four categories Daly produced an equation where they are each multiplied by the proportion of each social class in the population. Their sum is equal to total income divided by total population, i.e. the average per capita income for the entire population. Daly's disaggregation of this 'almost meaningless' average is far more informative.[2]

With this simple equation in hand Daly says that 'knowing the values of the four incomes per head, the rates of growth, the values and rates of changes [in the weights] would give us a vastly more complete, and yet succinct, description of an economy... Instead of the vague goal of "maximize GNP per head", we would be obliged to say something about composition', i.e. the relative sizes of the weights and their changes over time (ibid., p. 34). Daly also showed how concepts like development and over population can be better defined. In this approach development has two dimensions:

> one dimension is an increase in Y/P; the other is an increase in a_1 [the proportion owning property and controlling fertility] and a decrease in a_4 [the proportion relying on labour income and not controlling fertility, with the other two categories] serving as functional stages in the transfer.
>
> *ibid., p. 34*

Going further, Daly suggested that societies could be classified according to differing combinations of the four proportions. A 'primitive classless society' would be one where everyone is in the fourth class: all income goes to labour with no fertility control. The social and cultural characteristics of such societies could still vary of course, including the conventions by which labour income, defined broadly, is shared out. Another of Daly's examples is the 'future vision of a developed classless society' (ibid., p. 36), where everyone owns property (individually or collectively) and fertility control is widely practised. He explained that in 'dual' societies with great disparities in incomes and wealth and where the middle class is small or virtually non-existent, the population is divided between those totally dependent on labour income with little or no access to fertility control, and those who derive their much larger incomes from property ownership and have far fewer children.

Lastly, the much discussed and widely accepted 'demographic transition' where societies move from high birth and death rates to low birth and death rates as a result of economic improvement can be described quite conveniently using Daly's classification of social classes. Historically, declines in death rates due to improved living conditions and better health care have preceded reductions in birth rates. This has been typical of all high-income countries. In the period after the death rate has fallen and the birth rate has yet to follow the population expands. The greater the difference between the birth rate and death rate, the faster the growth in the population.

In terms of Daly's classification of social classes, a 'pre-demographic transition class society might correspond to the case where [the population is comprised of only the second and fourth classes]... Other possibilities are easily imagined' (ibid.). As to whether it makes sense to rely on the demographic transition 'induced by rising standard of living, as usually assumed...' Daly pointed out many years later that the overall burden on the environment could rise if a fall in the Indian fertility rate[3] to that of Sweden required a rise to Swedish living standards. 'Of course, indirect reduction in fertility by automatic correlation with rising living standard of living is politically easy, while direct fertility reduction is politically difficult. But what is politically easy may be environmentally ineffective' (Daly 2015c, pp. 107–108).

Daly considered his fourfold classification of a country or region's population to be an improvement on GNP (and GDP) per capita as an index of welfare yet he had doubts about using income for this purpose at all. Following Boulding, Daly wrote,

> the flow of income (or product) is the maintenance cost of the stock of wealth, and… it is the services of the capital stock, not the flow of additions and replacements, which serves human wants… a more meaningful ratio would seem to be capital stock to population – i.e. to redefine the Y's as wealth rather than income….
>
> *op cit., p. 35*

This is an important idea, though in practice the measurement of wealth is more difficult than the measurement of income. For example, Daly's suggestion that 'the wealth of the labouring class would be the capitalized value of the labour of himself and his household goods' (ibid.) is problematic, depending as it does on forecasts of incomes, choice of the discount rate, and whether the calculation should be based on gross income or net of the household's consumption. Following this logic, those not in the paid labour force and without property would be considered wealthless if not worthless.

By combining income and population in the same analytical framework Daly showed how Marxist and Malthusian explanations of poverty can complement each other. He combined the Marxist distinction between capitalists and workers and the social relations of property-owning employer and propertyless employee with the neo-Malthusian recognition that individual action to control fertility can increase their incomes. This synthesis may not appeal to those traditional Marxists who still look to the immiseration of the poor to spark a revolution; but to those who see other avenues for radical change, whether or not they are Marxists, Daly's analysis has much to offer.

Two controversies

Over the years Daly became embroiled in two controversies relating to population, both concerning measures to reduce the rate of population growth. The first concerned a proposal to reduce the birth rate through a system of tradeable birth licenses, and the second his advocacy of reduced immigration to the United States through lower numbers of immigrants and stricter enforcement of immigration laws. Each of these merits some discussion.

Transferable birth licenses

In a fascinating and still very relevant book entitled *The Meaning of the Twentieth Century*, Boulding wrote about three 'traps' that stood in the way of a great transition to a 'postcivilization'.[4] He named them the war trap, the population trap and the entropy trap. Writing in the 1960s, when the cold war between the USA and the USSR dominated politics, especially in the USA, nuclear war was seen as an ever-present threat. The arms race was a trap that offered no easy way out and still does not. The entropy trap, the one which Daly has given the most attention, concerned the depletion of low entropy resources and the threat it posed to continued economic growth. The population trap, the rapidly increasing number of humans on planet Earth, meant that unless population growth was brought to a halt and soon, improvements in the standard of living from economic growth would be short-lived.

Daly frequently acknowledged Boulding's influence on his own thinking, particularly on the steady-state economy and population. One idea that Daly adopted from Boulding on population was the use of transferable birth licenses. In the first and second editions of *Steady-State Economics* (Daly 1977a, 1991a) Daly described the plan 'as the best yet offered if the goal is to attain aggregate stability with a minimum sacrifice of individual freedom and variability' (ibid., pp. 50–51). In its most basic form 'the scale and distribution of the rights to bear children [are] determined by the community at large, but these rights [can] be traded in the free market' (Daly and Cobb 1989, 1994, p. 244). Daly considered a number of distribution schemes such as 1 license to every person, or perhaps 2 or 2.1 (allowing for infant and child mortality) to each woman. Under the scheme licenses could be transferred by sale or as a gift.

Daly anticipated several objections to this means of bringing the birth rate down to stabilize the population. He dismissed the argument that it is unjust because the rich have an advantage. 'Of course the rich *always* have an advantage, but is their advantage increased or decreased by this plan?' Daly's answer was that the effect of the plan is to make income distribution more equal because the licenses are distributed equally, and funds would likely flow from the rich to the poor. And if the rich end up having larger families than the poor it will reduce their family per-capita income and raise those of poorer families having fewer children. 'Whatever injustice there is in the plan stems from the prior existence of rich and poor not from Boulding's idea, which actually reduces the degree of injustice' (Daly 1977a, 1991a, p. 58). Daly thought people would be mistaken to think that the scheme adds to existing injustices.

A second objection to transferable birth licenses that Daly considered more reasonable was the problem of enforcement. 'Illegal' children could be put up for adoption and the adopting parents could be paid the market value for one of their licenses which would be retired to compensate for the child born without a license. Alternatively, parents of children for which they had no license could be punished. Despite Daly's suggestion that these transgressors are just like any others who break the law and should be treated no differently it is unlikely that most people would see it this way. A law which makes it illegal to have children without a license can hardly be seen as equivalent to one that makes it illegal to fish without a fishing license, to have dog without a dog license, or drive without a driving license

A third objection that neither Daly nor Boulding addressed is the scope of the market for transferable birth licenses. Is it to be local, national or international? Even at the national level there are problematic ethical questions and enforcement difficulties. These problems would only be magnified should the market be international and even if that were not the intention, surreptitious trades would be hard to avoid.

Unusually for Daly, who wrote so much about the ethical aspects of economics, he does not explore them in depth relating to transferable birth licenses. Instead, he quotes John Stuart Mill, 'one of the greatest champions of liberty who ever lived' (ibid., p. 61). Mill wrote:

> The laws which, in many countries on the continent, forbid marriage unless the parties can show that they have the means of supporting a family, do not exceed the legitimate powers of the state... they are not objectionable as violations of liberty... Yet the current ideas of liberty... would repel the... attempt to put any restraint upon his inclinations when the consequence of their indulgence is a life or lives of wretchedness and depravity to the offspring.
>
> *quoted in Daly ibid., p. 61*

Whether Mill would have favoured a market in birth licenses remains unknown. What he might very well have supported, and Daly certainly did, are better and more widely available sex education and family planning services to reduce unwanted pregnancies especially among adolescent girls, acceptance and support for couples who choose not to have children, and granting older people... 'the right to die on their own terms' (Daly and Cobb 1989, 1994, pp. 248–250).

Daly continued to propose a system of transferable birth licenses in *For the Common Good* (Daly and Cobb 1994), comparing it favourably with the Chinese policy of one-child families introduced in 1979. 'Our proposal is offered as an alternative to the Chinese approach for countries that are both in need of drastic action and capable of implementing it' (ibid., p. 246). And even then, with birth rates in decline, Daly said that 'we would hold in reserve the transferable birth quota plan... should present demographic trends reverse themselves' (Daly and Cobb 1994, p. 251).

After 1994 Daly wrote very little about transferable birth licenses and when he did it was as 'a paradigm for many sensible policies', such as pollution abatement where, as in the case of greenhouse gases, what matters most environmentally is total emissions. The allocation of tradeable emissions allowances among different sources is less important when emissions reduction at least cost is the objective. Tradeable emissions allowances are opposed by some who regard them as 'licenses to pollute' but that pales in comparison with the emotionally charged matter of the right to procreate which most people do not think should be for sale. Daly recognized this and turned his attention elsewhere but not before noticing that people now seem willing to accept much greater intrusions of the market into reproductive biology. He gives as examples the purchase and sale or donation of ova and sperm by young men, the now defunct 'Nobel Prize Sperm Bank', and commercial surrogate birthing where a women agrees, for payment, to bear a child for another woman unable to do so herself.

> Buying and selling a legal right to reproduce seems less an interference with biological reproduction than do these mentioned practices which also introduce an element of eugenics not present in the birth license plan. Perhaps any intervention that increases births is more acceptable than one reducing births.
>
> *Daly, personal communication*

It is not just the general lack of appeal of transferable birth licenses that has shifted Daly's attention away from them. Since Boulding and then Daly first made their proposals fertility rates have fallen significantly and in most rich countries, where the system had the best chance of success, the fertility rate is now at or below replacement rate. In the USA for example, data from the World Bank shows that births per woman fell from 3.6 in 1958 to 1.78 in 1978. It has fluctuated since but never exceeding the replacement rate of 2.1. In 2017 it was 1.8 and in decline.[5]

Countries in a similar position to the USA where the fertility rate is below replacement rate, but which still want population growth, must rely on immigration. Accordingly, in the past two decades or so Daly turned his attention to immigration where his views again proved controversial in some quarters.

BOX 11.1

Sometimes 'population' is used too sweepingly – but the choice to bring another rich world consumer into the world... is by far worse than a family in the global south having children (not least as that is the best insurance for old age support they will get). So there needs to be a much more nuanced discussion.

Katherine Trebeck

Immigration

Even when birth rates and death rates are equal a population will increase if net migration (immigrants minus emigrants) is positive. Such has been the case in the USA which, for over 100 years, has seen considerably more immigrants than emigrants. 'A stationary population is part of the definition of a steady-state economy, and in the U.S., Western Europe, and Canada, population growth is almost entirely due to net immigration' (Daly 2015d, p. 131). All countries try to control the number of immigrants they accept but the objectives of immigration policy can be confounded by illegal immigration. Since the turn of the century to 2018 the average number of legal immigrants to the USA exceeded 1 million per year. It was then reduced by discriminatory policies brought in by President Trump and virtually eliminated when immigration was halted when the Covid-19 pandemic began in 2020.

Emigration is less subject to control than immigration and often goes unmeasured. The USA for example does not keep records of emigrants but estimates of the number of US citizens living abroad show that the numbers are far below the number of immigrants, and even more so if illegal immigrants are included.

Daly began writing about immigration in the first edition of *For the Common Good* (Daly and Cobb 1989). The chapter was repeated without change in the second edition in 1994. Daly's position, shared by Cobb, was to 'favour continuing legal immigration close to its present volume of 600,000, although a gradual reduction may be necessary. But we also favour current efforts to gain control of our borders and to bring an end to illegal immigration.' Daly's contention, from which he has not wavered since, is that there are 'interests served by generous immigration or tolerance of illegal aliens. The supply of cheap, docile labor is increased. Unions are weakened. Wages decline and profits rise. Capital benefits from free migration just as it does from free trade. It is our laboring class that pays the bill for the generosity of the capitalists who want to let poor immigrants in (ibid., p. 247).

One reason that Daly gives for 'maintaining a generous quota for legal immigration is that the United States has a responsibility to refugees, especially if U.S. actions helped to create the situation from which people are fleeing' (ibid.). He gives the examples of U.S. foreign policy in El Salvador and Nicaragua:

> In Nicaragua [U.S.] national policy has been to take all possible actions to destabilize the Sandinista government. To this end the United States has done what it can to disrupt the economy so as to cause hardship and dissatisfaction with the government. It also recruits and finances rebels against the government.
>
> *ibid.*

Twenty-five years later the same policies were being used to destabilize the government of Venezuela, and the US military had intervened directly in Afghanistan, Iraq, Libya and Syria.

Global climate change has provided a new impetus for the large-scale relocation of people. As the country responsible for the largest contribution to the accumulated atmospheric greenhouse gases that cause climate change, the USA, following Daly's argument, has a special responsibility to admit large numbers of people whose living conditions have deteriorated as a result. However, Daly questions whether immigration to the USA is the best way to meet this obligation. He wonders if it would be better fulfilled though assistance along the lines of the collective post-WWII Marshall Plan, supplemented by individually sent CARE packages? He also notes that China, Japan and Russia are all significant contributors to global greenhouse gas emissions but are insignificant havens for refugees or economic immigrants. So, Daly contends, there is a political problem of international burden sharing (Daly, personal communication).

> Daly has identified '…three fundamental philosophical divergences that contribute to the difficulty of immigration policy… First, there are differing visions of world community.' He contrasts a view of open borders – believed by some to be the key to world peace – with his preferred one of a federation of interdependent nations 'as the road to true global community'.
>
> *Daly 2015d, p. 131*

Second is the ethical divergence between deontologists, who think that choices should be grounded in moral principles, and consequentialists, who think that choices should be judged by the state of affairs that result. Daly places himself in the second camp, though he recognizes the weaknesses of both.

> The difficulty with deontology is that in its devotion to absolute rules of rightness, it is sometimes blind to foreseeable evil consequences. The difficulty with consequentialism is that it can be wrong in its prediction of consequences, especially if they are complex or far in the future. It can also be too willing to employ bad means in the service of good ends.
>
> *ibid.*

Daly uses Garett Hardin's 'agonizing parable' of a lifeboat that only has room for a few more passengers but large numbers, swimming around the lifeboat, want to come aboard. Should those in the lifeboat allow all of them to come aboard which would sink the lifeboat, bringing 'complete justice, complete disaster', or should they keep them out and watch them drown as they save themselves which is 'unjust – survival of some at the sacrifice of others'? There is no obvious right answer to this question. For one thing, it raises other questions. Did everyone have an equal opportunity to get into the lifeboat? Do they all have an equal right to be there? Are all lives of equal value? Does this kind of circumstantial information matter? Daly avoids such questions by focusing instead on what can be learned from the parable in its starkest form:

> The lesson of the parable is that we must by all means avoid the trap in the first place – provide more lifeboats on the ship, take fewer passengers on each cruise, and go more slowly through safer passages. That is what a steady state economy seeks to do.
>
> *ibid.*

Daly's 'third philosophical divide is [between] the conception of people either as isolated individuals, or as persons-in-community. Are we independently defined atomistic individuals related to each other only externally? Or is our personal identity itself constituted by internal relations in community?' (ibid.). We considered this question in some detail in Chapter 8. Here it is sufficient to note, as Daly does, that individualists (most economists) concern themselves primarily with interactions among people mediated anonymously by the market. They pay little attention to 'relations in community, provided by families, places, nations, traditions, religious communities, language, culture – relations that largely define our identity as persons' (ibid., p. 132). These different conceptions of people lead to different positions on immigration. The individualistic conception emphasizes costs and benefits to migrants as individuals and the person-in-community conception favoured by Daly emphasizes costs and benefits to the communities from which immigrants come, such as the loss of educated and skilled workers, and where they settle, such as requirements for support and pressure on available services.

Daly provides no easy answers to the many dilemmas that large-scale immigration presents but he does outline a way forward. The first step is to distinguish 'legal from illegal immigrants, and economic migrants from refugees. Lumping them all together as "immigrants" and then referring to any policy of border control or selectivity as "nativism" or "bigotry" is unhelpful, unfair, and unrealistic' (ibid.). Next Daly says that we need to 'better distinguish winners and losers resulting from mass immigration, and to share the costs and benefits more equitably among them' (ibid.). Reflecting on the fourfold increase in population in his own lifetime Daly says we should:

> realize that the growth economy in the U.S. and other countries, protected by militarism, is contributing to wars over remaining resources and ecological space, and will increase the numbers of people displaced by war and by ecological destruction… Growth can no longer substitute for living within, and equitably sharing, the carrying capacity of Creation. And erasing national borders, while it has some appeal, unfortunately leads to the tragedy of open access commons, which is a bad way to share.
>
> *ibid., pp. 132–133*

In 2004 at the biennial conference in Montreal of the International Society for Ecological Economics which Daly had co-founded 15 years previously, a symposium was held on 'the global governance challenge of global economic integration and immigration: do environmental concerns have a role to play?' Daly was unable to attend the symposium, but he submitted two short papers that helped fuel the discussion. A special edition of *Ecological Economics* included papers presented at the symposium, one by Daly, and others from invited contributors. The editors of the special edition described Daly's paper as 'an important source of inspiration for the debate' (Muradian, Neumeyer, Røpke 2006, p. 185). In it, Daly set out his views on the relationships among population, migration and globalization along the lines covered in this chapter, and with respect to globalization and incomes, discussed more fully in the next. Inge Røpke, one of the organizers of the symposium and an editor of the special edition in *Ecological Economics*, agreed with much of what

Daly has to say about population and migration in the context of sustainability (Røpke 2004). However, she did not think that population policies in poor countries are influenced much by the possibilities for emigration and the expectation of remittances as Daly had suggested. Røpke has some sympathy for Daly's concern that, because of global competition, immigrants to rich countries reduce wages, working conditions and environmental standards, but she contends that these effects of immigration, as long as not of 'enormous proportions' can be counterbalanced by complementary measures to soften their impact. Where Daly thinks that 'the best option would be to roll back globalization – restrict capital flows, reduce migration, etc. – and return to a system with severe national restrictions' (Ropke 2004, p.193), Røpke thinks rolling back history will be very difficult and new forms of international governance have more promise. She also thinks that there is little support for turning back globalization, though with the experience of the Covid-19 pandemic where even the richest countries discovered the vulnerabilities of relying on long supply chains, this may change.

A more determined critique of Daly's position on immigration comes from Giorgos Kallis in his book *Degrowth* (2018). He begins by challenging Daly's use of the lifeboat analogy which he says is false, there being only one lifeboat, planet earth. What happens in each country's "boat" is irrelevant… [since]… immigration is simply a reallocation of the "cargo" to different parts of the ship' (ibid., p.184). Irrelevant to whom, Kallis does not say. Kallis rejects the idea that the USA 'or any other national "boat" is overloaded… [since] ecosystems do not follow national boundaries (ibid). Countries trade with other countries and their ecological footprints are distributed around the globe.

Kallis claims that, because of changes in values and expectations, the movement of people from high to low fertility countries will reduce population growth faster than relying on a demographic transition at home. But this change in values and expectations may also raise consumption levels such that overall environmental impacts increase.

These criticisms of Daly's use of the 'lifeboat analogy' overlook his main purpose in using the parable (Daly's term) which was to distinguish between deontology and consequentialism as ethical frameworks leading to different positions on immigration policy. Where Daly leans in the direction of consequentialism, Kallis leans towards deontology. They are leanings only because both Daly and Kallis draw on both philosophical positions in their discussions of immigration. For example, Kallis continues his criticism of Daly by saying that the USA and Europe, who benefited from colonial exploitation, should accommodate people from regions most affected by colonial exploitation. This is similar to Daly's view that the USA has a responsibility for refugees especially if they are fleeing from conditions that US actions helped create (Daly and Cobb 1989, 1994).

On the other hand, when Kallis says 'there is no evidence that immigration boosts an otherwise steady or contracting economy' and that 'new data… finds that refugees have little or no impact on the wages of average native workers and no large detrimental effects on less educated workers' he is criticizing Daly based on a disagreement about consequences. At the same time Kallis agrees with Daly on another consequence, that 'low-income groups may see their wages fall due to competition from migrants' (ibid., p.185). Despite Kallis's accusing Daly of 'gross oversimplification' they would both seem to agree that 'immigration might have good or bad social and economic consequences depending on the context' (ibid.).

We can dismiss Kallis's contention in his book (ibid, p. 186) that Daly confuses migrants with refugees in one of his papers (Daly 2015b). Daly was commenting on a response by the Swedish Minister of Immigration to a question about how many immigrants Sweden should accept and was making the point that Sweden, like other rich countries, looks to economic growth to solve any problems relating to immigration. The Minister did not distinguish between immigrants and refugees so neither did Daly in his comment, but further down the same page on which Kallis based his criticism Daly very clearly makes the distinction when it is germane to the point he is making (ibid., p. 130).

The practical differences on immigration policy between Kallis and Daly arising from these criticisms turn out to be less than might be expected. Kallis is not in favour of open borders, 'a nation, like any social system, needs to assert controls over the people or goods that enter its territory' (op cit., p. 186). He also cites Daly's view in support of the proposition that under capitalist competition, open borders 'may lead to a post-national feudalism' (ibid.). Kallis's real objection to Daly on immigration is to 'life-boat ethics' for which he claims 'there is no ecological-economic basis…' (ibid., p. 187). For the rest, his own description, perhaps made in haste, is that they are 'my personal, half-baked thoughts and questions on a topic that I have not researched' (ibid.).

It is a moot point whether Daly's use of the life-boat parable to differentiate between deontology and consequentialism is an endorsement of Hardin's 'life-boat ethics'. Lifeboat ethics is not a term that Daly uses in his discussion of ethics in relation to immigration or of any other issues which have concerned him over the years. It comes from Garret Hardin, a biologist who wrote the highly influential paper 'The Tragedy of the Commons' (Hardin 1968). Hardin explored a class of problems for which there are no technical solutions which require 'a change only in the techniques of the natural sciences, demanding little or nothing in the way of change in human values or ideas of morality' (ibid., p. 1243). He discussed many such problems but the one most relevant here and the one of greatest interest to him was population. He argued that the result of people choosing how many children to have based only on the benefits and costs to them personally and without regard to the depletion and destruction of shared resources of all kinds, would result inevitably and tragically in overpopulation. It was in this context that Hardin introduced the idea of lifeboat ethics in a subsequent paper where he argued against helping the poor on the grounds that doing so will only make matters worse (Hardin 1974).

Even though Daly admired Hardin who was a personal friend, there is no evidence in any of Daly's writings that he endorsed Hardin's lifeboat ethics. He used the lifeboat parable to raise ethical questions that arise from excessive population, for which there are no easy answers, and which he believed are best avoided by moving towards a steady-state economy and stable population. When Daly referred to the tragedy of the commons, as he often did, he made it clear that he was talking about open access commons (e.g. Daly 2013a). If access to the commons is restricted by any of a wide variety of social arrangements, not just private property and markets, the tragedy that so exercised Hardin can be avoided.

Another matter about which there is some confusion is the significance of Daly's membership of the Advisory Board of the Carrying Capacity Network (CCN), an NGO originally founded by Hardin when it was known by a different name. Daly joined the Board at Hardin's invitation but subsequently resigned because he objected to the director's anonymity on the CCN's website and his awareness, drawn to his attention by a letter from the Degrowth Research group, of political views with which he did not agree.

None of this would be very important except for the fact that Kallis and others in the degrowth movement who admire Daly's critique of economic growth and proposals for a steady-state economy, have criticized Daly for adopting lifeboat ethics and for his involvement with CNN. This has placed a wedge between Daly and themselves that should be removed.

Laudato Si'

In 2015 Pope Francis released *Laudato Si'*, an encyclical letter 'On the Care for Our Common Home' (Pope Francis 2015). It is a remarkable document whether seen through the eyes of Catholics or those of anyone else who cares for planet Earth and our place within it. In *Laudato Si'*, the Pope covers many of the topics that Daly has written on for decades: economic growth, over consumption, inequality, environmental degradation, and he does so in a way that reveals Daly's influence. 'Francis… skates fairly close to the idea of steady-state economics, of qualitative development without quantitative growth in the physical scale of the economy…' (Daly 2019c, p. 77). In his commentary on Laudato Si' Daly describes himself as a 'Christian of Protestant persuasion' and so he is quite comfortable talking about 'God's creation'. He says of Pope Francis: 'At a minimum he has given us a more truthful, informed, and courageous analysis of the environmental and moral crisis than have our secular political leaders' (ibid., p. 76).

The one area where Daly thinks that *Laudato Si'* falls short is in the lack of 'attention given to population and its consequences' (ibid.). To make his case, Daly draws on all that he had previously written on population – of humans and artifacts. He points out that

> growth of the human footprint that displaces other parts of God's creation is measured by total resource use… which of course is the product of per capita

resource use times population. Creation care, limiting the human footprint, requires serious attention to both factors.

ibid., p. 78

This contrasts with *Laudato Si'* which states,

> to blame population growth instead of extreme selective consumerism on the part of some… is an attempt to legitimize the present model of distribution, where a minority believes that it has the right to consume in a way which can never be universalized, since the planet could not even contain the wastes of such consumption.

op cit., p. 36

The fault that Daly would no doubt find with this rebuttal is that it is not a question of population growth *instead* of a more equal distribution of incomes and wealth, both are important, both require attention.

Daly does not set himself the task of resolving the Catholic church's position on fertility control, but he does offer a suggestion

> that might help fit population into *Laudato Si* [which is] to proclaim at the outset that more people (and other creatures) are better than fewer – but only if they are not all alive at the same time! Sustainability means longevity for creation – more people, in the company of other creatures, enjoying a sufficient level of consumption for a good life over more generations – not more simultaneously living high-consuming people elbowing each other, and God's other creatures, off the planet.

ibid., p. 83

Conclusion

Population and immigration are difficult topics. They touch on human rights, ethics, morality and values, as well as on biology, ecology, politics and economics. It is no wonder they are contentious and prone to sharply different views. Many people shy away from discussing them, fearing charges of racism and privilege. Others may admit that population and immigration are problems but insist that economic growth provides a more than adequate solution. Daly is different. He understands that population size multiplied by affluence measured as average income per capita, multiplied by environmental impacts per dollar of income determines the impacts of humans on the environment. This is a simple mathematical identity that must be true.[6] It is useful for examining the relative contributions of changes in population, affluence and technology on impacts. These contributions can differ over time and space. Increases in affluence rather than population have come to dominate the environmental impacts of rich countries. In poorer countries faster growing population rather than increasing affluence tends to dominate. Daly contends that all three variables warrant attention: population, affluence and technology as do their interdependencies. It is a mistake, as Daly tells us, to neglect any of them.

Notes

1 This relationship can be expressed in a very simple equation:

where:

r = annual per capita rate of growth of real output

s = annual savings rate (capital formation) as a percentage of annual output

k = marginal capital-output ratio

p = annual rate of population growthIn the Harrod-Domar framework total income growth rate is given by s/k, the savings rate divided by the marginal capital-output ratio.

2 $$\frac{Y}{P} = a1\left(\frac{Yp}{Pc}\right) + a2\left(\frac{Yp}{Pn}\right) + a3\left(\frac{Yw}{Pc}\right) + a4\left(\frac{Yw}{Pn}\right)$$

where:

$a1, a2, a3, a4$ are the proportions of each social class defined by income and fertility control in the population.

Daly also considered using the proportion of total income going to each class in place of the proportion of the population.

3 The birth rate is the number of births in a year per 1,000 of the population. The fertility rate is the number of live births in a year per woman of reproductive age in a population.

4 Interest in a great transition is even stronger today. See for example the online forum The Great Transition Initiative (greattransition.org)

5 www.macrotrends.net/countries/USA/united-states/fertility-rate

6 It is known as the IPAT equation: $I = P \times A \times T$

where: I = human impact on the environment

P = population

A = Affluence (income per capita)

T = Technology (impact per $ of income)

12

MONEY AND BANKING

> Although real wealth cannot grow exponentially for long, our cultural symbol and measure of wealth, money, may indeed grow both exponentially and indefinitely. This lack of symmetry in behavior between the reality measured and the measuring rod has serious consequences.
>
> *Daly*

When the global financial crisis struck in 2007 most people thought it had something to do with the banks, the US banks in particular, and they were right. The deregulation of the banking system in the USA and other countries had broadened the range of activities in which banks could engage and reduced a number of regulatory requirements which allowed them to take on greater risks. Instability in the financial sector created instability and a deep recession in the rest of the economy in ways rather like those described by economist Hyman Minsky in *Stabilizing and Unstable Economy* (Minsky 1986). Was this a 'Minsky Moment' the pundits wondered, when markets collapse after a prolonged period of reckless speculation? In the USA the housing sector was especially hard hit – one risk factor being careless mortgage lending – which was bad for business and even worse for the many families who lost their homes.

Very few economists forecast the financial crisis and its outfall in terms of unemployment, displacement of people and widespread damage to whole communities. The few who did had in common an appreciation of the intimate relation between finance, accounting and economics (Bezemer 2009). This was in contrast to the vast majority of economists who did not, who were relying on 'equilibrium models ubiquitous in mainstream policy and research' (ibid., p. 1). All too many of these economists were working with models of economies incorporating little or no representation of the financial sector even though its influential presence in the real world had already become quite apparent. Daly was not one of these

DOI: 10.4324/9781003094746-12

economists. It was not his style to work with the kind of macroeconomic models favoured by econometricians or to rely heavily on mainstream macroeconomic theory, both of which lacked adequate links to the environment and disregarded the first and second laws of thermodynamics.

Yet for some considerable time before the financial crisis of 2007/08 one aspect of the macroeconomy that had attracted Daly's attention was money: what it is, how it is created, and why the banking system needed to be subject to much greater control to avoid excessive risk-taking causing crises. He also thought that stricter lending requirements would bring the added benefit of supporting local communities because more consideration would be given to their specific circumstances when extending credit.

The strange case of Professor Soddy

Daly's interest in money and banking was sparked by the work of Frederick Soddy, a British Nobel Laureate in Chemistry in 1921, who turned his attention to economics. One day in the mid-1970s Daly was browsing in the stacks at the Louisiana State University library. As he cast his eye along the shelves of books by authors with names beginning with 'S' he came across a slender volume entitled *Cartesian Economics, The Bearing of Physical Science upon State Stewardship* (Soddy 1922). The book consisted of two lectures given by Soddy in 1921 to students at Birkbeck College and the London School of Economics. Economists did not respond well to being told by an outsider what was wrong with their understanding of economies and Soddy's attempt to do so met with disdain from the outset. Henry Higgs who reviewed *Cartesian Economics* in *The Economic Journal* likened Soddy the chemist to Robert Louis Stevenson's Dr. Jekyll and Soddy the economist to his Mr. Hyde: 'It is sad to see so distinguished a physicist transformed into a pitiable purveyor of economic fallacies' (Higgs 1923, p. 101). Reviewers of Soddy's later work on economics, with the notable exception of Frank Knight, were no less dismissive which explains why a new generation of economists, Daly included, had never heard of Soddy.

Daly's reaction to Soddy's foray into economics was quite different. At the time of his serendipitous discovery, Daly was one of very few economists who thought that the laws of thermodynamics were relevant to economics, something he had learned from Georgescu-Roegen and Boulding in the 1960s. You can imagine his surprise when he opened the pages of *Cartesian Economics* and read, 'The starting point of *Cartesian economics* is thus the well-known laws of the conservation and transformation of energy, usually referred to as the first and second laws of thermodynamics' (op cit., pp. 4–5). How well known these laws were to economists in the 1920s is unclear. What we do know is that then as now (with a few exceptions) they did not see the relevance of these laws to economics. Soddy's aim was to put that right.

Coming to economics from chemistry Soddy located its subject matter between 'the electron and the soul… concerned rather with the interaction, with the middle world of life of these two end worlds of physics and mind' (ibid., p. 6). This is highly reminiscent of Daly's Ends-Means Spectrum that framed so much of his work and was bound to resonate with Daly who had come upon it quite independently. Soddy also anticipated Shroedinger's explanation of life in thermo-dynamic terms as discussed by Daly in his 1968 paper 'On Economics as Life Science' (see Chapter 4).

> Life derives the whole of its physical energy or power, not from anything self-contained in living matter, and still less from an external deity, but solely from the inanimate world. It is dependent for all the necessities of its physical continuance primarily upon the [thermodynamic] principles of the steam-engine. The principles and ethics of human law and convention must not run counter to those of thermodynamics.
>
> *Soddy 1922, quoted in Daly 1980c, p. 473*

Daly found the last sentence of this quote especially significant 'because it provides the basis for many of Soddy's criticisms of the economy as a presumed perpetual motion machine' (ibid.) exactly as he himself had begun to do before he discovered Soddy and forever afterwards. Soddy's account of economics in thermo-dynamic terms anticipated those of Georgescu-Roegen and Boulding by over 40 years. Apparently, neither of them was aware of Soddy's work on economics although Boulding did attend Soddy's lectures on chemistry at Oxford but, as he confessed to Daly, had slept through them.

What deeply concerned Soddy and what brought him to the study of money and banking, was the observation that unlike the physical world which is governed by the laws of thermodynamics, in the economic world – money and debt in particular –

> are subject to the laws of mathematics rather than physics… Unlike wealth… debts do not rot with old age and are not consumed in the process of living. On the contrary, they grow at so much percent per annum, by the well-known mathematical laws of simple and compound interest.
>
> *Soddy 1926, p. 70*

Soddy traced many of the ills of society to the tensions that arise from an economic system in which financial claims on current and future production in the form of debt rise exponentially, outpacing any conceivable increases in physical production. In Daly's words,

> Debt can endure forever; wealth cannot, because its physical dimension is subject to the destructive force of entropy. 'Since [real] wealth cannot continually grow as fast as debt, the one-to-one relation between the two will at some point be broken – i.e. there must be some repudiation or cancellation of debt. The positive feedback of compound interest must be offset by

counteracting forces of debt repudiation, such as inflation, bankruptcy, or confiscatory taxation, all of which breed violence.

Daly, op cit., pp. 475–476

The idea that financial claims on the real output of an economy can exceed, by some considerable amount, the value of that output at current prices was of profound concern to Soddy and to Daly. Neither of them brought empirical data to bear on the disconnect between financial claims and what an economy is capable of producing. While it is not entirely clear how that can best be done, one way is to compare the total annual income in a country with the total value of all financial assets owned by individuals and institutions. That would show that the value of all the financial assets created by the financial system far exceed the value of the annual output of the economy.[1] According to Soddy, and Daly, the financial assets represent a claim on real output that could not be met if all those who own these assets tried to liquidate and spend them at the same time. This is cause for concern, even allowing for the fact that some of these financial assets, such as pensions, are a claim on future rather than current output.

Another perspective on the issue raised by Soddy of financial claims on output far exceeding what is available was amplified by Daly with a quote from Nobel Laureate economist James Tobin. Tobin said that

the community's wealth now has two components: the real goods accumulated through past real investment and fiduciary or paper "goods" manufactured by the government from thin air. Of course, the non-human wealth of such a nation really consists only of its tangible capital.

Tobin 1965, quoted in Daly ibid., p. 479

This is made crystal clear from an examination of a country's national balance sheet which shows that total financial assets equal total financial liabilities so that net financial assets in the economy are zero. The country's net worth from the standpoint of its national accounts reside exclusively in its non-financial assets or real capital, comprised of those that have been produced and those such as land and natural resources that are considered gifts of nature. (See, for example, Statistics Canada 2020 which shows Canada's net financial assets equal to zero.)

It was the neglect of physical principles in economists' understanding of an economy and the unrecognized clash between what was actually possible with the human conventions that guide its production, distribution, and use that led Soddy to money, 'the topic which most occupied Soddy's attention' (Daly ibid.) and in whose footsteps Daly would follow. Indeed, Daly was so impressed with Soddy's work that he wrote a paper on the 'Economic Thought of Frederick Soddy' (Daly 1980c) in an attempt to make Soddy's ideas more widely known. Initially, the paper was rejected by the journal *History of Political Economy*. In response Daly wrote to the editor objecting to the referee's reasons that Soddy's ideas were inconsistent with orthodox monetary theory. Daly explained that he was only trying to present the thought of Frederick Soddy and was not trying to degrade or disprove him on the basis of what current economists think. Happily, for Daly, the editor reconsidered, and the paper was published.

Forty years later Daly co-authored a paper with economists Ansel Renner and Kozo Mayumi on 'the dual nature of money' (Renner, Daly and Mayumi 2021). The dual nature of money refers to the idea derived from Soddy that 'money can be understood both as a form of wealth from an individual perspective and as a source of biophysical debt from a communal perspective (ibid., p. 6). They distinguish between the functional and structural components of material objects. With the inevitable decay of the structure of material objects their capacity to perform their function also decays. Money, they emphasize, is different. '*A remarkable peculiarity of money derives from the fact that the functional component of money can survive independently of its structural constituent* [which may be nothing more that bytes in a computer's memory]' (ibid., p. 11, italics in original). Consequently, there is a disconnect between financial debt on which interest is paid and which endures and the physical goods that the debt represents which inevitably deteriorate. The result is an accumulation of lasting personal wealth from the perspective of those that hold the debt and an accumulation of biophysical debt from the perspective of the community to be repaid with real economic output which can be both inequitable and destabilizing. New institutional arrangements for money are required to avoid this problem.

Full reserve banking

Daly's interest in money had begun before he discovered Soddy. One of his first publications was about the shortage of change – coins and notes of low denomination – in Northeast Brazil. Daly attributed it to a combination of inflation which had reduced the value of metal coins so that people switched to low denomination paper money which wore out with frequent use and, in rural areas especially, were not replaced fast enough with new ones (Daly 1970a).

Even earlier, Daly had written about the proliferation of banks in Uruguay during a time in the late 1950s and early 1960s of economic contraction (Daly 1967b). He explained this phenomenon through a combination of 'legally sanctioned collusion in restraint of price competition [among banks]… the maintenance of the loan rate above the rate of inflation' and opportunities for the banks to make profitable investments in real estate (ibid., pp. 92–93). He noted the less obvious effect of the Uruguayan banking system in redistributing 'economic power from the public to the private sector' and concluded that 'If Uruguay is to recover from its economic ills, and particularly if it is to do so through government planning, as it intends, then the money doctors must prescribe some effective treatment for the financial tumor' (ibid., p. 94).

One prescription that Daly came to favour, not just for Uruguay but for all countries with private banking systems, was the introduction of full reserve banking. This would mean a legal requirement that commercial banks maintain 100% reserves preventing them from lending demand deposits and only allowing loans from time deposits. This is in contrast to fractional reserve banking where commercial banks maintain only a small fraction of reserves as cash. Soddy had proposed full reserve banking in 1926 (Soddy 1926), as had the influential American economist Irving Fisher in 1936, though without acknowledging Soddy. It is not widely understood that the coins and paper money that are manufactured for the central bank (i.e., the Federal Reserve in the USA and the Bank of England in the UK) and put into circulation represent only a minor fraction of all the money in a modern economy. As the Bank of England explains:

> most of the money in the economy is created not by printing presses at the central bank, but by [commercial] banks when they provide loans... if you borrow £100 from the bank, and it credits your account with the amount, "new money" has been created. It didn't exist until it was credited to your account.
>
> *Bank of England, n.d.*

The point here is that banks do not hold cash in reserve sufficient to cover the money they create through loans. Far from it. In some countries, the USA for example, large banks are required to have a cash reserve of 10% of their deposits. (The reserve requirement is lower for smaller banks.) As Daly and Farley explain, if someone deposits $100 cash in a single monopoly bank that is just meeting its 10% reserve requirement, the bank can increase its loans by up to $900. The bank would now have deposits totalling $1,000 backed by $100 cash as reserves. The same expansionary result happens with many banks as with one monopoly bank, due to a chain of relending and redepositing excess reserves (Daly and Farley 2011, p. 293).

Since all banks operate in much the same way, the result is that if everyone went to their bank to withdraw all the funds in their accounts in cash, they would be gravely disappointed.[2] It follows from this that

> if banks had to keep 100% reserves against the demand deposits they create, then there will be no creation of money. Therefore, the reform called for by Soddy, Fisher, Knight, and others was for a 100% reserve requirement on demand deposits.
>
> *ibid., pp. 292*

Daly and Cobb endorsed full reserve banking together with Soddy's two other proposals: a policy of maintaining a constant price level [by careful control of the money supply by government], and freely fluctuating international exchange rates (op cit., p. 428). Soddy thought that the money supply should be strictly controlled to ensure that the value of the currency would remain fixed, paralleling the measurement units used in the physical sciences. Today central banks typically 'aim to make sure the amount of money creation in the economy is consistent with low and stable inflation' (McLeay, Radia and Thomas 2014, p. 14), which is a step towards the second proposal. Freely fluctuating exchange rates have become common in the post-Bretton Woods era which was Soddy's third proposal.

When Daly and Cobb decided to write a second edition of *For the Common Good* they did so for two reasons. One was to provide a much more detailed treatment of the Index of Sustainable Economic Welfare (see Chapter 6) and the other was to add an Afterword on Money, Debt and Wealth, a subject that was omitted from the first edition. The Afterword, written by Daly, made the case that, 'Without fundamental changes in the current thinking about finance, and in the institutions that embody that thinking, the other changes we propose may prove insufficient' (op cit., p. 407). Much of the analysis in the Afterword is based squarely on Soddy though Daly makes clear that he does not agree with Soddy that 'nearly all economic problems would yield to reform of finance' (ibid., p. 419). Many other reforms, as detailed in *For the Common Good* and discussed in various chapters of this book, are also required.

Another aspect where Daly does echo Soddy is in the relationships between community and finance. Daly's view, borrowed from Soddy, is that:

> Money should not bear interest as a condition of its existence, but only when genuinely lent by an owner who gives up its use while it is in the possession of the borrower. When the commercial banking system lends money it gives up nothing, creating the deposits *ex nihilo* up to the limit set by the reserve requirement. [or where there is no legislated reserve requirement as in Canada, then to the limit set by the bank's lending policy]. Unlike an individual, when a bank lends money it does not abstain from spending that money for the duration of the loan. The burden of abstinence is shifted on to the public.'
>
> *ibid., p. 426*

It is through this mechanism of money creation that the community is the real lender rather than the bank that made the loan since 'the community has abstained from the use of real assets, making them available to the bank's borrower in exchange for the money created by the bank and loaned to the borrower' (ibid.).

'Seigniorage' is the difference between the face value of money and the costs of creating it. Another concern of Daly's relating to fractional banking is the transference of seigniorage from the government (historically the monarch) to commercial banks which now create most of the money in the economy. It costs them virtually nothing to create it, just a few keystrokes on a computer, and yet they are able to charge interest for its use.

> Private bankers are able to lend over 90% of the virtual wealth of the community, which does not really belong to them, and earn interest on it, which

does not belong to them... It is important to recognize that money and virtual wealth[3] are social phenomena that arise not from the mere aggregating of atomistic individuals, but from the community consensus and resulting practical general willingness to accept the agreed-upon token as money... the essence of money is that it be generally accepted as such within a community, and acceptance of the same monetary standard becomes one of the defining bonds of community.

ibid, p. 422

Proposals to end, or seriously curtail, the ability of commercial banks to create money have been widely discussed in recent years. For example, the aim of the International Movement of Monetary Reform with membership of non-profit organizations in some 30 countries is 'Money creation free of debt by a public institution in the public interest'. It is a clear expression of Daly's position. Even more telling is the paper by Benes and Kumhof, two economists at the International Monetary Fund (IMF) who examined the claims made by Fisher (1936) for full reserve banking: '(1) Much better control of a major source of business cycle fluctuations, sudden increases and contractions of bank credit and of the supply of bank-created money. (2) Complete elimination of bank runs. (3) Dramatic reduction of the (net) public debt. (4) Dramatic reduction of private debt, as money creation no longer requires simultaneous debt creation' (Benes and Kumhof 2012, p. 1). Using a simulation model of the U.S. economy Benes and Kumhof found support for all of Fisher's claims though this finding has been contested by other economists.

BOX 12.1

I strongly agree with and support everything Herman has written about except his writings on money, although this is a more recent view of mine having totally agreed with Herman in the past. I have come to believe that it is the level of spending, not the size of the money supply, that needs to be in keeping with the sustainable productive capacity of the economy.

Philip Lawn

The most direct approach to strengthening communities through a reform of the system of money and banking is the creation of local currencies. The essential idea of a local currency, which complements rather than replaces the national currency, is that it stays within the community that created it, stimulating local employment and business. Local currencies were mentioned in a footnote in *For the Common Good* but received more attention in Daly's textbook with Farley though more for expository purposes than as a key component of Daly's proposals for monetary reform. Of greater interest to Daly is the 'special threat [to the common good] from the international nature of finance' (op cit., p. 437). The fact that 'money flows around the world far more freely than labor, or means of production or even products... [makes it impossible] to develop stable regional or national economies if regions or nations have no control over the movement of money' (ibid.). Daly sees this as especially threatening to Third World countries that became indebted to banks in the USA and other Northern Hemisphere countries in the 1970s.

> When it became clear in the 1980s that earnings from these "investments" would not suffice for repayment, creditors began to pressure multilateral development banks to speed up their lending to the South in order to provide them with the foreign currency necessary to pay their debts to northern commercial banks.
>
> *ibid., pp. 438–439*

Daly, who had an insider's view of the World Bank for six years, says this is one of the reasons for the shift from project-based lending to policy-based lending. Project-based lending requires careful pre-approval assessment which takes time. Policy-based lending is contingent on the adoption by the borrowing country of 'structural adjustment' policies of market liberalization, reduced taxation and privatization is less scrutinized and faster. 'The World Bank should focus more on investing in natural capital restoration, in resource efficiency, and in domestic production of basic goods and services for local use under local control' (ibid., p. 441). The kind of national banking system most amenable to this kind of approach, and not just in developing countries, would be one with separate institutions for 'specifically banking functions of safekeeping, checking, and clearing, from the investment function [and for] ... small investment trusts as complements to the banks' (ibid.), as proposed by Henry C. Simons, founder of the Chicago School of Economics. Daly bemoaned the fact that Simons' ideas, which expressed 'our own idea of the financial requirements of an economics for community, [were] forgotten in the subsequent rise of extreme individualism associated with the current Chicago School' (ibid., p. 442).

Daly's proposals for reforming money and banking reform, for full reserve banking in particular, have not attracted much attention even among ecological economists. One exception is Inge Røpke at the University of Aalborg in Denmark. Røpke considered it paradoxical that full reserve banking finds favour with ecological economists, yet they seldom refer to Daly's basic arguments which go back to Frederick Soddy, notably the incongruity between limits on the rate of economic growth due to the laws of thermodynamics and the capacity of monetary debt to increase ad infinitum (Røpke 2017).

In her assessment of the contribution that full reserve banking could make to sustainability Røpke recognized that full reserve banking 'is used as a heading for several proposals'. Since the 'sovereign money system' proposed by Positive Money, a UK non-governmental organization, is popular among environmentalists Røpke chose to focus her attention on it including criticisms it has received from

Post-Keynesian economics. Although the Positive Money proposal shares with full reserve banking the aim of preventing commercial banks from creating money when they make loans against reserves, the method is different. It has to do with the continuance of central bank money and commercial bank money in full reserve banking, and a single state-created means of payment under Positive Money's sovereign money proposal (Van Lerven 2016). It is unfortunate, therefore, that Røpke chose to assess the case for and against full reserve banking using a proposal that its originators consider to be something different. Nonetheless, her conclusions are interesting since they relate to Daly's thinking on full reserve banking even though Røpke based them on a somewhat different approach.

Røpke agrees with Daly that the availability of credit should be constrained and that banks should hold much larger reserves at the central bank. She also agrees that reserve requirements should be based on social rather than private risk, which implies high requirements in relation to loans for the purchase of existing assets, to limit speculation. Røpke also considers structural reforms to the banking system as important, such as a regulated separation between retail and investment banking and within investment banking as well. She writes that

> smaller and more localized banks may be useful for local economies... and from a Sustainability perspective, it could be useful to improve the conditions for values-based financial institutions, to establish green investment banks and funds, to strengthen sustainability in the mandates of pension funds and sovereign wealth funds, and to apply guided lending and quantitative regulations.
>
> *ibid., p. 188*

These are all supply side measures intended to reduce the availability of credit and/or direct it towards more environmentally and socially worthwhile uses. To reduce the demand for loans Røpke mentions the removal of tax incentives to borrow and to reduce the bias in favour of debt finance. Reducing the attractiveness of speculation in various assets is also important.

These proposals, which Røpke describes in more detail, reveal a greater affinity to Daly's proposals than perhaps Røpke realizes. One of her main conclusions is that 'financial reform must be seen in relation to the phases of capitalist development and potential new models' (ibid., p. 189). This is very similar to Daly's view that 'our national institutions governing money and finance are embedded in a culture which has come to accept exponential growth as the norm' (op cit., p. 408). He refers to Marx's historical analysis of the transformation from exchange based on barter to the exchange of commodities mediated by money. Both forms of exchange increase use value. This was followed by a further transformation from a system that increases use value to 'capitalist circulation' which increases exchange value.

> An initial capital... is used to hire labor and buy raw materials, which are then turned into a commodity..., which in turn is sold for a greater amount of money. The shift of focus from use value to exchange value is crucial.

Commodity accumulation and use values... are self-limiting... There is nothing to limit how much abstract exchange value one can own... In our own time this historical process of abstracting farther away from use value has perhaps been carried to the limit in the so-called 'paper economy' [now the digital economy]... the direct conversion of money into more money without reference to commodities.

ibid., pp. 409–410

At the level of specific policy proposals, the increased requirements for reserves which Post-Keynesians and Røpke support, point in the direction of full reserve banking even though they do not go as far. But this is not as different from Daly's position as it may appear. 'Obviously, the 100% mark could be approached gradually overtime by raising the reserve requirement a few percent a year' (Daly and Cobb, p. 434). This would allow learning to take place as reserve requirements are increased and the process could be halted if the warnings of the Post-Keynesians that it can become counter-productive turn out to be correct.

Daly is clearly in agreement with Røpke's proposals for: structural reform of the banking system, the separation of banking functions, smaller local banks that are better informed about the communities they serve, and the redirection of investment to strengthen communities, protect the environment and promote sustainability. Daly has discussed and promoted them all. Perhaps what is most telling in terms of Daly's influence on the debate about full reserve banking is Røpke's statement that,

Since the effectiveness of measures to reduce speculative credit creation depends on whether capital is mobile across borders, other measures such as taxing cross-border capital flows and constraining the use of tax havens would be necessary complements to reduce this mobility.

op cit., p. 189

No one has called more loudly and more often than Daly for reducing the international mobility of capital.

Conclusion

Daly and Cobb closed their Afterword on money, debt and wealth by describing their work 'not as a technical blueprint for detailed reform, but rather a few basic policy directions'. They invited 'experts to review our proposals... not only to point out errors, but to come up with something better' (op cit., p. 442). Daly can take some satisfaction from the many initiatives led by civil society towards monetary reform, such as Positive Money. He shares the concerns of all who are troubled by the continuing financialization of the economy – 'the increasing role of financial motives, financial markets, financial actors and financial institutions in the operation of the domestic and international economies' (Epstein quoted in Røpke 2017,

p. 181), and the rise of cryptocurrencies such as Bitcoin that are subject to no government control. It is clear that a great deal of work remains to be done to redesign the banking system so that it facilitates rather than impedes a transition to a steady-state economy.

Notes

1 Data for Canada gives the ratio of financial assets to GDP in 2019 at about 16:1.
2 The central bank is there to lend funds to any commercial bank that finds itself short of cash reserves. As long as the central bank can be relied upon to provide sufficient reserves, commercial banks can increase their loans knowing that reserves will be available to meet their reserve requirement.
3 The concept of virtual wealth introduced by Soddy is 'the total value of real assets that the community voluntarily abstains from holding in order to hold money instead. Since individuals can easily convert their money into real assets, they count their money holdings as wealth. Yet the community as a whole cannot convert money into wealth because someone has to end up holding the money. Money wealth is therefore "virtual"' (Daly and Farley op cit., p. 494).

13

GLOBALIZATION, INTERNATIONALIZATION AND FREE TRADE

> To make an omelette, you have to break some eggs. The dis-integration of the national egg is necessary to integrate the global omelette.
>
> *Daly*

Globalization has a long history, longer than most people realize. Historian Valerie Hansen says it started 1,000 years ago with the arrival of the Vikings in north-eastern Canada (Hansen 2020). Others say it started before then. For Europeans, globalization took a major leap forward when, in the Late Middle Ages, they scrambled to find long-distance trade routes to the Moluccan Spice Islands to supply the rising demand for spices. Later, in the sixteenth century, Dutch, British and French entrepreneurs broke the Portuguese monopoly in the spice trade by smuggling out seeds and growing spices in the tropics. There followed an aggressive expansion of international trade in other commodities and European colonization of the Americas, Africa and India with devastating impacts on indigenous peoples and their cultures.

Through successive scientific, technical, economic and political innovations the world contracted. Globalization historian Alex MacGillivray singles out five decades in which 'global contraction [has] been immediate and unmistakable': *Contraction 1. The Iberian carve-up (1490–1500)* when the shape and size of the globe finally became known and the Pope divided the globe between Spain and Portugal. *Contraction 2. The Britannic meridian (1880–90)* when the British Empire, the leading imperial power, was at the height of its global power and influence. *Contraction 3. Sputnik World (1955–65)* marked by conflict and competition between the USA (capitalism) and the USSR (communism) and symbolized by the space race and myriad technological spin-offs. *4. The global supply chain (1995-2005)* when supply chains became dominated by huge multi-national corporations trading east-west rather than north-south. *5. Thermo-globalization.* Writing in 2005 MacGillivray

DOI: 10.4324/9781003094746-13

could only describe this fifth 'planet-shrinking decade... [that]... could be another 20 or more years in coming' (MacGillivray 2005, pp. 18–21). Fifteen years on we can see how prescient he was:

> The economic cake will continue to grow, but not as fast as the mounting appetite of the world's people. In this decade, the first crumbs will be visibly picked from the plates of rich nations, and they won't like it. Social networks will reach the greatest feasible extent, and virtual consumerism will supplement diminishing commodities. The dominant fault-line will be drawn not by explorers, traders or diplomats but by climate change... The first four planetary contractions were largely unpredicted. What is different this time is that we already know that global contractions on a heating planet will present huge economic, social, cultural and environmental challenges.
>
> *ibid., p. 21*

It is in the context of this last contraction, Daly's full world, that his ideas about globalization are best appreciated. But what is globalization? Again, MacGillivray is helpful. He distinguishes between narrow economic definitions of globalization and broader social definitions. An example of the former comes from Paul Krugman, who says globalization is 'a catchall phrase for growing world trade, the growing linkages between financial markets in different countries, and the many other ways in which the world is becoming a smaller place' (quoted in ibid., p. 5). Joseph Stiglitz is even more specific: globalization entails 'the removal of barriers to free trade and the closer integration of national economies' (quoted in ibid.).

Finding these economically focused definitions of globalization too narrow, others, such as Manfred Steger offer a broader perspective: globalization is 'a multidimensional set of social processes that create, multiply, stretch, and intensify worldwide social interdependencies and exchanges...' (quoted in ibid.).

As we shall see, Daly's discussion and analysis of globalization and the related issues of free trade and capital mobility, fall somewhere between a strictly economic definition, though he has much to say about economics, and a broader definition that encompasses social and environmental dimensions.

One aspect of globalization that neither the narrow economic nor broader social definitions bring out is the difference between globalization and internationalization. Daly explains the distinction as follows:

> Globalization, considered by many to be the inevitable wave of the future, is frequently confused with internationalization, but is in fact something totally different. Internationalization refers to the increasing importance of international trade, international relations, treaties, alliances, etc. Inter-national, of course, means between or among nations. The basic unit remains the nation, even as relations among nations become increasingly necessary and important. Globalization refers to global economic integration of many formerly national economies into one global economy, mainly by free trade

and free capital mobility, but also by easy or uncontrolled migration. It is the effective erasure of national boundaries for economic purposes. International trade (governed by comparative advantage) becomes interregional trade (governed by absolute advantage). What was many becomes one.

Daly 1999c

Unlike others who call for a global government to govern a global economic system in the name and interests of a single global community, Daly is adamant that:

Global community must be built up from below as the federated community of interdependent local and national communities. It cannot be some single, integrated, top-down, ahistorical, abstract global club. Free trade, free capital mobility, and free migration do not create a global community, they simply destroy national community. Globalization is just neoclassical atomistic individualism writ large. Such globalization destroys the local historical relations in community by which persons produce for and take care of each other, and from which we might step-by-step federate into a global community of communities. ...

Daly 2014a, p. 160

As Daly observes in many of his writings, the main international institutions established at the end of World War II were founded on the idea of internationalization not globalization. This includes the United Nations, the International Monetary Fund (IMF), The World Bank, the General Agreement on Tariffs and Trade (GATT) and its successor the World Trade Organization (WTO). It was only later that the IMF, the World Bank and the WTO reinterpreted their mission to become vehicles of globalization. One feature of this expanded mission that deeply concerns Daly is the removal of constraints on the mobility of capital that were part of the original Bretton Woods agreement. Daly's contention is that the removal of these constraints undermines the very basis on which the mutual gains from international trade is based. If Daly is correct, then the presumed mutual gains from international trade on which much of the case for globalization rests are without foundation.

Comparative advantage and the gains from trade

'No economic doctrine is more widely accepted among economists than that of free trade based on comparative advantage.' So begins Chapter 11 in both editions of Daly and Cobb's *For the Common Good* (Daly and Cobb 1989, 1994, p. 209). Statements of this sort from other economists are easy to find. Steven Suranovic opens his overview of the theory of comparative advantage in much the same way. 'The theory of comparative advantage is perhaps the most important concept in international trade theory' (Suranovic 2010). He also recounts the story of the celebrated neoclassical economist Paul Samuelson who, when asked for an example of a non-trivial result from economics, immediately replied 'comparative advantage' (ibid.). Suranovic continues by saying that comparative advantage 'is also one of the most commonly misunderstood principles', a view shared by Paul Krugman, 'anyone who becomes involved in discussions of international trade beyond the narrow circle of academic economists quickly realizes that [comparative advantage] must be, in some sense, a very difficult concept indeed' (Krugman 1998, p. 22).

BOX 13.1

I was always impressed with his knowledge of the history of economics, his ability to draw on basic trade theory when economists were in denial of their own theory, and his ability to write on community. One cannot help but greatly admire Herman for his brilliance and tenacity.

Richard Norgaard

The first and still most widely used explanation of comparative advantage was by David Ricardo in 1817. He gave the highly simplified example of two countries, Portugal and England, each able to produce just two commodities, cloth and wine, using only labour. Ricardo assumed that in England the production of a unit of cloth requires 100 hours of labour and a unit of wine 120 hours of labour. He assumed that in Portugal a unit of cloth requires 90 hours of labour and a unit of wine 80 hours. In the absence of trade between Portugal and England, cloth and wine would be traded within each country according to the relative amounts of labour required to produce them. (Ricardo subscribed to the labour theory of value which assumed that prices were based on the labour required for production.) Since the relative amounts of labour required for producing cloth and wine is different in Portugal and England their relative prices would also be different in the absence of international trade. In England the price of 1 unit of wine before trade would be 1.2 units of cloth since these are the quantities that can be produced with 100 hours of labour. In Portugal the price of 1 unit of wine would be 1.125 units of cloth for the same reason.

When trade between the two countries is permitted things change. Instead of having to give up 1.2 units of cloth to get 1 unit of wine (by transferring 100 hours of labour from cloth production to wine production), England could offer Portugal anything greater than 1.125 units of cloth but less than 1.2 in exchange for 1 unit of wine. Such an exchange would obviously be to England's advantage. It would also be to Portugal's advantage to accept such an offer because on its own, Portugal can only obtain 1.125 units of cloth if it produces 1 unit of wine less and England is offering more than that.

The point which Ricardo makes with this example is that even though Portugal can produce both cloth and wine more cheaply in terms of labour hours than England it pays both countries to specialize in the production of a single commodity and trade for the other rather than to produce both commodities for themselves. The commodity in which each should specialize is determined by the one that it is most efficient in producing, wine in the case of Portugal, or the one it is the least inefficient in producing, cloth in the case of England. Portugal has an absolute advantage in the production of both commodities and a comparative advantage in the production of wine. England has an absolute disadvantage in the production of both commodities and a comparative advantage in the production of cloth. International trade said Ricardo is determined by comparative advantage, not absolute advantage, and it results in both countries gaining from trade.

'This argument is ingenious and, given the conditions it presupposes, no doubt true. In the real world, however, these conditions are not always met' (Daly and Cobb ibid., p. 213; Daly and Farley 2011, pp. 355–363). Daly is well aware that a similar statement can be made about any theoretical model. By their nature all models are simplifications of the phenomena they seek to describe and explain. This is true not just in economics but in the natural sciences too. Problems arise when a specific simplification crucial to the results obtained is at significant variance with reality. One simplification made by Ricardo that is essential for the principle of comparative advantage is that capital is immobile between countries. If it were not and capitalists in England could use their capital to employ labour in Portugal rather than in England, then they would do so if it increased their profits. 'Everything that differentiates domestic from international trade depends, for Ricardo, explicitly on the international immobility of capital (labour immobility between nations was taken for granted)' (op cit., p. 214). Ricardo understood this.

> The difference in this respect, between a single country and many, is easily accounted for, by considering the difficulty with which capital moves from one country to another, to seek a more profitable employment, and the activity with which it invariably passes from one province to another in the same country.
>
> *Ricardo 1817, p. 154*

If it were otherwise and

> capital freely flowed towards those countries where it could be most profitably employed, there could be no difference in the rate of profit, and no other difference in the real or labour price of commodities, than the additional quantity of labour required to convey them to the various markets where they were to be sold
>
> *ibid., p. 155*

It is interesting that Ricardo did not regret this difference between domestic and international circumstances. On the contrary, he says he would 'be sorry to see weakened… these feelings [that] induce most men of property to be satisfied with a low rate of profits in their own country, rather than seek a more advantageous employment for their wealth in foreign nations' (ibid.).

In Ricardo's day capital was much less mobile between countries than within them, so the assumption of capital immobility was not an unreasonable simplification. In the Bretton Woods era after World War II, capital controls were the norm as a matter of national policy. They included limits on the purchase or sale of currencies and financial assets, and government approval for foreign investment. Capital controls began to be loosened before the dissolution of the Bretton Woods system when in 1968 President Nixon temporarily suspended the convertibility of the dollar into gold in 1968, making it permanent in 1973. The system of fixed exchange rates (the price of one country's currency in terms of another's) and capital controls gave way to floating exchange rates and increasingly mobile capital. By the 1990s capital controls were largely abolished by advanced and emerging economies. At the same time globalization had become widely accepted, and one of the key arguments put forward in its favour was free trade.

The problem that has exercised Daly and which has yet to receive an adequate response from mainstream economists, is that the case for free trade is based heavily on comparative advantage, but comparative advantage is based on the immobility of capital and capital is now anything but immobile. 'The free flow of capital and goods (instead of goods only) means that investment is governed by absolute profitability and not by comparative advantage' (op cit., p. 214). Daly commends Ricardo and Adam Smith, who also wrote about international trade, for being 'realists who paid close attention to actual conditions. Their intellectual progeny, however, are idealists, ideologues, and logicians. Academic economists have become so enamored of the logical argument for comparative advantage... [that] they have suppressed recognition of the fact that the empirical cornerstone of the whole classical free trade argument, capital immobility, has crumbled into loose gravel' (ibid., 216). Daly concludes that 'it will not do to advocate every move towards setting aside national boundaries in the economic order in the name of a principle (comparative advantage)... that itself depends on the functioning of these very boundaries' (ibid., p. 218).

Daly's objection to basing arguments for free trade on comparative advantage in a world where capital is mobile would be of little consequence were the results of free trade indisputably positive but they are not. Daly finds fault with the theory underlying the case for free trade stemming from the false assumption of capital immobility. He also argues that the inclusion of mobile capital under the heading of free trade has negative economic, environmental and social impacts that can be avoided by appropriate changes to the rules governing international trade and finance.

Economic effects of globalization

The theory of international trade moved on after Ricardo although the foundational role of comparative advantage remained. One key development which Daly notes is the contribution of economist Bertil Ohlin (Ohlin 1933). It is easy to understand why the prices of commodities traded internationally would tend to be the same, allowing for transport costs. What Ohlin showed was that the same tendency applies to the prices of the factors of production even when they do not move across national boundaries. This reflects the tendency of nations to export commodities made from their most abundant factors. For example, Canada, with its huge forests, exports a lot of wood products and China, with its large, well-trained labour force, exports vast quantities of manufactured goods. Ohlin explained that the addition of foreign demand to the domestic demand for these commodities increases the total domestic demand for the factors from which they are made which puts upward pressure on their prices. At the same time the import of commodities that are more costly to produce domestically reduces the total domestic demand for the scarcer factors from which they are made, lowering their prices. However, as with the theory of comparative advantage, there are many conditions that must be met for complete factor price equalization: zero transport costs, pure competition, identical constant-returns-to-scale production functions – none of which hold in practice. However, the point remains that 'free trade, even with factor immobility, pushes us, however incompletely, in the direction of international equalization of factor prices – and factor prices include wage rates' (op cit., p. 219).

Daly and co-author Goodland[1] recognized that

> major welfare gains can be achieved through international trade, and trade-induced competition can lower costs by increasing efficiency as well as, unfortunately, by lowering standards so that any regulation of trade for environmental purposes needs to be approached with all due caution.
>
> *Daly and Goodland 1994, p. 74*

One aspect of international trade that concerns Daly is that these welfare gains, and the specialization in production that generates them, comes with a reduction in the choice of occupations in each country and insecurity of supply. These are sources of welfare reduction that are too often overlooked when there is an over-emphasis on incomes and consumption.

> Most people's enjoyment of life depends at least as much on how they earn their living as on how they spend their earnings. For example, a country like Uruguay, with a clear comparative advantage in cattle and sheep ranching would afford a citizen the choice of being either a cowboy or a shepherd, if it adhered strictly to the rule of specialization and trade; however, to sustain a viable national community Uruguayans have felt that they need their own legal, financial, medical, insurance, and educational services, as well as basic agriculture and industry.
>
> *Daly 1996, p. 146*

An effect of specialization is that it increases a country's dependence on long supply chains over which it has no control for food, energy, medical supplies and other inputs that can become critical in a time of crisis. 'A corollary of specialization is interdependence and vulnerability to disruption of trade' (Daly and Goodland, pp. 87–88).

One of Daly and Goodland's greatest concerns was that 'global economic integration via free trade will favour a privileged minority at the expense of the majority in both industrial and developing countries' (ibid.). This can happen for two reasons. The first is that 'International trade, despite the name, is not trade between nations, but trade between individuals [people and corporations] that crosses national boundaries... Mutual advantage between individuals in different countries does not guarantee mutual advantage for the two countries.' (Daly and Cobb op cit., p. 221). This means that the gains of parties to an international transaction can come at the expense of others in the countries where the trading parties are located. This is an especially serious problem when transnational corporations

> understate the price of commodities to evade taxes in the producing country and raise profits when they sell at full price in the importing country. This is done to transfer profits to the country with lowest taxes... This usually means in practice the underpricing of raw material exports.
>
> *Daly and Goodland op cit., p. 85. See also Goodland and Daly 1996*

BOX 13.2

He is brilliant at using metaphors to explain difficult concepts, and has a clarity of thought rarely seen among economists. Though extremely kind, he also has a rapier wit – I remember reading one review of a book in which the author had described starving Africans searching through feces for an undigested kernel of corn. He said he felt the same way searching through the book for a kernel of wisdom.

Elizabeth Trout

Environmental and social effects of globalization

Daly and Goodland consider multiple issues to support their case that 'deregulation of trade would be environmentally and socially harmful'. They deliberately use the term 'deregulation of trade' rather than free trade which is a 'rhetorically persuasive label for "deregulated international commerce". Who can be against freedom?' (ibid., p. 75). Their case about the harmful environmental and social effects of deregulation of trade is organized under three basic policy goals: allocative efficiency, distributional equity and ecologically sustainable scale. These are derived directly from Daly's principles of allocation, distribution and scale (see Chapter 5). The common link in the four 'mostly allocative issues' concerns external costs. An external cost is a cost not included, or imperfectly included, in the financial cost and price of a traded commodity. Adverse environmental impacts of production and consumption are examples of external costs. If a country, through regulation or an economic instrument such as an emissions charge, imposes additional costs on its domestic producers to protect the environment they will be at a competitive disadvantage against producers in other countries not facing similar regulations or charges.

> Therefore, national protection of a basic policy of internalization of environmental costs constitutes a clear justification for tariffs on imports from a country which does not internalize its environmental costs… There is a clear inconsistency between a national policy of cost internalization and an international policy of deregulated trade with non-cost internalizing nations.
>
> *ibid., p. 78*

A related issue raised by Daly and Goodland concerns labour standards. The GATT (the General Agreement on Tariffs and Trade, the predecessor of the WTO), included a provision allowing governments to restrict imports relating to products made by prison labour. This was to eliminate the competitive advantage obtained by artificially low labour costs. Writing in the GATT era Daly and Goodland said 'if GATT admits prison labour as an exception, can they continue to exclude similar issues such as child labour, debt peonage, chattel slavery, or jobs with high mortality/morbidity rates, especially where the job can be made tolerably safe?' (ibid., p. 79). The difference, if there is one, is that prison labour would be paid the market wage if performed outside prison, so the lower labour costs are regarded as a subsidy. These other concerns are humanitarian rather than direct subsidies, though they have economic causes and consequences, and are more complicated. For one thing, some developing countries hold the view that the campaign to bring labour issues into the WTO is a protectionist move by industrial countries 'to undermine the comparative advantage of lower wage trading partners' (WTO 2020). As it turns out, under the WTO the exception for prison labour has been side-lined. WTO agreements do not deal with labour standards as such, and member countries having decided to leave the negotiation of labour standards to the weaker International Labour Organization (ibid.).

One area of trade that especially concerned Daly and Goodland was international trade in toxic wastes. They ask: 'Should 'toxics be traded on the principle of comparative advantage? Should toxins be exported by the polluting nation or firm to "underpolluted" or cheap lands, likely to be low wage nations in the Third World?'[2] Daly and Goodland give several reasons why toxic wastes should not be exported that are still very relevant. The first is that the waste importer has less information about the risks than the exporter which places them at a disadvantage. Second, there is a mismatch between the distribution of the benefits from the trade to exporters and the inadequate compensation of the people in the importing country taking the risks. Third, proper management and oversight of toxic waste movement, treatment (if any) and disposal require long-term institutional stability that may be absent in the importing country. Fourth, since the volume of toxic wastes is increasing, the likelihood of serious accidents from the transportation of these wastes is also increasing and should be avoided if possible. Finally, the 'problem with toxics trade is that it removes the powerful and dynamic incentive for producers to internalize waste disposal' (op cit., p. 80). If it is easier and cheaper for those who generate the toxic wastes to ship them abroad rather than to reduce their production in the first place or treat and dispose of them at home then that is what they will most likely do. Building on the principle that the cost of toxic waste should be internalized to the firm that produces them, Daly and Goodland consider it a small extension to also internalize them to the country under whose laws the firm was allowed to generate toxic wastes, rather than allow externalization by exporting them.

The mainly distribution issues that Daly and Goodland raise in their case against deregulated 'free' trade follow directly from the mobility of capital.

> When capital flows abroad, the opportunity for new domestic employment diminishes, which drives down the price for domestic labor. Even if free trade and capital mobility raise wages in low wage countries (and that tendency is thwarted by overpopulation and rapid population growth), they do so at the expense of labor in the high-wage countries.
>
> *Daly 1993, p. 54*

In support of this proposition Daly reported that the 'real wages of the 80 per cent of the US labour force classified as "nonsupervisory employees"… have fallen 17 per cent between 1973 and 1990, in significant part because of trade liberalization' (ibid.).[3] If the reduced incomes of workers in the USA were matched by increases in the incomes of workers in poorer countries there would be some justification for the transfer. But Daly does not think that low-wage countries necessarily gain from free trade. He gives the example of the effects of NAFTA (the North American Free Trade Agreement), suggesting that it will 'ruin Mexican peasants when

> 'inexpensive' U.S. corn (subsidized by depleting topsoil, aquifers, oil wells and the federal treasury) can be freely imported. Displaced peasants will bid down wages. Their land will be bought cheaply by agribusinesses to produce fancy vegetables and cut flowers for the US market.
>
> *ibid.*

Daly wrote these words in 1993 in anticipation of the effects of NAFTA which was signed by the USA, Canada and Mexico in 1994. Twenty years later Laura Carlsen reported that,

> As heavily subsidized U.S. corn and other staples poured into Mexico, producer prices dropped, and small farmers found themselves unable to make a living. Some two million have been forced to leave their farms since Nafta.

At the same time, consumer food prices rose, notably the cost of the omni-present tortilla.

Carlsen 2013

The effects of capital mobility on the income distribution of labour in different countries is only one aspect of the distributional impacts of trade governed by absolute advantage. The other more troublesome one, as Daly noted in 1994, is that the main beneficiaries of capital mobility are the owners of capital who benefit from access to cheap foreign labour and the downward pressure it places on domestic wages. His concern was that high wages, reasonable working hours, social insurance and other benefits obtained by the working class in developed economies would be competed away in competition from low paid works in developing countries if capital is relocated.

> Through unregulated trade, Northern capitalists share the wages of Northern laborers with Southern laborers – although the Southern elites may also gain. This levelling of wages will be overwhelmingly downward due to the vast number and rapid growth rate of under employed populations in the Third World.
>
> *Daly and Goodland 1994, p. 83*

The net result, according to Daly and Goodland, is an increase in the share of national income going to capital. Thomas Piketty documented such a shift in shares which he attributed to differences between the relative rates of economic growth and the return to capital in advanced economies. Piketty did not include the particular dynamic described by Daly and Goodland which would contribute to a higher return to capital, though it is consistent with Piketty's explanation of increasing inequality.

Daly and Goodland contend that lower labour costs are not the only cost savings that promise increased returns to capital once it can move across national borders:

> International capital mobility, coupled with free trade of products, stimulates an international standards-lowering competition to attract capital: wages can be lowered, as can health insurance, worker safety standards, environmental standards etc. – all in the name of reducing costs. But reducing costs by increasing efficiency, and reducing costs by lowering standards are two very different things. Avoiding standards-lowering competition requires more than 'free trade'.
>
> *ibid., p. 83*

Not everyone agrees that free trade and capital mobility depresses wages. On the contrary,

> the dogma of global economic integration… is based on the assumption that wages can be levelled upward rather than downward. It assumes that the

whole world and many future generations can consume resources at the per capita levels current in today's high wage countries without inducing ecological collapse.

ibid., p. 86

Daly and Goodland dispute this assumption:

Free trade sins against the criterion of sustainable scale… Trade between nations or regions offers a way of loosening local constraints by importing environmental services, including waste absorption, from elsewhere. Within limits this can be quite reasonable and justifiable. But carried to extremes in the name of free trade it becomes destructive… it converts a set of problems, some of which are manageable at the national level, into one big, integrated, unmanageable, global problem.

ibid., pp. 86–87

The case for globalization and free trade

Where does this leave the case for globalization and free trade? Daly understands that the argument for free trade could be based on specialization according to capital mobility and absolute advantage rather than comparative advantage. For one thing, world output is greater with capital mobility than without it. But trade based on comparative advantage with immobile capital promises mutual gains. Trade based on absolute advantage with mobile capital does not. It is quite possible for residents of one country to reap all the gains as capital flows in at the expense of residents in another country as capital flows out. As Daly and Goodland observe, all countries have a comparative advantage in something but there is no reason to expect them all to have an absolute advantage in anything. This means that potential losses to some countries could be very great when capital is mobile.

Daly and Goodland conclude that the default presumption of most economists that free trade based on comparative advantage is good and restrictions require justification is faulty. They think the default presumption should be that national production for national markets is good and that international trade should be justified. They are in good company. In the midst of the Great Depression John Maynard Keynes wrote an article in which he admitted to changing his mind about free trade:

I was brought up, like most Englishmen, to respect free trade not only as an economic doctrine which a rational and instructed person could not doubt, but almost as a part of the moral law. I regarded ordinary departures from it as being at the same time an imbecility and an outrage.

Keynes 1933, p. 755

Only later did Keynes recognize that the 'flight of capital' limits the cope for domestic policies. He was particularly concerned about the divorce between ownership and management, a rather new phenomenon in his day, which when applied internationally,

> is, in times of stress, intolerable… experience is accumulating that remoteness between ownership and operation is an evil in the relations among men, likely or certain in the long run to set up strains and enmities which will bring to nought the financial calculation [used to justify the investment of capital abroad].
>
> *ibid., p. 759*

Daly describes the following quote as his favourite from Keynes:

> I sympathize, therefore, with those who would minimize, rather than with those who would maximize, economic entanglement among nations. Ideas, knowledge, science, hospitality, travel – these are the things which should of their nature be international. But let goods be homespun whenever it is reasonably and conveniently possible, and, above all, let finance be primarily national.
>
> *Keynes quoted in Daly 1996, p. 236*

Such a view of the proper relationship among trading nations is by no means common, especially among economists who have not given an adequate response to Daly's critique of the misapplication of comparative advantage as a justification for free trade. In 1993, *Scientific American* considered Daly's assessment of the 'perils of free trade' important enough to publish an article by him under that title. In the article Daly presents his argument that capital mobility undermines the case for gains from trade based on comparative advantage. He also makes many of the arguments against free trade based on the environmental and social impacts described above, adding that, to the extent that free trade promotes economic growth, it also induces an increase in throughput with all the associated, negative environmental consequences.

To provide a counter view to Daly's, *Scientific American* published in the same issue an article entitled 'The Case for Free Trade' written by Jagdish Bhagwati an expert on international trade. Although the title suggested a comprehensive response to Daly's critique, Bhagwati limited his argument to the claim that 'environmentalists are in error when they fear that trade, through growth, will necessarily increase pollution' (1993, p. 43). This is clearly a difference between Bhagwati and Daly and an important one but Bhagwati failed to address Daly's fundamental argument that the case for free trade cannot be based on comparative advantage in a world of capital mobility.

A very different assessment of globalization and free trade is offered by Joseph Stiglitz whom we encountered earlier in his dispute with Daly over economic growth (see Chapter 5). Stiglitz held several key positions that gave him an insider's view of globalization to complement his expertise as an academic economist of great standing. He was the Chair of the U.S. Council of Economic Advisors from 1995 to 1997 under President Clinton and Chief Economist of the World Bank from 1997 to 2000. After returning to academia in 2000 Stiglitz wrote *Globalization and its Discontents*, in which he described the many adverse effects of globalization, especially on developing countries (Stiglitz 2002). The book was deservedly successful, selling more than 1 million copies, which indicates the extent to which his account of globalization in practice found favour, especially in developing countries. In 2018 Stiglitz published a second edition of *Globalization and its Discontents* further documenting the many negative effects of globalization which he attributed largely to its mismanagement by the key international organizations: the IMF, the World Bank and the WTO. Stiglitz (2018) explained this mismanagement by pointing to the undue influence of corporations, especially American ones, on the policies and priorities of all three organizations.

There are several features of Stiglitz' second edition that stand out in relation to Daly's much earlier critique of globalization. Stiglitz provides extensive, documented evidence validating Daly's concerns about the effects of globalization: increased unemployment; increased risk to nations of becoming too dependent on others for essential supplies of energy, food or other essential items; the ambiguous desirability of free trade when there is imperfect competition giving large corporations substantial market power; the capture of government by corporate interests; the shift of production from the United States to China induced by cheap labour, with no likelihood of the production returning; advantages to companies accruing from low environmental and work place standards; and more than 100 per cent of the gains from trade, accruing to corporations, with labour becoming worse off. Stiglitz concludes: 'We should be asking for the minimal level of harmonization required to make the global system work, not for the minimal level of regulation – the level of regulation that maximizes corporate profits.' Reflecting on recent trade agreements Stiglitz says,

> there was a broader, more invidious agenda: to develop a system of globalization under which countries competed in every way possible to attract business – lower wages, weaker regulations, and reduced taxation… Globalization has become a race to the bottom, where corporations are the only winners and the rest of society, in both the developed and developing world, is the loser.
>
> *ibid., p. 28*

Conclusion

The close resemblance of Stiglitz's account of globalization in practice to Daly's critique written three decades earlier is striking. So too is the similarity between their recommendations for improving the situation. In his farewell speech marking his departure from the World Bank Daly offered four prescriptions for how the Bank could better serve the goal of environmentally sustainable development:

1. Stop counting the consumption of natural capital as income…
2. Tax labor and income less, and tax resource throughput more…
3. Maximize the productivity of natural capital in the short run, and invest in increasing its supply in the long run…
4. Move away from the ideology of global economic integration by free trade. free capital mobility. and export led growth-and toward a more nationalist orientation that seeks to develop domestic production for internal markets as the first option. having recourse to international trade only when clearly much more efficient…

Daly 1994a, pp. 62–68

Stiglitz lays out a plan for a fairer globalization, 'globalization with a human face' that would be more effective for raising living standards by changing institutions and the mind-set around globalization. Although he does not distinguish between globalization and internationalization the way that Daly does, much of what he is calling for – the more equitable treatment of nations to assure mutual gains and more sovereign control over vital interests – is similar to the internationalization that Daly favours.

Of course, there are differences between them on specifics and in the depth of their investigations into issues they both agree are important. (Stiglitz's book is 472 pages long and is one of several he has written on globalization.) The one difference that stands out is their views on capital mobility and its implications. They agree, to paraphrase Stiglitz, that if not well managed, freeing up capital can lead to greater instability and more inequality (ibid p. 30) What Stiglitz fails to consider is that capital mobility undermines the main assumption on which comparative advantage is based, something that Daly has emphasized for over three decades. Perhaps this is because Stiglitz's interpretation of Ricardo's example of Portugal and England is that comparative advantage is based on weather (ibid., p. 408), even though Ricardo made it clear that it was the immobility of capital between the two countries that was key – he did not even mention weather. However, Stiglitz did acknowledge that in a modern economy 'with capital, and even skilled labor, relatively mobile… comparative advantage resides in other aspects of production which are not mobile, such as the legal institutions' (ibid.).

Daly does not mention other possible determinants of comparative advantage. He might have noted, for example, the uneven distribution of deposits of raw materials which are inherently immobile until traded. Hence, Daly may have overstated the extent to which comparative advantage depends on immobile capital, but then he was discussing Ricardo's original formulation and Ricardo confined his discussion to that very feature. A more general point is that comparative advantage can arise from any immobile aspect of production. The more such aspects there are the greater will be the potential gains from trade. The removal of a really significant one – the immobility of capital – reduces the potential for mutually beneficial trade and expands the scope for absolute advantage with its unbalanced outcomes, to determine the scope and pattern of trade. This raises a question about the extent to which the mismanagement and its adverse effects that Stiglitz describes are the inevitable result of capital mobility, just as Daly warned and which has yet to be adequately acknowledged or proven wrong.

Notes

1 Environment Advisor to the World Bank, responsible for developing environmental assessment criteria for the Bank, and incidentally for hiring Daly when he was head of the Environmental Review Department for Latin America. He was the instigator and co-editor with Daly and El Serafy of *Building on Brundtland* (1991). Goodland, a fine man, was frequently referred to as the 'environmental conscience of the World Bank'. Unfortunately, the Bank frequently ignored its conscience. Over their years together at the Bank Goodland and Daly became close friends.
2 When economist Larry Summers was Vice-President of Development Economics and Chief Economist of the World Bank, he signed a memo stating 'the economic logic behind dumping a load of toxic waste in the lowest wage country is impeccable and we should face up to that... I've always thought that under-populated countries in Africa are vastly underpolluted.'
3 More recent data shows that average real (inflation adjusted) weekly earning of production and nonsupervisory employees in the US continued to decline to the mid-1990s and then started to rise again reaching the 1979 level in 2019 but still well short of the level in 1973 (Robertson 2019).

EPILOGUE

Most of this book was written during the first full year of the Covid-19 pandemic, from which some 2.7 million people died – one quarter of whom lived in the United States, the richest country in the world. Vaccines were developed in record time and were rolling out unevenly around the world. There was much discussion about which changes wrought by the pandemic would continue and which would be soon be forgotten. Would working from home become normal for more people? Would stronger community and neighbourhood ties continue? Would the popularity of cycling continue to grow? Would public transport fully recover from the loss of passengers? Would social distancing become a thing of the past? Would hugs and handshaking return?

On the economic front there was some realization that globalization had gone too far. The massive international movement of goods and peoples was cited as a reason that the Covid-19 virus had spread for far and so fast. Countries that considered themselves 'advanced' discovered that they could not manufacture urgently required personal protection equipment let alone vaccines. People became nervous, hoarding anything and everything from flour to toilet paper. Concern was expressed about supply lines that had become too long, too prone to disruption. Disparities between rich and poor increased. While many blue collar and other frontline workers bore the brunt of the pandemic, white collar workers who thought their jobs were secure discovered what precarious employment was like. Some businesses thrived, others collapsed, their vulnerability to the pandemic being so different and seemingly arbitrary. Income and wealth inequality increased. Government, which had fallen into disrepute in many countries in the wake of neoliberalism, had regained some grudging appreciation in the public mind for what it could do that the private sector could not, supporting the most vulnerable for example and planning mass vaccination.

There was also plenty of discussion about what kind of economic recovery would follow the pandemic. To the alphabet soup of economic recoveries the letter K was added, describing how some parts of the economy grow while others languish. What would this mean for monetary and fiscal policy? Would employment levels return to pre-pandemic levels? Balancing the budget became less of a concern or at least the lesser of evils.

Running through all this was the call to 'get back on track', as if the track we were on before the pandemic was the right one. 'Building back better' was also a common refrain, though it was not always clear what this meant – perhaps some sort of green new deal? What was notably absent was an insufficient appreciation that ideas that might have made sense in an empty world – that economic growth was essential for prosperity and virtually synonymous with progress – no longer applied now that the world was full. Fortunately, if we want to know what an economics for a full world looks like we can turn to Daly.

There are several themes in Herman's economics for a full world that run throughout his career of more than six decades. His End–Means Spectrum located economics within a physical and ethical framework which is reflected in his novel approach to economics. Herman's understanding that the economy is a sub-system of the biosphere came very early as did his appreciation of the relevance of the first and second laws of thermodynamics to economic processes. In this he followed the lead of Georgescu-Roegen and Boulding. There was already sufficient evidence in the 1960s to convince Herman that the empty world prevailing when economics was first developed was being transformed into a full world by the destruction and desecration of nature to support the historically unprecedented and remarkably unequal material growth of the human economy and population. He saw that nature was rapidly becoming scarce while capital and labour were comparatively abundant, yet these changing circumstances were having little impact on mainstream economics and on governments which had adopted growth as their primary macroeconomic objective.

This full-world perspective prompted Herman to build on the steady-state economics of John Stuart Mill that had been neglected by economists for over a century. Herman proposed a hierarchical relationship among scale, distribution and allocation and insisted on the distinction between qualitative development and quantitative growth. Rich economies, he said, can and should develop without growing, especially since there were already strong indications that growth had become uneconomic in the sense that the costs of growth were exceeding its benefits. In economies with high incomes and wealth Herman said redistribution rather than growth is the answer to poverty. He argued that the excessive use by advanced economies of the Earth's capacity to support life – human and non-human – was depriving those living in poorer countries and future generations everywhere of the opportunity to enjoy a good life based on sufficiency. All countries, Herman said, should seek qualitative development, but only poor ones should pursue quantitative economic growth while every effort is made by all to stay within planetary boundaries for resources and ecological systems.

Throughout his long career Herman did not allow himself to be constrained by the pre-analytic vision, presuppositions and narrow scope of the neoclassical economics in which he was trained. He demanded freedom to think outside the intellectual boundaries of his discipline and beyond the growth paradigm, believing it to be outmoded and counterproductive. The world had become full and it called for an economics of a different kind. It has been Herman's life's work to provide it. He has done much to point the way, sometimes to lead the way, and his influence has spread far and wide, to the chagrin, no doubt, of the naysayers. Now it is time for a new generation of enlightened economists to build on the foundations that Herman has provided and continue the task of building an economics for a full world.

BOX 14.1

Herman Daly's work has had a huge influence on me – more than any other academic. If he had not written his books and articles, I would probably be doing something very different right now.

Dan O'Neill

BOX 14.2

As a doyen of ecological economics Dr Daly stands tall as a leader in the field and his ideas have touched many a student, practitioner, researcher and policy maker.

Andrew Spezoka

BOX 14.3

As an undergraduate student a lecturer… told me about Herman's work and recommended his book *Valuing the Earth*. I read it in one go unlike anything else study-related before. It changed my life, as from this moment I knew I wanted to know more. His work has influenced me ever since in my career.

Christian Kerschner

BOX 14.4

I first encountered Herman Daly as an undergraduate studying economics. His work presented the relationship between humans and our ecosphere so clearly, and so logically that I wondered why this was missing from the conventional economics textbooks in my courses. Reading Herman Daly set me on a path to study ecological economics.

Brett Dolter

BOX 14.5

So much of Herman's work has had a profound influence on my thinking and is one of the reasons that I have focused much of my energy in the last 15 years on being an advocate for economic system change. Not only is his work totally fundamental but he has communicated this in such powerful and simple ways.

Stewart Wallis

BOX 14.6

Upon encountering Daly's scholarship during my Ph.D. research, I immediately recognized it as having the most potent policy implications I had ever seen for environmental and economic policy. As I came to know Herman personally, I also found him to be one of the most principled, conscientious scholars I'd known. Truly a gentle giant in the field.

Brian Czech

BOX 14.7

Intellectually, Herman has had more influence on me than anyone else. I consider myself to be a Daly disciple. Daly's work should be read by everyone studying economics and everyone who is not. He is a giant with countless people standing on his shoulders and countless more who should be.

Philip Lawn

REFERENCES

Adriaanse, A. et al. (1997), *Resource Flows: The Material Basis of Industrial Economies*. Washington, DC: World Resources Institute.

Alvaredo F. et al. (2020), *World Inequality Report 2018*, WID. World.

Aron (2011), 'Difference between Religion and Theology', Difference Between.com.

Arrow, K.J. (1951), *Social Choice and Individual Values*, New Jersey: John Wiley.

Arrow, K.J. and Debreu, G. (1954), 'Existence of an equilibrium for a competitive economy', *Econometrica*, 22(3), 265–290.

Arrow, K.J. et al (2004), 'Are we Consuming too Much?', *The Journal of Economic Perspectives*, 18(3), 147–117.

Arrow, K.J. et al (2007), 'Consumption, Investment, and Future Well-Being: Reply to Daly et al', *Conservation Biology*, 21(5), 1363–1365.

Awe, Y.A. (2012), *Toward a green, clean, and resilient world for all: a World Bank Group environment strategy 2012–2022 (English)*. Washington, DC: World Bank Group.

Ayres, R.U. and Kneese A.V. (1969), 'Production, Consumption and Externalities', *The American Economic Review*, 59(3), 282–297.

Ayres, R.U. and Warr, B. (2010), *The Economic Growth Engine*, Cheltenham: Edward Elgar.

Bank of England (nd), 'How is Money Created?'

Bates, M. (1960), *The Forest and the Sea*. New York: Random House.

Bao Hong, T. (2008), 'Cobb-Douglas Production Function', https://studylib.net/doc/8182897/cobb-douglas-production-function.

Barnett, H. and Morse C. (1963), *Scarcity and Growth*, Resources for the Future, Baltimore: The Johns Hopkins Press.

Becker, G. and Tomes, N. (1979), 'An Equilibrium Theory of the Distribution of Income and Intergenerational Mobility', *Journal of Political Economy*, 87(6), pp. 1153–1189.

Beckerman, W. (1974), *In Defence of Economic Growth*, London: Jonathan Cape.

Benes, J. and Kumhof, M. (2012), 'The Chicago Plan Revisited', IMF Working Paper WP/12/202.

Bezemer, D.J. (2009), '"No One Saw This Coming": Understanding Financial Crisis Through Accounting Models', *MPRA Paper* 15892.

Bhagwati, J. (1993), 'The Case for Free Trade', *Scientific American*, 269(5), 42–49.

Birch, C. (1991), NSW, Australia: *On Purpose*, New South Wales University Press.

Birch, C. and Cobb, J. (1981), *The Liberation of Life: from the Cell to the Community*, New York: Cambridge University Press.

Bonaiuti, M. (2011), *From Bioeconomics to Degrowth*, London: Routledge.

Boulding K.E. (1966), 'Economics of the Coming Spaceship Earth'. *In* Environmental Quality in a Growing Economy: Essays from the Sixth RFF Forum. Jarrett H., Ed.: 3–14, Baltimore: Johns Hopkins Press.

Boulding K.E. (1973), 'The Shadow of the Stationary State' in *The No-Growth Society*, Olson M. and Landsberg, H.H. eds., 89–101, New York: W.W. Norton.

Boulding, K.E. (1964), *The Meaning of the Twentieth Century*, New York: Harper and Row.

Burkett P. (2004), 'Marx's reproduction schemes and the environment', *Ecological Economics*, 49, 457–446.

Burness, S. et al, (1980), 'Thermodynamic and Economic Concepts as Related to Resource-Use Policies', *Land Economics*, 56(1), 1–9.

Burness, H.S. and Cummings, R.G. (1986), 'Thermodynamic and Economic Concepts as Related to Resource-Use Policies: Reply', *Land Economics*, 62(3), pp. 323–324.

Campbell, R.B. (2108), *Gone to Texas. A History of the Lone Star State*, 3rd edition, New York: Oxford University Press.

Carlsen, L. (2013), 'Under NAFTA, Mexico suffered, and United States Felt its P', *New York Times*, 24 November.

Carson, R, (1962), *Silent Spring,* Houghton-Miffin.

Carter, S. (2011), 'On the Cobb—Douglas and all that…: the Solow—Simon correspondence over the aggregate neoclassical production function', *Journal of Post Keynesian Economics*, 34(2), 255–273.

Clark, J.B. (1899), *The Distribution of Wealth: A Theory of Wages, Interest and Profits*, New York: The Macmillan Company.

Cobb Jr., J. and Daly, H.E. (1990), 'Free Trade versus Community: Social and Environmental Consequences of Free Trade in a World with Capital Mobility and Overpopulated Regions', *Population and Environment*, 11(3), 175–191.

Cobb, C.W. and Douglas, P.H. (1928), 'A Theory of Production'. *American Economic Review*, 18 (Supplement), 139–165.

Cohen, A.J. and Harcourt, G.C. (2003), 'Whatever Happened to the Cambridge Capital Theory Controversies?', *The Journal of Economic Perspectives*, 17(1), 199–214.

Commission on Growth and Development (2008), *The Growth Report: Strategies for Sustained Growth and Inclusive Development*, Washington, DC: The World Bank,.

Commoner, B. (1971), *The Closing Circle*, London: John Cape.

Costanza, R. and Daly, H.E. (1987), 'Towards an Ecological Economics', *Economic Modelling*, 38, 1–7.

Costanza, R. and Daly, H.E. (1991), 'Goals, Agenda, and Policy Recommendations for Ecological Economics', Chapter 1 in Costanza, R. (ed.), *Ecological Economics. The Science and Management of Sustainability*, New York: Columbia University Press.

Couix, Q. (2019), 'Natural Resources in the theory of production: the Georgescu-Roegen/ Daly versus Solo/Stiglitz controversy', *The European Journal of the History of Economic Thought*, 26(6), 1341–1378.

Cumberland, J.H. (1966), 'A Regional Inter-industry Model for Analysis of Development Objectives', *Regional science Association Papers*, 17, 65–95.

D'Alisa, G., Demaria, F. and Kallis, G. (2015), *Degrowth. A Vocabulary for a New Era*, New York/ London: Routledge.

Dale, G. (2012), 'The growth paradigm: a critique', *International Socialism*, Issue: 134.

Daly, H.E. (1965), 'The Uruguayan Economy: Its Basic Nature and Current Problems', *Journal of Inter-American Studies*, 7(3), 316–330.

Daly, H.E. (1966), 'An Historical Question and Three Hypotheses Concerning the Uruguayan Economy', *Inter-American Economic Affairs*, xx(1).

Daly, H.E. (1967a), 'A Brief Analysis of Recent Uruguayan Trade Control Systems', *Economic Development and Cultural Change*, 15(3), 286–296.

Daly, H.E. (1967b), 'A Note on the Pathological Growth of the Uruguayan Banking Sector', *Economic Development and Cultural Change*, 16(1), 91–96.

Daly, H.E. (1968a), 'Economics as a Life Science', *Journal of Political Economy*, 76, 392–406.

Daly, H.E. (1968b), 'Desenvolvimento economico e o problema demografico no Nordeste Brasileiro', *Revista Brasileira de Economia*, 22(4).

Daly, H.E. (1969), 'The Population Question in Northeast Brazil: Its Economic and Ideological Dimensions', Centre Discussion Paper 75, Economic Growth Center, Yale University.

Daly, H.E. (1970a), 'The Population Question in Northeast Brazil: Its Economic and Ideological Dimensions', *Economic Development and Cultural Change*, 18(4), Part 1, 536–574.

Daly, H.E. (1970b), 'The Canary has Fallen', *The New York Times*, October 14, 47.

Daly, H.E. (1970c), 'Some Observations on the Causes and Consequences of the Shortage of Change in Northeast Brazil', *Journal of Political Economy*, 78(1), 181–184.

Daly, H.E. (1971a), 'A Marxian-Malthusian View of Poverty and Development', March, 25–37.

Daly, H.E. (1971b), 'The Stationary-State Economy', Distinguished Lecture Series 2, Department of Economics, University of Alabama.

Daly, H.E. (1972), 'In Defense of a Steady-State Economy', *American Journal of Agricultural Economics*, Dec, 945–954.

Daly, H.E. (1973a) ed., *Toward a Steady-State Economy*, San Francisco: W.H. Freeman.

Daly, H.E. (1973b), Introduction to *Toward a Steady State-Economy*, (1973a).

Daly, H.E. (1973c), Electric Power, Employment, and Economic Growth: A Case Study in Growthmania. Chapter 13 in Daly 1973a.

Daly, H.E. (1974), 'The Economics of the Steady State', *American Economic Review*, 64, Papers and Proceedings of the 86th Annual Meeting of the American Economic Association, 15–21.

Daly, H.E. (1976), Entropy, Growth, and the Political Economy of Scarcity', presented at the RFF Conference on Natural Resource Scarcity, October 19. Reprinted in *Scarcity and Growth Reconsidered* (1979) V.K. Smith ed., *Resources for the Future*, Baltimore and London: The Johns Hopkins Press.

Daly, H.E. (1977a), *Steady-State Economics: The Economics of Biophysical Equilibrium and Moral Growth*, San Francisco: W. H. Freeman.

Daly, H.E. (1977b), 'Steady-State and Thermodynamics', Letters, *BioScience*, 27(12), 770–771.

Daly, H.E. (1979), 'Entropy, Growth, and the Political Economy of Scarcity', in Smith, K. ed., *Scarcity and Growth Reconsidered*, chapter 3, pp. 67–94.

Daly, H.E. (1980a), *Economics, Ecology, Ethics. Essays Towards a Steady-State Economy*, San Francisco: W.H. Freeman.

Daly, H.E. (1980b), 'Growth Economics and the Fallacy of Misplaced Concreteness', *American Behavioral Scientist*, 24(1), 79–105.

Daly, H.E. (1980c), The Economic Thought of Frederick Soddy', *History of Political Economy*, 12(4), 469–488.

Daly, H.E. (1980d), 'The Ecological and Moral Necessity for Limiting Economic Growth', in Roger Shinn, ed., *Faith and Science in an Unjust World*, I, Plenary Presentations, World Council of Churches, Geneva, 212–220.

Daly, H.E. (1982), 'Chicago School Individualism versus Sexual Reproduction: A Critique of Becker and Tomes', *Journal of Economic Issues*, 16(1), 307–312.

Daly, H.E. (1985), 'Marx and Malthus in North-east Brazil: A Note on the World's Largest Class Difference in Fertility and its Recent Trends', *Population Studies*, 39(2), 329–338.

Daly, H.E. (1986), 'Thermodynamic and Economic Concepts as Related to Resource-Use Policies: Comment', *Land Economics*, 62(3), 319–322.

Daly, H.E. (1987), The Economic Growth Debate: What Some Economists Have Learned But Many Have Not', *Journal of Environmental Economics and Management*, 14, 323–336.

Daly, H.E. (1989), 'Toward a Measure of Sustainable Social Net National Product', Chapter 2 in Ahmad, El Serafy, Lutz eds (1989), *Environmental Accounting for Sustainable Development*, Washington DC: The World Bank.

Daly, H.E. (1990), 'Towards Some Operational Principles of Sustainable Development', *Ecological Economics*, 2, 1–6.

Daly, H.E. (1991a), *Steady-State Economics, Second Edition,* Washington DC: Island Press.

Daly, H.E. (1991b), 'From empty-world to full-world economics: Recognizing an historical turning point in economic development', in Goodland et al. (eds) 1991, 17–27.

Daly, H.E. (1992a), 'Is the Entropy Law Relevant to the Economics of Natural Resource Scarcity – Yes, of Course it is!, *Journal of Environmental Economics and Management*, 23, 91–95.

Daly, H.E. (1992b), 'Allocation, distribution, and scale: towards an economics that is efficient, just, and sustainable', *Ecological Economics*, 6, 185–193.

Daly, H.E. (1993), 'The Perils of Free Trade', *Scientific American*, 269(5), 50–57.

Daly, H.E. (1994a), 'Farewell Lecture to the World Bank', Speech, in Daly (1999b), *Ecological Economics and the Ecology of Economics. Essays in Criticism*, 60–68.

Daly, H.E. (1994b), Reply, *Ecological Economics*, 10, 90–91.

Daly, H.E. (1995a), 'Against free trade: neoclassical and steady-state perspectives', *Journal of Evolutionary Economics*, 5, 131–326.

Daly, H.E. (1995b), 'Reply to Mark Sagoff's "Carrying Capacity and Ecological Economics", *BioScience*, 45(9), 621–624.

Daly H.E. (1995c), 'Consumption and Welfare: Two Views of Value Added', *The Social Economics of Environmental Issues*, 53(4), 451–473.

Daly, H.E. (1995d), 'On Nicholas Georgescu-Roegen's contributions to economics: an obituary essay', *Ecological Economics*, 13, 149–154.

Daly, H.E. (1996), *Beyond Growth*, Boston: Beacon Press.

Daly, H.E. (1997a), 'Georgescu-Roegen versus Solow/Stiglitz', *Ecological Economics*, 22, 261–266.

Daly, H.E. (1997b), 'Reply to Solow/Stigitz', Ecological Economics, 22, 271–273).

Daly, H.E. (1999a), 'Reply to Marcus Stewen', *Ecological Economics*, 30, 1–3.

Daly, H.E. (1999b), *Ecological Economics and the Ecology of Economics. Essays in Criticism*, 27–33, Cheltenham: Edwards Elgar.

Daly, H.E. (1999c), Globalization Versus internationalization', *Global Policy Forum*.

Daly, H.E. (1999d), Unpublished letter to M. Bonauiti, 22 October.

Daly, H.E. (2001), 'A Prayer of gratitude for the life of Donella Meadows' (unpublished).

Daly, H.E. (2002), Policy, Possibilities and Purpose', *Worldviews*, 6(2),183–197.

Daly, H.E. (2004a), *Statement before Senate Democratic Policy Committee, March 5,* 'Off-Shoring in the Context of Globalization'.

Daly, H.E. (2004b), 'Globalization and National Defense', as presented to the Office of Strategic Services, 16 April.

Daly, H.E. (2005), 'Economics in a Full World', *Scientific American*, 293, 100–107.

Daly, H.E. (2006), 'Population, migration, and globalization', *Ecological Economics*, 59(2), 187–190, previously published by Worldwatch Institute, *World Watch Magazine*, 17(5), 2004.

Daly, H.E. (2007), *Ecological Economics and Sustainable Development. Selected Essays of Herman Daly*. Cheltenham: Edward Elgar Publishing.

Daly, H.E. (2008a), 'Growth and Development. Critique of a Credo', *Population and Development Review*, 34(3), 511–518.

Daly, H.E. (2008b), 'Towards a Steady-State Economy'. Essay commissioned by the Sustainable Development Commission, UK (April 24).

Daly, H.E. (2010), 'The Operative Word is Somehow', *Real World Economics Review*, issue 54, 103.

Daly, H.E. (2012), 'Growth and Free Trade: Brain-Dead Dogmas Still Kicking Hard', *The Daly News, Casse,* 6 February.

Daly, H.E. (2013a), 'Open Borders and the Tragedy of Open Access Commons', *The Daly News*, Center for the Advancement of the Steady-State Economy, June 3.

Daly, H.E. (2013b), 'Top 10 Policies for a Steady-State Economy', *The Daly News*, CASSE.

Daly, H.E. (2014a), *From Uneconomic Growth to a Steady-State Economy*, Cheltenham: Edward Elgar.

Daly, H.E. (2014b) 'An Economics Fit for Purpose', *The Daly News*, 2014, 1.

Daly, H.E. (2015a), 'Economics for a Full World', *Great Transition Initiative*.

Daly, H.E. (2015b), 'Author's response to GTI Roundtable "Full-World Economics"', *Great Transition Initiative.*

Daly, H.E. (2015c), 'A population perspective on the steady state economy', *Real-World Economics Review*, 70, 106–109.

Daly, H.E. (2015d), 'Mass Migration and Border Policy', Real-World Economics Review, Issue, 73, 130–133.

Daly, H.E. (2015e), Foreward to *God? Very Probably*, R.H. Nelson, Eugene: Wipf and Stock Publishers.

Daly, H.E. (2016), email to Elke Pirgmaier (unpublished).

Daly, H.E. (2017), 'A new economics for our full world', Chapter 8 in *The Handbook on Growth and Sustainability* (eds P.A. Victor and B. Dolter), Cheltenham: Edward Elgar.

Daly, H.E. (2018), 'Do Red and Green Mix?: A Roundtable,' *Great Transition Initiative*, December.

Daly, H.E. (2019a), 'Draft of an Incomplete Memoire' (unpublished).

Daly, H.E. (2019b), 'Some overlaps in the first and second thirty years of ecological economics', *Ecological Economics*, 164, 1–3.

Daly, H.E. (2019c), 'Laudato Si and Population' in *Laudato Si and the Environment: Pope Francis' Green Encyclical*, R. McKim (ed), Routledge.

Daly, H.E. (2020a), 'A Note in Defense of the Concept of Natural Capital', *Ecosystem Services*, 41, 1–3.

Daly, H.E. (2020b), 'Reply to Troy Vetesse's "Against steady state economics"', Steady State Herald, *Center for the Advancement of the Steady State Economy*, March 18.

Daly, H.E. and Cobb, J. (1989), *For the Common Good. Redirecting the economy toward community, and a sustainable future*, Boston: Beacon Press.

Daly, H.E. and Cobb, J. (1994), *For the Common Good. Redirecting the economy toward community, and a sustainable future*, 2nd edition, Beacon Press: Boston.

Daly, H.E. and Farley, J. (2004), *Ecological Economics: Principles and Applications*, Washington DC: Island Press.

Daly, H.E. and Farley, J. (2011), *Ecological Economics. Principles and Applications*, 2nd edition, Washington: Island Press.

Daly, H.E. and Goodland, R. (1994), 'An ecological-economic assessment of deregulation of international commerce under GATT', *Ecological Economics*, 9, 73–92.

Daly, H.E. and Townsend, K.N. (1993), *Valuing the Earth*, Cambridge, MA: MIT Press.

Daly, H.E., et al. (2007), 'Are We Consuming Too Much – for What?', *Conservation Biology*, 21(5), 1359–1362.

Darwin, C. (1868), *The Variation of Animals and Plants under Domestication*, London: John Murray.

Dawkins, R. (1995), *River out of Eden: A Darwinian View of Life*, New York: Basic Books.

Dietz, S. and E. Neumayer, (2006), 'Some constructive criticisms of the Index of Sustainable Economic Welfare', Chapter 9, 186–206 in *Sustainable Development Indicators in Ecological Economics*, edited by P. Lawn, Cheltenham: Elgar Publishing.

Easterlin, R.A. (1974), 'Does economic growth improve the human lot?' in *Nations and Households in Economic Growth*, edited by P. David and M. Reder, Academic Press, pp. 89–125.

Easterlin, R.A. (2016), 'Paradox Lost?, USC Dornsife Institute for New Economic Thinking, Working paper 16-02.

Eddington, A.S. (1928), *The Nature of the Physical World*, The Macmillan Company, [Everyman's Library Edition 1935].

Elhacham, E., et al. (2020), 'Global human-made mass exceeds all living biomass', *Nature* 588, 17 December, 442–444.

Ehrlich, P. (1968), *The Population Bomb*, New York: Ballantyne Books.

Ehrlich, P., Holdren, J.P. and Ehrlich A. (1970, 1972, 1977), San Francisco: W.H Freeman.

El Serafy, S. (1989), 'The Proper Calculation of Income from Depletable Natural Resources', *Environmental Accounting for Sustainable Development*, Y.J. Ahmad, S. El Serafy and E. Lytz (eds), Washington DC: The World Bank.

Ellen McArthur Foundation (2013), Towards the Circular Economy, Volume Ellen McArthur Foundation (2013), Towards the Circular Economy, Volume 1.

Eurostat (2018), *Economy-wide material flow accounts HANDBOOK* 2018 edition, European Union.

Farley, J. and Washington, H. (2018), 'Circular Firing Squads: A Response to 'The Neoclassical Trojan Horse of Steady-State Economics' by Pirgmaier', *Ecological Economics*, 147, 442–449.

Felipe, J. and McCombie, J. (2011–12), 'On Herbert Simon's criticisms of the Cobb—Douglas and the CES production functions', *Journal of Post Keynesian Economics*, 34(2), 275–293.

Feynman. R.P. (2005). *The Meaning of it All: Thoughts of a Citizen-Scientist*, New York: Basic Books.

Fisher, I. (1892), *Mathematical investigations in the theory of value and prices, and appreciation and interest*, New York: A.M. Kelly.

Fisher, I. (1906), *The Nature of Capital and Income*, London: Macmillan.

Fisher, I. (1936), '100% Money and the Public Debt', *Economic Forum*, Spring Number, 406–420.

Foster, J.B. (2018), 'Do Red and Green Mix?: A Roundtable', *Great Transition Initiative*, December.

Foster, J.B. (2000), *Marx's Ecology: Materialism and Nature*, New York: Monthly Review Press.

Foster, J.B. (2020), *The Return of Nature, Socialism and Ecology*, New York: Monthly Review Press.

Foster, J.B. and Clark, B. (2009), 'The Paradox of Wealth: Capitalism and Ecological Destruction', *Monthly Review*, Nov 1.

Galbraith, J.K. (1958), *The Affluent Society*, Houghton. Mills, Harcourt.

George A. (2017), ed. *How Evolution Explains Everything about Life*, New Scientist, John Murray Learning.

Georgescu-Roegen, N. (1966), *Analytical Economics*, Harvard University Press.

Georgescu-Roegen, N. (1970), 'The Economics of Production', *American Economic Review*, 60(2), 1–9.

Georgescu-Roegen, N. (1971), *The Entropy Law and The Economic Process*, Harvard University Press.

Georgescu-Roegen, N. (1975), 'Energy and Economic Myths', *Southern Economic Journal*, 41(3), 347–381.

Georgescu-Roegen, N. (1977a), 'The Steady State and Ecological Salvation: A Thermodynamic Analysis ', *BioScience*, 27(4), 266–270.

Georgescu-Roegen, N. (1977b), 'Author's Reply', Letters, *BioScience*, 27(12), 771.

Georgescu-Roegen, N. (1979), Comments on the Papers by Daly and Stiglitz', in *Scarcity and Growth Reconsidered*, ed V. Kerry Smith, Resources for the Future (1979).

Germain, M. (2019), 'Georgescu-Roegen versus Solow/Stiglitz: Back to a controversy, *Ecological Economics*, 160–182.

Global Footprint Network (2020), *Ecological Footprint*. https://www.footprintnetwork.org/our-work/ecological-footprint.

Goodland, R. and Daly, H.E. (1996), 'If tropical log export bans are so perverse, why are there so many?', *Ecological Economics*, 18, 189–196.

Goodland, R., Daly, H.E., El Serafy, S. and von Droste, B. (1991), *Environmentally Sustainable Economic Development: Building On Brundtand*, Paris: UNESCO.

Gould, S.J. (1999), *Rocks of Ages: Science and Religion in the Fullness of Life*, New York: Ballantine Books.

Gowdy, J. (2016), Review of Mauro Bonaiuti (ed.): 'From bioeconomics to degrowth: Georgescu-Roegen's "New Economics" in eight essays', *Journal of Bioeconomics*, 18, 79–85.

Haas, W. et al, (2020), 'Spaceship earth's odyssey to a circular economy – a century long perspective', *Resources, Conservation & Recycling*, 163, 1–10.

Haberl H., Erb, K. and Krausmann, F. (2007), Human appropriation of net primary production (HANPP). *Internet Encyclopedia of Ecological Economics*. Neumeyer E., Ed. International Society for Ecological Economics.

Haldane, J.B.S. (1926), 'On Being the Right Size' *Harper's Magazine,* March.

Hansen, V. (2020), *The Year 1000: When Explorers Connected the World – and Globalization Began*, Scribener.

Hardin, G. (1968), 'The Tragedy of the Commons', *Science*, 162(3859), 1243–1248.

Hardin, G. (1974), 'Lifeboat Ethics: The Case Against Helping the Poor', *Psychology Today*, September.

Harris, S. (2012), *Free Will*, New York: Free Press.

Heilbroner, R. (1953), *The Worldy Philosphers*, New York, Simon & Schuster.

Helliwell, J., Layard, R., Sachs, J. and De Neve J-E., eds. (2020), *World Happiness Report 2020*. New York: Sustainable Development Solutions Network.

Harris, S. (2012), *Free Will*, New York: Free Press.

Higgs, H. (1923), 'Review: Cartesian Economics: The Bearing of Physical Science upon State Stewardship', *The Economic Journal*, 33(129), 100–101.

Hirsch, F. (1976), *Social Limits to Growth*, Cambridge MA: Harvard University Press.

Holland, A. (2002), 'Evolution and Purpose: A Response to Herman Daly', *Worldviews*, 6(2), 198–206.

Hubbert M.K. (1956), 'Nuclear Energy and the Fossil Fuels', Publication 95, Houston: Shell Development Corporation.

Hubbert M.K. (1974), 'M. King Hubbert on the Nature of Growth', Testimony to Hearing on the National Energy Conservation Policy Act of 1974, hearings before the Subcommittee on the Environment of the committee on Interior and Insular Affairs House of Representatives. June 6, Resilience.org.

Huesemann, M. and Huesemann, J. (2011), *TechNo-Fix*, Gabriola Island: New Society Publishers.

Isard, W. (1969), 'Some Notes on the Linkage of the Ecologic and Economic Systems', paper delivered to the Regional Science and Landscape Analysis Project, Department of Landscape Architecture, Harvard University and the Regional Science Research Institute.

Jackson, T. (2015), 'Contribution to GTI Roundtable "Full-World Economics,"' *Great Transition Initiative*, June, www.greattransition.org/commentary/tim-jackson-economics-for-a-full-world-herman- daly.

Jackson, T. (2017), *Prosperity without Growth*, 2nd edition, London: Earthscan.

Kåberger, T. and Månsson, B. (2001), Entropy and economic processes – physics perspectives', *Ecological Economics*, 36, 165–179.

Kahneman, D. (2011), *Thinking Fast and Slow*, London: Penguin Books.

Kallis, G. (2011), 'In defence of degrowth', *Ecological Economics*, 70, 873–880.

Kallis, G. (2018), *Degrowth*, Newcastle: Agenda Publishing.

Kallis, G., Kerschner, C. and Martinez-Alier, J. (2012), 'The economics of degrowth', *Ecological Economic*, 84, 172–180.

Keen, S. (2020), 'The macroeconomics of degrowth: can planned economic contraction be stable?', Bravenew Europe.com.

Kennedy, R. (1968), Remarks at the University of Kansas. https://www.jfklibrary.org/learn/about-jfk/the-kennedy-family/robert-f-kennedy/robert-f-kennedy-speeches/remarks-at-the-university-of-kansas-march-18-1968.

Kerschner, C. (2010), 'Economic de-growth vs. steady-state economy', *Journal of Cleaner Production*, 18, 544–551.

Keynes, J.M. (1930), 'Economic Possibilities for our Grandchildren', *Essays in Persuasion*. New York: W.W. Norton.

Keynes, J.M. (1933), 'National Self-Sufficiency', *The Yale Review*, 22(4), 755–769.

Kitzes, J. (2013), 'An Introduction to Environmentally– Extended Input-Output Analysis', *Resources*, 2, 489–503.

Klee, R.J. and T.E. Graedel (2004), 'Elemental cycles: A status report on human or natural dominance', *Annual Review of Environment and Resources*, 29, 69–107.

Krausmann, F., Gingrich, S., Eisenmenger, N., Erb, K.-H., Haberl, H. and Fischer-Kowalski, M. (2009), 'Growth in global materials use, GDP and population during the 20th century', *Ecological Economics*, 68(10), 696–2705.

Krausmann, F. et al. (2017), 'Global socioeconomic material stocks rise 23-fold over the 20th century and require half of annual resource use', *PNAS*, 114(8), 1880–1885.

Krausmann, F. et al. (2018), 'From resource extraction to outflows of wastes and emissions: The socioeconomic metabolism of the global economy, 1900–2015'. *Global Environmental Change*, 52, 131–140.

Krugman, P. (1998), 'Ricardo's Difficult Idea', chapter 3 in *The Economics and Politics of International Trade*, G. Cook (ed), London: Routledge.

KU Natural History Museum (2020), Investigating VIST Evolutionary Principles, KU Natural History.

Kubiszewski, I. et al. (2013), 'Beyond GDP: Measuring and achieving global genuine progress', *Ecological Economics*, 93, 57–68.

Kuhn, T. (1962), *The Structure of Scientific Revolutions*, Chicago: University of Chicago Press.

Kunkel, B. (2018), 'Ecologies of Scale', *New Left Review*, 109, 3–29.

Latouche, S. (2010), 'Degrowth', *The Journal of Cleaner Production*, 18, 519–522.

Leontief, W. (1936), 'Quantitative Input and Output Relations in the Economic Systems of the United States', *The Review of Economics and Statistics*, 18(3), pp. 105–125.

Leontief, W. (1970), 'Environmental Repercussions and the Economic Structure: An Input-Output Approach', *Review of Economics and Statistics*, II, 262–271.

Lewis, W. Arthur, (1955), *The Theory of Economic Growth*, Routledge.

Linder, S.B. (1970), *The Harried Leisure Class*, New York: Columbia University Press.

Löwy, M. (2018), 'Why Ecosocialism: For a Red-Green Future', *Great Transition Initiative*, December.

Lozado, G.L. (1991), 'A defense of Nicholas Georgescu-Roegen's Paradigm', *Ecological Economics*, 3, 157–160.

MacGillivray, A. (2005), *A Brief History of Globalization*, London: Robinson.

MacNeill, J. (2006), 'The forgotten imperative of sustainable development', *Environmental Policy and Law*, 16, 167–170.

Maddison, A. (2006), *The World Economy*, Paris: OECD.

Malthus, T. (1798), *An Essay on the Principle of Population* (1st edn). London: J. Johnson.

Marshall, A. (1920), *Principles of Economics*. 8th edition. London: Macmillan, Papermac 16 1966.

Mathews, E. et al. (2012), *The Weight of Nations*, Washington DC: World Resources Institute.

McKean, R.N. (1973), 'Growth vs. No Growth: An Evaluation', in *The No-Growth Society*, 207–227 in M. Olson and H.L Landsberg eds, New York: W.W. Norton.

McLeay, M., Radia, A. and Thomas, R. (2014), 'Money creation in the modern economy', *Quarterly Bulletin* Q1, 14–22.

Meadows, D.H. et al. (1972), *The Limits to Growth: A Report for the Club of Rome's Project on the Predicament of Mankind*, New York: Universe Books.

Midgely, M. (2009), 'Purpose, Meaning and Darwinism', *Philosophy Now*, 71.

Mill, J.S. (1848), *Principles of Political Economy: With some of their applications to social philosophy*. LondonL John W. Parker.

Miller, R.E. and Blair. P.D. (2009) second edition, *Input-Output Analysis*, Cambridge University Press.

Minsky, H. (1986), *Stabilizing and Unstable Economy*, Yale University Press.

Mirowski, P. (1988), 'Energy and energetics in economic theory: A review essay'. *Journal of Economic Issues*, XXII, 811–830.

Mirowski, P. (1989), *More Heat than Light*, Cambridge University Press.

Mirowski, P. (1991), 'The When, the How and the Why of Mathematical Expression in the History of Economic Analysis', *Journal of Economic Perspectives*, 5(1), 145–157.

Muradian, R., Neumeyer, E. and Røpke, I., (2006), 'Migration, globalization and the environment – introduction to the special issue', *Ecological Economics*, 59, 185–186.

Niemietz, K. (2012), 'A Critique of 'Steady-State Economics', *Economic Affairs Student and Teacher Supplement*, 32, issue s1, Spring, 4.

Nordhaus, W. and Tobin, J. (1972), 'Is Growth Obsolete?', *Economic Growth*, Fifteenth Anniversary Colloquium V, National Bureau of Economic research. New York: Columbia University Press.

O'Neill, D.W. (2012), 'Measuring progress in the degrowth transition to a steady state economy', *Ecological Economics*, 84, 221–231.

OECD (2001), Glossary of Statistical Terms: Gross Domestic Product (GDP).

OECD (2008), *Measuring Material Flows and Resource Productivity Volume I. The OECD Guide*. Paris: OECD.

Ohlin, B. (1933), *Interregional and International Trade*, Harvard University Press.

Okun, A.M. (1975), *Equality and Efficiency: The Big Tradeoff*, Washington, DC: Brookings Institution Oxford Reference (2017), Oxfordreference.com.

Parrique, T. (2019), *The political economy of degrowth*. Economics and Finance. Université Clermont.

Passell, P. et al. (1972), 'Review of *Limits to Growth*', New York Times Book Review, April 2.

Pen, J. (1971), *Income Distribution Facts, Theories, Policies*, Praeger Publishers.

Persky, J. (1995), 'The Theology of *Homo Economicus*', *Journal of Economic Perspectives*, 9(2), 221–231.

Pigou, A.C. (1920), *The Economics of Welfare*, London: Macmillan. All references to this book are to the fourth edition published in 1962.

Piketty, T. (2014), *Capital in the Twenty-First Century* (English edition), Harvard University Press.

Pirgmaier, E. (2017), 'The Neoclassical Trojan Horse of Steady-State Economics', *Ecological Economics*, 133, 52–61.

Plimsoll, S. (1873), *Our Seamen: An Appeal*, London: K.Mason.

Polanyi, K. (1944), *The Great Transformation*, Farrar & Rinehart.

Polanyi, M. (1958), *Personal Knowledge*, London: Routledge.

Pope Francis (2015), '"Encyclical Letter Laudato Si" Of The Holy Father Francis On Care For Our Common Home (official English-language text of encyclical)'.

Prakash, A. and Gupta, A.K. (1994), 'Are efficiency, equity, and scale independent?', *Ecological Economics*, 10, 89–91.

Rawls, J. (2005), *A Theory of Justice*, Cambridge: Harvard University Press.

Raworth, K.R. (2017), *Doughnut Economics*, Vermont: Chelsea Green Publishing.

Renner, A., Daly, H.E. and Mayumi, K. (2021), The Dual Nature of Money: A Note on Finance in an Equitable Bioeconomy, Environmental Economics and Policy Studies, (accepted for publication).

Ricardo, D. (1817), *Principles of Political Economy and Taxation*, Pelican Classics, Harmondsworth: Penguin Books 1971.

Robbins, L. (1932), *An Essay on the Nature and Significance of Economic Science*, London: Macmillan.

Robertson, L. (2019), 'Are Wages Rising or Flat?', FactCheck.org

Rockström, J. et al. (2009), 'A safe operating space for humanity', *Nature*, 461, 472–475.

Røpke, I. (2004), Migration and sustainability—compatible or contradictory?, *Ecological Economics*, 59, 191–194.

Røpke, I. (2017), 'Sustainability and the Governance of the Financial System: What role for full reserve banking?', *Environmental Policy and Governance*, 27, 177–192.

Røpke, I. (2020), 'Econ 101 – In need of a sustainability transition', *Ecological Economics*, 169.

Sagoff, M. (1995), 'Carrying Capacity and Ecological Economics', *BioScience*, 45(9), 610–620.

Sagoff, M. (2012), 'The Rise and Fall of Ecological Economics', *The Breakthrough Institute*, January 13. https://thebreakthrough.org/journal/issue-2/the-rise-and-fall-of-ecological-economics.

Sandelin, B. (1976), 'On the origin of the Cobb-Douglas production function', *Economy and History*, 19(2), 117–123, DOI: 10.1080/00708852.1976.10418933

Sanders, T.G. (1984), 'Family planning and population policy in Brazil', Abstract, *UFSI Rep.* (16), 1–7.

Schneider, E. and D. Sagan (2005), *Into the Cool. Energy Flow, Thermodynamics, and Life*, The University of Chicago Press.

Schrödinger, E. (1944), *What is Life?* Cambridge University Press.

Simon H. (1979), 'On Parsimonious Explanations of Production Relations', *The Scandinavian Journal of Economics*, 81(4), 459–474.

Smith, V.K. (1979), *Scarcity and Growth Reconsidered*, RFF, Johns Hopkins University Press.

Smith, R. (2010), 'Beyond growth or beyond capitalism?', *Real-World Economics Review*, Issue 53, 28–42.

Soddy, F. (1922), *Cartesian Economics, The Bearing of Physical Science upon State Stewardship*, London: Hendersons.

Soddy, F. (1926), *Wealth, virtual wealth and debt: The solution of the economic paradox*, London: Allen and Unwin.

Solow, R. (1966), Review of Capital and Growth. *American Economic Review*, 56(5), pp. 1257–1260.

Solow, R.M. (1973), 'Is the End of the World at Hand?', *Challenge*, 16(1), 39–50.

Solow, R. (1974), 'The Economics of Resources or the Resources of Economics', *The American Economic Review*, 64(2), Papers and Proceedings of the Eighty-sixth Annual Meeting of the American Economic Association, May, 1–14.

Solow, R. (1978), 'Resources and Economic Growth', *The American Economist*, 61(1), 52–60.

Solow, R. (1992), 'An almost practical step toward sustainability'. Washington, DC: Resources for the Future.

Solow, R.M. (1997), 'Georgescu-Roegen versus Solow/Stiglitz', *Ecological Economics*, 22, pp. 267–268.

Spash, C.L. (2013), 'The shallow or the deep ecological economics movement?', *Ecological Economics*, 93, 351–362.

Spash, C.L. (2015), 'The Future Post-Growth Society', Review Essay, *Development and Change*, 46(2), 366–380.

Spash, C.L. (2020), 'A tale of three paradigms: Realising the revolutionary potential of ecological economics', *Ecological Economics*, 169, 2–14.

Statistics Canada (2020), 'Table 36-10-0580-01 National Balance Sheet Accounts'.

Steffen, W. et al. (2015), 'Planetary boundaries: Guiding human development on a changing planet', *Science*, 347, Issue 6223, 1259855.

Stewen, M. (1998), 'The interdependence of allocation, distribution, scale and stability – A comment on Herman E. Daly's vision of an economics that is efficient, just and sustainable', *Ecological Economics*, 27, 119–130.

Stiglitz, J. (1974), 'Growth with Exhaustible Natural Resources: Efficient and Optimal Growth Paths', *The Review of Economic Studies*, Symposium on the Economics of Exhaustible Resources, 41, 123–137.

Stiglitz, J. (1997), 'Georgescu-Roegen versus Solow/Stiglitz', *Ecological Economics*, 22, 269–270.

Stiglitz, J. (2002), *Globalization and its Discontents*, New York: W.W. Norton.

Stiglitz, J.E., Sen, A. and Fitoussi, J-P. (2009), *Report by the Commission on the Measurement of Economic Performance and Social Progress*.

Stiglitz, J. (2018), *Globalization and its Discontents*, 2nd ed, New York: W.W. Norton.

Stoll, S. (2008), 'Fear of Fallowing', *Harper's Magazine*, March, 88–94.

Summers, K. (1991), Internal Memo, 2 December, The Whirled Bank Group.

Suranovic, S. (2010), 'The Theory of Comparative Advantage – Overview', *International Trade Theory and Policy*.

Tang, A.M., Wesfield, F.M. and Worley, J.S. (1976), *Evolution, Welfare and Time in Economics. Essays in Honour of Nicolas Georgescu-Roegen,* Lexington, MA: D.C. Heath & Co.

The JBH Foundation, (1996), 'The Short History of Race-Based Affirmative Action at Rice University.', *The Journal of Blacks in Higher Education*, 13, 36–38.

Tinbergen, J. (1952), *On the Theory of Economic Policy*, Amsterdam: North-Holland Publishing Co.

Tobin, J. (1965), 'Money and Economic Growth', *Econometrica* 33, Oct.

Townsend, K. (1992), 'Is the Entropy Law Relevant to the Economics of Natural Resource Scarcity?', *Journal of Environmental Economics and Management*, 23, 96–100.

Turner, G.M. (2012), 'On the Cusp of Global Collapse? Updated Comparison of The Limits to Growth with Historical Data', *GAIA*, 21(2), 116–124.

UN (2014), *System of Environmental-Economic Accounting 2012*. Central Framework, New York: United Nations.

UN (2020), *Material Flow Accounts*, System of Environmental Economic Accounting.

UNEP (2016), *Global Material Flows and Resource Productivity. An Assessment Study of the UNEP International Resource Panel*. H. Schandl et al., Paris, United Nations Environment Programme.

Valdes-Viera, O. (2017), 'The Borrowed Science of Neoclassical Economics', *Economic Questions*.

Van Den Bergh, J. (2011), 'Environment versus growth – A criticism of "degrowth" and a plea for "a-growth"', *Ecological Economics*, 70, 881–890.

Van Lerven, F. (2016), 'Setting the Record Straight: Sovereign Money is not Full-Reserve Banking', *Positive Money*.

Vettese, T. (2020), 'Against steady-state economics', *The Ecological Citizen*, 3, 35–46.

Victor, P.A. (1972), *Pollution: Economy and Environment*, London: George Allen & Unwin, republished in 2018 by London and New York: Routledge.

Victor, P.A. (1979), 'Economics and the challenge of environmental issues', in W. Leiss (ed.), *Politics and Ecology in Canada*. Toronto: University of Toronto Press. Reprinted in H.E. Daly, ed., *Economics, Ecology, Ethics*. W.H. Freeman, 1980.

Victor, P.A. (1991), 'Indicators of sustainable development: some lessons from capital theory', *Ecological Economics*, 4, 191–213.

Victor, P.A. (2016), 'The Steady-State Economy', chapter 16 in *Beyond Uneconomic Growth, Economics, Equity and the Ecological Predicament*, J. Farley and D. Malghan (eds), Advances in Ecological Economics series, Cheltenham: Edward Elgar.

Victor, P.A. (2019), *Managing without Growth. Slower by Design, not Disaster*, 2nd edition, Cheltenham: Edward Elgar Publishing.

Victor, P.A. (2020), 'Cents and nonsense: A critical appraisal of the monetary valuation of nature', *Ecosystem Services*, 42.

Vitousek, P.M., Ehrlich, P.R., Ehrlich, A.H. and Matson, P. (1986), 'Human appropriation of the products of photosynthesis', *BioScience*, 36, 368–373.

Wackernagel M. and Rees W.E. (1996), Our Ecological Footprint: Reducing Human Impact on the Earth. Gabriola Island, Canada and Philadelphia: New Society Publishers.

Wallich, H.C. (1972), 'Zero Growth', *Newsweek*, January 24.

Ward, B., and Dubos, R. (1972), *Only One Earth; the Care and Maintenance of a Small Planet*, New York: Norton.

Whitehead, A.N. (1925), *Science and the Modern World*, The Macmillan Company.

World Commission on Environment and Development (1987), *Our Common Future*, Oxford; New York, NY: Oxford University Press.

WTO (2020), 'Labour standards: consensus, coherence and controversy'.

Young, J.T. (1994), 'Entropy and Natural Resource Scarcity: A Reply to the Critics', *Journal of Environmental Economics and Management*, 26(2), pp. 210–213.

Zimmer, C. (2006), *Evolution. The Triumph of an Idea*, New York: Harper.

Zolotas, X. (1981), *Economic Growth and Declining Economic Welfare*, New York: New York University Press.

INDEX